For my father

'Sets out the twentieth-century history of the use of "psychedelic" substances with clarity, insight and humour . . . I hope that this book will be widely read. It is, in its gentle way, an important contribution to psychedelic research' Galen Strawson, *The Times Literary Supplement*

'There seems little doubt these drugs have far more to offer psychiatry than has been acknowledged. Pollan's exposition will spur exploration into this strange and captivating field' Oliver Thring, *Sunday Times*

'Truly astonishing' *Guardian*

'Challenging and fascinating . . . The book was having an after-effect. I was changing my mind . . . the claims are plausible. There really could be something to it' David Aaronovitch, *The Times*

'A compulsive book' Tim Adams, *Observer*

'Both beautiful and personal . . . shines new light on the revitalized field of psychedelic medicine' Benjamin Bell, *Salon*

'Pollan's deeply researched chronicle will enlighten those who think of psychedelics chiefly as a kind of punchline to a joke about the Woodstock generation and hearten the growing number who view them as a potential antidote to our often stubbornly narrow minds . . . engaging and informative' *Boston Globe*

'A deep dive into the history of psychedelics . . . Deliciously trippy' *New York Post*

'Journalist Michael Pollan explored psychoactive plants in *The Botany of Desire*. In this bold, intriguing study, he delves further . . . Pollan even "shakes the snow globe" himself, chemically self-experimenting in the spirit of psychologist William James, who speculated about the wilder shores of consciousness more than a century ago' *Nature*

'Pollan's complexly elucidating and enthralling inquiry combines fascinating and significant history with daring and resonant reportage and memoir' *Booklist*

'This nuanced and sophisti[...] questions about meaning-making and sp[...] is thought-provoking and eminently readable' *Publishers Weekly*

'A trip well worth taking, eye-opening and even mind-blowing' *Kirkus*

'Known for his writing on plants and food, Michael Pollan ... brings all the curiosity and skepticism for which he is well known to a decidedly different topic ... *How to Change Your Mind* beautifully updates and synthesizes the science of psychedelics, with a highly personalized touch' *Science Magazine*

'I've never regretted my adolescent use of LSD, but reading this fascinating, lucid, wise and hopeful book did make me wonder if those drug experiences weren't another example of youth wasted on the young. Michael Pollan, who waited until he was a grownup to experiment, is the perfect guide to today's dawning psychedelic renaissance' Kurt Andersen, author of *Fantasyland*

'Michael Pollan masterfully guides us through the highs, lows, and highs again of psychedelic drugs' Daniel Goleman, co-author of *Altered Traits*

'Michael walks the tight-rope between an objective "reporter" and a spiritual pilgrim ... It's an extraordinary achievement, and no matter what you may think you know about psychedelics, if you even know the word, you should read this book' Peter Coyote, author and Zen Buddhist priest

'After 50 years underground, psychedelics are back. We are incredibly fortunate to have Michael Pollan be our travel guide for their renaissance' Thomas R. Insel, former director of National Institute of Mental Health

ABOUT THE AUTHOR

Michael Pollan is an award-winning author, activist and journalist. His No.1 international bestselling books about the way we live today – including *The Omnivore's Dilemma*, *In Defence of Food* and *Cooked* (also a successful Netflix series) – combine meticulous reporting with anthropology, philosophy, culture, health and natural history. *Time* magazine has named him one of the hundred most influential people in the world. He lives in the Bay Area of California with his wife and son.

MICHAEL POLLAN

How to Change Your Mind

The New Science of Psychedelics

PENGUIN BOOKS

First published in the United States of America by Penguin Press,
an imprint of Penguin Random House LLC 2018
First published in Great Britain by Allen Lane 2018
Published in Penguin Books 2019
003

Images on the inside front cover from 'Homological scaffolds of brain
functional networks,' by G. Petri, P. Expert, F. Turkheimer, R. Carhart-Harris,
D. Nutt, P. J. Hellyer, and F. Vaccarino, *Journal of the Royal Society Interface*, 2014.

NOTE: This book relates the author's investigative reporting on, and related self-
experimentation with, psilocybin mushrooms, the drug lysergic acid diethylamide (or,
as it is more commonly known, LSD), and the drug 5-methoxy-N,N-dimethyltryptamine
(more commonly known as 5-MeO-DMT or The Toad). It is a criminal offence in the
United States and in many other countries, punishable by imprisonment and/or fines,
to manufacture, possess, or supply LSD, psilocybin mushrooms, and/or the drug
5-MeO-DMT, except in connection with government-sanctioned research. You should
therefore understand that this book is intended to convey the author's experiences and to
provide an understanding of the background and current state of research into these
substances. It is not intended to encourage you to break the law and no attempt should be
made to use these substances for any purpose except in a legally sanctioned clinical trial. The
author and the publisher expressly disclaim any liability, loss, or risk, personal or otherwise,
that is incurred as a consequence, directly or indirectly, of the contents of this book.

Certain names and locations have been changed in order to protect the author and others.

Printed and bound in Great Britain by Clays Ltd, Elcograf S.p.A.

A CIP catalogue record for this book is available from the British Library

ISBN: 978-0-141-98513-8

www.greenpenguin.co.uk

MIX
Paper from
responsible sources
FSC
www.fsc.org   FSC® C018179

Penguin Random House is committed to a
sustainable future for our business, our readers
and our planet. This book is made from Forest
Stewardship Council® certified paper.

The soul should always stand ajar.

—EMILY DICKINSON

Contents

HOW TO CHANGE YOUR MIND

A New Door

MIDWAY THROUGH the twentieth century, two unusual new molecules, organic compounds with a striking family resemblance, exploded upon the West. In time, they would change the course of social, political, and cultural history, as well as the personal histories of the millions of people who would eventually introduce them to their brains. As it happened, the arrival of these disruptive chemistries coincided with another world historical explosion—that of the atomic bomb. There were people who compared the two events and made much of the cosmic synchronicity. Extraordinary new energies had been loosed upon the world; things would never be quite the same.

The first of these molecules was an accidental invention of science. Lysergic acid diethylamide, commonly known as LSD, was first synthesized by Albert Hofmann in 1938, shortly before physicists split an atom of uranium for the first time. Hofmann, who worked for the Swiss pharmaceutical firm Sandoz, had been looking for a drug to stimulate circulation, not a psychoactive compound. It wasn't until five years later when he accidentally ingested a minuscule quantity of the new chemical that he realized he had created something powerful, at once terrifying and wondrous.

The second molecule had been around for thousands of years, though no one in the developed world was aware of it. Produced not by a chemist but by an inconspicuous little brown mushroom, this molecule, which would come to be known as psilocybin, had been used by the indigenous peoples of Mexico and Central America for hundreds of years as a sacrament. Called *teonanácatl* by the Aztecs, or "flesh of the gods," the mushroom was brutally suppressed by the Roman Catholic Church after the Spanish conquest and driven underground. In 1955, twelve years after Albert Hofmann's discovery of LSD, a Manhattan banker and amateur mycologist named R. Gordon Wasson sampled the magic mushroom in the town of Huautla de Jiménez in the southern Mexican state of Oaxaca. Two years later, he published a fifteen-page account of the "mushrooms that cause strange visions" in *Life* magazine, marking the moment when news of a new form of consciousness first reached the general public. (In 1957, knowledge of LSD was mostly confined to the community of researchers and mental health professionals.) People would not realize the magnitude of what had happened for several more years, but history in the West had shifted.

The impact of these two molecules is hard to overestimate. The advent of LSD can be linked to the revolution in brain science that begins in the 1950s, when scientists discovered the role of neurotransmitters in the brain. That quantities of LSD measured in micrograms could produce symptoms resembling psychosis inspired brain scientists to search for the neurochemical basis of mental disorders previously believed to be psychological in origin. At the same time, psychedelics found their way into psychotherapy, where they were used to treat a variety of disorders, including alcoholism, anxiety, and depression. For most of the 1950s and early 1960s, many in the psychiatric establishment regarded LSD and psilocybin as miracle drugs.

The arrival of these two compounds is also linked to the rise of the counterculture during the 1960s and, perhaps especially, to its particular tone and style. For the first time in history, the young had a rite of passage all their own: the "acid trip." Instead of folding the young into the adult world, as rites of passage have always done, this one landed them in a country of the mind few adults had any idea even existed. The effect on society was, to put it mildly, disruptive.

Yet by the end of the 1960s, the social and political shock waves unleashed by these molecules seemed to dissipate. The dark side of psychedelics began to receive tremendous amounts of publicity— bad trips, psychotic breaks, flashbacks, suicides—and beginning in 1965 the exuberance surrounding these new drugs gave way to moral panic. As quickly as the culture and the scientific establishment had embraced psychedelics, they now turned sharply against them. By the end of the decade, psychedelic drugs—which had been legal in most places—were outlawed and forced underground. At least one of the twentieth century's two bombs appeared to have been defused.

Then something unexpected and telling happened. Beginning in the 1990s, well out of view of most of us, a small group of scientists, psychotherapists, and so-called psychonauts, believing that something precious had been lost from both science and culture, resolved to recover it.

Today, after several decades of suppression and neglect, psychedelics are having a renaissance. A new generation of scientists, many of them inspired by their own personal experience of the compounds, are testing their potential to heal mental illnesses such as depression, anxiety, trauma, and addiction. Other scientists are using psychedelics in conjunction with new brain-imaging tools to explore the links between brain and mind, hoping to unravel some of the mysteries of consciousness.

One good way to understand a complex system is to disturb it and then see what happens. By smashing atoms, a particle accelerator forces them to yield their secrets. By administering psychedelics in carefully calibrated doses, neuroscientists can profoundly disturb the normal waking consciousness of volunteers, dissolving the structures of the self and occasioning what can be described as a mystical experience. While this is happening, imaging tools can observe the changes in the brain's activity and patterns of connection. Already this work is yielding surprising insights into the "neural correlates" of the sense of self and spiritual experience. The hoary 1960s platitude that psychedelics offered a key to understanding—and "expanding"—consciousness no longer looks quite so preposterous.

How to Change Your Mind is the story of this renaissance. Although it didn't start out that way, it is a very personal as well as public history. Perhaps this was inevitable. Everything I was learning about the third-person history of psychedelic research made me want to explore this novel landscape of the mind in the first person too—to see how the changes in consciousness these molecules wrought actually feel and what, if anything, they had to teach me about *my* mind and might contribute to my life.

o o o

THIS WAS, FOR ME, a completely unexpected turn of events. The history of psychedelics I've summarized here is not a history I lived. I was born in 1955, halfway through the decade that psychedelics first burst onto the American scene, but it wasn't until the prospect of turning sixty had drifted into view that I seriously considered trying LSD for the first time. Coming from a baby boomer, that might sound improbable, a dereliction of generational duty. But I was only twelve years old in 1967, too young to have been more than dimly aware of the Summer of Love or the San Francisco Acid Tests. At fourteen, the only way I was going to get to Woodstock was if my parents drove me. Much of the 1960s I experienced through the pages of *Time* magazine. By the time the idea of trying or not trying LSD swam into my conscious awareness, it had already completed its speedy media arc from psychiatric wonder drug to counterculture sacrament to destroyer of young minds.

I must have been in junior high school when a scientist reported (mistakenly, as it turned out) that LSD scrambled your chromosomes; the entire media, as well as my health-ed teacher, made sure we heard all about it. A couple of years later, the television personality Art Linkletter began campaigning against LSD, which he blamed for the fact his daughter had jumped out of an apartment window, killing herself. LSD supposedly had something to do with the Manson murders too. By the early 1970s, when I went to college, everything you heard about LSD seemed calculated to terrify. It worked on me: I'm less a child of the psychedelic 1960s than of the moral panic that psychedelics provoked.

I also had my own personal reason for steering clear of psychedelics: a painfully anxious adolescence that left me (and at least one psychiatrist) doubting my grip on sanity. By the time I got to college, I was feeling sturdier, but the idea of rolling the mental dice with a psychedelic drug still seemed like a bad idea.

Years later, in my late twenties and feeling more settled, I did try magic mushrooms two or three times. A friend had given me a Mason jar full of dried, gnarly *Psilocybes*, and on a couple of memorable occasions my partner (now wife), Judith, and I choked down two or three of them, endured a brief wave of nausea, and then sailed off on four or five interesting hours in the company of each other and what felt like a wonderfully italicized version of the familiar reality.

Psychedelic aficionados would probably categorize what we had as a low-dose "aesthetic experience," rather than a full-blown ego-disintegrating trip. We certainly didn't take leave of the known universe or have what anyone would call a mystical experience. But it was *really* interesting. What I particularly remember was the preternatural vividness of the greens in the woods, and in particular the velvety chartreuse softness of the ferns. I was gripped by a powerful compulsion to be outdoors, undressed, and as far from anything made of metal or plastic as it was possible to get. Because we were alone in the country, this was all doable. I don't recall much about a follow-up trip on a Saturday in Riverside Park in Manhattan except that it was considerably less enjoyable and unselfconscious, with too much time spent wondering if other people could tell that we were high.

I didn't know it at the time, but the difference between these two experiences of the same drug demonstrated something important, and special, about psychedelics: the critical influence of "set" and "setting." Set is the mind-set or expectation one brings to the experience, and setting is the environment in which it takes place. Compared with other drugs, psychedelics seldom affect people the same way twice, because they tend to magnify whatever's already going on both inside and outside one's head.

After those two brief trips, the mushroom jar lived in the back of our pantry for years, untouched. The thought of giving over a whole day to a psychedelic experience had come to seem inconceivable. We were working long hours at new careers, and those vast swaths of unallocated time that college (or unemployment) affords had become a memory. Now another, very different kind of drug was available, one that was considerably easier to weave into the fabric of a Manhattan career: cocaine. The snowy-white powder made the wrinkled brown mushrooms seem dowdy, unpredictable, and overly demanding. Cleaning out the kitchen cabinets one weekend, we stumbled upon the forgotten jar and tossed it in the trash, along with the exhausted spices and expired packages of food.

Fast-forward three decades, and I really wish I hadn't done that. I'd give a lot to have a whole jar of magic mushrooms now. I've begun to wonder if perhaps these remarkable molecules might be wasted on the young, that they may have more to offer us later in life, after the cement of our mental habits and everyday behaviors has set. Carl Jung once wrote that it is not the young but people in middle age who need to have an "experience of the numinous" to help them negotiate the second half of their lives.

By the time I arrived safely in my fifties, life seemed to be running along a few deep but comfortable grooves: a long and happy marriage alongside an equally long and gratifying career. As we do, I had developed a set of fairly dependable mental algorithms for navigating whatever life threw at me, whether at home or at work. What was missing from my life? Nothing I could think of—until, that is, word of the new research into psychedelics began to find its way to me, making me wonder if perhaps I had failed to recognize the potential of these molecules as a tool for both understanding the mind and, potentially, changing it.

o   o   o

HERE ARE THE THREE DATA POINTS that persuaded me this was the case.

In the spring of 2010, a front-page story appeared in the *New York Times* headlined "Hallucinogens Have Doctors Tuning In Again." It reported that researchers had been giving large doses of psilocybin—

the active compound in magic mushrooms—to terminal cancer patients as a way to help them deal with their "existential distress" at the approach of death.

These experiments, which were taking place simultaneously at Johns Hopkins, UCLA, and New York University, seemed not just improbable but crazy. Faced with a terminal diagnosis, the very *last* thing I would want to do is take a psychedelic drug—that is, surrender control of my mind and then in that psychologically vulnerable state stare straight into the abyss. But many of the volunteers reported that over the course of a single guided psychedelic "journey" they reconceived how they viewed their cancer and the prospect of dying. Several of them said they had lost their fear of death completely. The reasons offered for this transformation were intriguing but also somewhat elusive. "Individuals transcend their primary identification with their bodies and experience ego-free states," one of the researchers was quoted as saying. They "return with a new perspective and profound acceptance."

I filed that story away, until a year or two later, when Judith and I found ourselves at a dinner party at a big house in the Berkeley Hills, seated at a long table with a dozen or so people, when a woman at the far end of the table began talking about her acid trips. She looked to be about my age and, I learned, was a prominent psychologist. I was engrossed in a different conversation at the time, but as soon as the phonemes L-S-D drifted down to my end of the table, I couldn't help but cup my ear (literally) and try to tune in.

At first, I assumed she was dredging up some well-polished anecdote from her college days. Not the case. It soon became clear that the acid trip in question had taken place only days or weeks before, and in fact was one of her first. The assembled eyebrows rose. She and her husband, a retired software engineer, had found the occasional use of LSD both intellectually stimulating and of value to their work. Specifically, the psychologist felt that LSD gave her insight into how young children perceive the world. Kids' perceptions are not mediated by expectations and conventions in the been-there, done-that way that adult perception is; as adults, she explained, our minds don't simply take in the world as it is so much as they make educated guesses about it. Relying on these guesses, which are based on past experience, saves the mind time and energy, as when, say, it's trying to figure out what that fractal pattern of green dots in its visual field might be. (The leaves on a tree, probably.) LSD appears to disable such conventionalized, shorthand modes of perception and, by doing so, restores a childlike immediacy, and sense of wonder, to our experience of reality, as if we were seeing everything for the first time. (*Leaves!*)

I piped up to ask if she had any plans to write about these ideas, which riveted everyone at the table. She laughed and gave me a look that I took to say, *How naive can you be?* LSD is a schedule I substance, meaning the government regards it as a drug of abuse with no accepted medical use. Surely it would be foolhardy for someone in her position to suggest, in print, that psychedelics might have anything to contribute to philosophy or psychology—that they might actually be a valuable tool for exploring the mysteries of human consciousness. Serious research into psychedelics had been more or less purged from the university fifty years ago, soon after Timothy Leary's Harvard Psilocybin Project crashed and burned in 1963. Not even Berkeley, it seemed, was ready to go there again, at least not yet.

Third data point: The dinner table conversation jogged a vague memory that a few years before somebody had e-mailed me a scientific paper about psilocybin research. Busy with other things at the time, I hadn't even opened it, but a quick search of the term "psilocybin" instantly fished the paper out of the virtual pile of discarded e-mail on my computer. The paper had been sent to me by one of its co-authors, a man I didn't know by the name of Bob Jesse; perhaps he had read something I'd written about psychoactive plants and thought I might be interested. The article, which was written by the same team at Hopkins that was giving psilocybin to cancer patients, had just been published in the journal *Psychopharmacology*. For a peer-reviewed scientific paper, it had a most unusual title: "Psilocybin Can Occasion Mystical-Type Experiences Having Substantial and Sustained Personal Meaning and Spiritual Significance."

Never mind the word "psilocybin"; it was the words "mystical" and "spiritual" and "meaning" that leaped out from the pages of a pharmacology journal. The title hinted at an intriguing frontier of research, one that seemed to straddle two worlds we've grown accustomed to think are irreconcilable: science and spirituality.

Now I fell on the Hopkins paper, fascinated. Thirty volunteers who had never before used psychedelics had been given a pill containing either a synthetic version of psilocybin or an "active placebo"—methylphenidate, or Ritalin—to fool them into thinking they had received the psychedelic. They then lay down on a couch wearing eyeshades and listening to music through headphones, attended the whole time by two therapists. (The eyeshades and headphones encourage a more inward-focused journey.) After about thirty minutes, extraordinary things began to happen in the minds of the people who had gotten the psilocybin pill.

The study demonstrated that a high dose of psilocybin could be used to safely and reliably "occasion" a mystical experience—typically described as the dissolution of one's ego followed by a sense of merging with nature or the universe. This might not come as news to people who take psychedelic drugs or to the researchers who first studied them back in the 1950s and 1960s. But it wasn't at all obvious to modern science, or to me, in 2006, when the paper was published.

What was most remarkable about the results reported in the article is that participants ranked their psilocybin experience as one of the most meaningful in their lives, comparable "to the birth of a first child or death of a parent." Two-thirds of the participants rated the session among the top five "most spiritually significant experiences" of their lives; one-third ranked it *the* most significant such experience in their lives. Fourteen months later, these ratings had slipped only slightly. The volunteers reported significant improvements in their "personal well-being, life satisfaction and positive behavior change," changes that were confirmed by their family members and friends.

Though no one knew it at the time, the renaissance of psychedelic research now under way began in earnest with the publication of that paper. It led directly to a series of trials—at Hopkins and several other universities—using psilocybin to treat a variety of indications, including anxiety and depression in cancer patients, addiction to nicotine and alcohol, obsessive-compulsive disorder, depression, and eating disorders. What is striking about this whole line of clinical research is the premise that it is not the pharmacological effect of the drug itself but the kind of mental experience it occasions—involving the temporary dissolution of one's ego—that may be the key to changing one's mind.

o   o   o

AS SOMEONE not at all sure he has ever had a single "spiritually significant" experience, much less enough of them to make a ranking, I found that the 2006 paper piqued my curiosity but also my skepticism. Many of the volunteers described being given access to an alternative reality, a "beyond" where the usual physical laws don't apply and various manifestations of cosmic consciousness or divinity present themselves as unmistakably real.

All this I found both a little hard to take (couldn't this be just a drug-induced hallucination?) and yet at the same time intriguing;

part of me wanted it to be true, whatever exactly "it" was. This surprised me, because I have never thought of myself as a particularly spiritual, much less mystical, person. This is partly a function of worldview, I suppose, and partly of neglect: I've never devoted much time to exploring spiritual paths and did not have a religious upbringing. My default perspective is that of the philosophical materialist, who believes that matter is the fundamental substance of the world and the physical laws it obeys should be able to explain everything that happens. I start from the assumption that nature is all that there is and gravitate toward scientific explanations of phenomena. That said, I'm also sensitive to the limitations of the scientific-materialist perspective and believe that nature (including the human mind) still holds deep mysteries toward which science can sometimes seem arrogant and unjustifiably dismissive.

Was it possible that a single psychedelic experience—something that turned on nothing more than the ingestion of a pill or square of blotter paper—could put a big dent in such a worldview? Shift how one thought about mortality? Actually change one's mind in enduring ways?

The idea took hold of me. It was a little like being shown a door in a familiar room—the room of your own mind—that you had somehow never noticed before and being told by people you trusted (scientists!) that a whole other way of thinking—of being!—lay waiting on the other side. All you had to do was turn the knob and enter. Who *wouldn't* be curious? I might not have been looking to change my life, but the idea of learning something new about it, and of shining a fresh light on this old world, began to occupy my thoughts. Maybe there *was* something missing from my life, something I just hadn't named.

Now, I already knew something about such doors, having written about psychoactive plants earlier in my career. In *The Botany of De-*

sire, I explored at some length what I had been surprised to discover is a universal human desire to change consciousness. There is not a culture on earth (well, one*) that doesn't make use of certain plants to change the contents of the mind, whether as a matter of healing, habit, or spiritual practice. That such a curious and seemingly maladaptive desire should exist alongside our desires for nourishment and beauty and sex—all of which make much more obvious evolutionary sense—cried out for an explanation. The simplest was that these substances help relieve pain and boredom. Yet the powerful feelings and elaborate taboos and rituals that surround many of these psychoactive species suggest there must be something more to it.

For our species, I learned, plants and fungi with the power to radically alter consciousness have long and widely been used as tools for healing the mind, for facilitating rites of passage, and for serving as a medium for communicating with supernatural realms, or spirit worlds. These uses were ancient and venerable in a great many cultures, but I ventured one other application: to enrich the collective imagination—the culture—with the novel ideas and visions that a select few people bring back from wherever it is they go.

o   o   o

NOW THAT I HAD DEVELOPED an intellectual appreciation for the potential value of these psychoactive substances, you might think I would have been more eager to try them. I'm not sure what I was waiting for: courage, maybe, or the right opportunity, which a busy life lived mainly on the right side of the law never quite seemed to afford. But when I began to weigh the potential benefits I was hearing about against the risks, I was surprised to learn that psychedelics

---

* The Inuit appear to be the exception that proves the rule, but only because nothing psychoactive grows where they live. (At least not yet.)

are far more frightening to people than they are dangerous. Many of the most notorious perils are either exaggerated or mythical. It is virtually impossible to die from an overdose of LSD or psilocybin, for example, and neither drug is addictive. After trying them once, animals will not seek a second dose, and repeated use by people robs the drugs of their effect.* It is true that the terrifying experiences some people have on psychedelics can risk flipping those at risk into psychosis, so no one with a family history or predisposition to mental illness should ever take them. But emergency room admissions involving psychedelics are exceedingly rare, and many of the cases doctors diagnose as psychotic breaks turn out to be merely short-lived panic attacks.

It is also the case that people on psychedelics are liable to do stupid and dangerous things: walk out into traffic, fall from high places, and, on rare occasions, kill themselves. "Bad trips" are very real and can be one of "the most challenging experiences of [a] lifetime," according to a large survey of psychedelic users asked about their experiences.† But it's important to distinguish what can happen when these drugs are used in uncontrolled situations, without attention to set and setting, from what happens under clinical conditions, after careful screening and under supervision. Since the revival of sanctioned psychedelic research beginning in the 1990s, nearly a thou-

---

* David J. Nutt, *Drugs Without the Hot Air: Minimising the Harms of Legal and Illegal Drugs* (Cambridge, U.K.: UIT, 2012). This is why people "microdosing" on psychedelics never take them on consecutive days.

† Theresa M. Carbonaro et al., "Survey Study of Challenging Experiences After Ingesting Psilocybin Mushrooms: Acute and Enduring Positive and Negative Consequences," *Journal of Psychopharmacology* (2016): 1268–78. The survey found that 7.6 percent of respondents sought treatment for "one or more psychological symptoms they attributed to their challenging psilocybin experience."

sand volunteers have been dosed, and not a single serious adverse event has been reported.

o o o

IT WAS AT THIS POINT that the idea of "shaking the snow globe," as one neuroscientist described the psychedelic experience, came to seem more attractive to me than frightening, though it was still that too.

After more than half a century of its more or less constant companionship, one's self—this ever-present voice in the head, this ceaselessly commenting, interpreting, labeling, defending I—becomes perhaps a little *too* familiar. I'm not talking about anything as deep as self-knowledge here. No, just about how, over time, we tend to optimize and conventionalize our responses to whatever life brings. Each of us develops our shorthand ways of slotting and processing everyday experiences and solving problems, and while this is no doubt adaptive—it helps us get the job done with a minimum of fuss—eventually it becomes rote. It dulls us. The muscles of attention atrophy.

Habits are undeniably useful tools, relieving us of the need to run a complex mental operation every time we're confronted with a new task or situation. Yet they also relieve us of the need to stay awake to the world: to attend, feel, think, and then act in a deliberate manner. (That is, from freedom rather than compulsion.) If you need to be reminded how completely mental habit blinds us to experience, just take a trip to an unfamiliar country. Suddenly you wake up! And the algorithms of everyday life all but start over, as if from scratch. This is why the various travel metaphors for the psychedelic experience are so apt.

The efficiencies of the adult mind, useful as they are, blind us to the present moment. We're constantly jumping ahead to the next thing. We approach experience much as an artificial intelligence (AI) program does, with our brains continually translating the data of the present into the terms of the past, reaching back in time for the relevant experience, and then using that to make its best guess as to how to predict and navigate the future.

One of the things that commends travel, art, nature, work, and certain drugs to us is the way these experiences, at their best, block every mental path forward and back, immersing us in the flow of a present that is literally wonderful—wonder being the by-product of precisely the kind of unencumbered first sight, or virginal noticing, to which the adult brain has closed itself. (It's so inefficient!) Alas, most of the time I inhabit a near-future tense, my psychic thermostat set to a low simmer of anticipation and, too often, worry. The good thing is I'm seldom surprised. The bad thing is I'm seldom surprised.

What I am struggling to describe here is what I think of as my default mode of consciousness. It works well enough, certainly gets the job done, but what if it isn't the only, or necessarily the best, way to go through life? The premise of psychedelic research is that this special group of molecules can give us access to other modes of consciousness that might offer us specific benefits, whether therapeutic, spiritual, or creative. Psychedelics are certainly not the only door to these other forms of consciousness—and I explore some non-pharmacological alternatives in these pages—but they do seem to be one of the easier knobs to take hold of and turn.

The whole idea of expanding our repertoire of conscious states is not an entirely new idea: Hinduism and Buddhism are steeped in it, and there are intriguing precedents even in Western science. William James, the pioneering American psychologist and author of *The Varieties of Religious Experience*, ventured into these realms more than a century ago. He returned with the conviction that our everyday waking consciousness "is but one special type of consciousness, whilst all about it, parted from it by the filmiest of screens, there lie potential forms of consciousness entirely different."

James is speaking, I realized, of the unopened door in our minds. For him, the "touch" that could throw open the door and disclose these realms on the other side was nitrous oxide. (Mescaline, the psychedelic compound derived from the peyote cactus, was available to researchers at the time, but James was apparently too fearful to try it.)

"No account of the universe in its totality can be final which leaves these other forms of consciousness quite disregarded.

"At any rate," James concluded, these other states, the existence of which he believed was as real as the ink on this page, "forbid a premature closing of our accounts with reality."

The first time I read that sentence, I realized James had my number: as a staunch materialist, and as an adult of a certain age, I had pretty much closed my accounts with reality. Perhaps this had been premature.

Well, here was an invitation to reopen them.

o   o   o

IF EVERYDAY WAKING CONSCIOUSNESS is but one of several possible ways to construct a world, then perhaps there is value in cultivating a greater amount of what I've come to think of as neural diversity. With that in mind, *How to Change Your Mind* approaches its subject from several different perspectives, employing several different narrative modes: social and scientific history; natural history; memoir; science journalism; and case studies of volunteers and patients. In the middle of the journey, I also offer an account of my own firsthand research (or perhaps I should say search) in the form of a kind of mental travelogue.

In telling the story of psychedelic research, past and present, I do not attempt to be comprehensive. The subject of psychedelics, as a matter of both science and social history, is too vast to squeeze between the covers of a single book. Rather than try to introduce readers to the entire cast of characters responsible for the psychedelic renaissance, my narrative follows a small number of pioneers who constitute a particular scientific lineage, with the inevitable result that the contributions of many others have received short shrift. Also in the interest of narrative coherence, I've focused on certain drugs to the exclusion of others. There is, for example, little here about MDMA (also known as Ecstasy), which is showing great promise in the treatment of post-traumatic stress disorder. Some researchers count MDMA among the psychedelics, but most do not, and I follow their lead. MDMA operates through a different set of pathways in the brain and has a substantially different social history from that of the so-called classic psychedelics. Of these, I focus primarily on the ones that are receiving the most attention from scientists—psilocybin and LSD—which means that other psychedelics that are equally interesting and powerful but more difficult to bring into the laboratory—such as ayahuasca—receive less attention.

A final word on nomenclature. The class of molecules to which psilocybin and LSD (and mescaline, DMT, and a handful of others) belong has been called by many names in the decades since they have come to our attention. Initially, they were called hallucinogens. But they do so many other things (and in fact full-blown hallucinations are fairly uncommon) that researchers soon went looking for more precise and comprehensive terms, a quest chronicled in chapter three. The term "psychedelics," which I will mainly use here, does have its downside. Embraced in the 1960s, the term carries a lot of countercultural baggage. Hoping to escape those associations and underscore the spiritual dimensions of these drugs, some researchers have proposed they instead be called "entheogens"—from the Greek for "the divine within." This strikes me as too emphatic. Despite the 1960s trappings, the term "psychedelic," coined in 1956, is etymologically accurate. Drawn from the Greek, it means simply "mind manifesting," which is precisely what these extraordinary molecules hold the power to do.

CHAPTER ONE

A Renaissance

IF THE START of the modern renaissance of psychedelic research can be dated with any precision, one good place to do it would be the year 2006. Not that this was obvious to many people at the time. There was no law passed or regulation lifted or discovery announced to mark the historical shift. But as three unrelated events unfolded during the course of that year—the first in Basel, Switzerland, the second in Washington, D.C., and the third in Baltimore, Maryland— sensitive ears could make out the sound of ice beginning to crack.

The first event, which looked back but also forward like a kind of historical hinge, was the centennial of the birth of Albert Hofmann, the Swiss chemist who, in 1943, accidentally found that he had discovered (five years earlier) the psychoactive molecule that came to be known as LSD. This was an unusual centennial in that the man being feted was very much in attendance. Entering his second century, Hofmann appeared in remarkably good shape, physically spry and mentally sharp, and he was able to take an active part in the festivities, which included a birthday ceremony followed by a three-day symposium. The symposium's opening ceremony was on January 13, two days after Hofmann's 100th birthday (he would live to be 102). Two thousand people packed the hall at the Basel Congress Center, rising to applaud as a stooped stick of a man in a dark suit and a necktie, barely five feet tall, slowly crossed the stage and took his seat.

Two hundred journalists from around the world were in attendance, along with more than a thousand healers, seekers, mystics, psychiatrists, pharmacologists, consciousness researchers, and neuroscientists, most of them people whose lives had been profoundly altered by the remarkable molecule that this man had derived from a fungus half a century before. They had come to celebrate him and what his friend the Swiss poet and physician Walter Vogt called "the only joyous invention of the twentieth century." Among the people in the hall, this did not qualify as hyperbole. According to one of the American scientists in attendance, many had come "to worship" Albert Hofmann, and indeed the event bore many of the hallmarks of a religious observance.

Although virtually every person in that hall knew the story of LSD's discovery by heart, Hofmann was asked to recite the creation myth one more time. (He tells the story, memorably, in his 1979 memoir, *LSD, My Problem Child*.) As a young chemist working in a unit of Sandoz Laboratories charged with isolating the compounds in medicinal plants to find new drugs, Hofmann had been tasked with synthesizing, one by one, the molecules in the alkaloids produced by ergot. Ergot is a fungus that can infect grain, often rye, occasionally causing those who consume bread made from it to appear mad or possessed. (One theory of the Salem witch trials blames ergot poisoning for the behavior of the women accused.) But midwives had long used ergot to induce labor and stanch bleeding postpartum, so

Sandoz was hoping to isolate a marketable drug from the fungus's alkaloids. In the fall of 1938, Hofmann made the twenty-fifth molecule in this series, naming it lysergic acid diethylamide, or LSD-25 for short. Preliminary testing of the compound on animals did not show much promise (they became restless, but that was about it), so the formula for LSD-25 was put on the shelf.

And there it remained for five years, until one April day in 1943, in the middle of the war, when Hofmann had "a peculiar presentiment" that LSD-25 deserved a second look. Here his account takes a slightly mystical turn. Normally, when a compound showing no promise was discarded, he explained, it was discarded for good. But Hofmann "liked the chemical structure of the LSD molecule," and something about it told him that "this substance could possess properties other than those established in the first investigations." Another mysterious anomaly occurred when he synthesized LSD-25 for the second time. Despite the meticulous precautions he always took when working with a substance as toxic as ergot, Hofmann must somehow have absorbed a bit of the chemical through his skin, because he "was interrupted in my work by unusual sensations."

Hofmann went home, lay down on a couch, and "in a dreamlike state, with eyes closed . . . I perceived an uninterrupted stream of fantastic pictures, extraordinary shapes with intense, kaleidoscopic play of colors." Thus unfolds the world's first LSD trip, in neutral Switzerland during the darkest days of World War II. It is also the only LSD trip ever taken that was entirely innocent of expectation.

Intrigued, Hofmann decided a few days later to conduct an experiment on himself—not an uncommon practice at the time. Proceeding with what he thought was extreme caution, he ingested 0.25 milligrams—a milligram is one-thousandth of a gram—of LSD dissolved in a glass of water. This would represent a minuscule dose of any other drug, but LSD, it turns out, is one of the most potent psychoactive compounds ever discovered, active at doses measured in micrograms—that is, one thousandth of a milligram. This surprising fact would soon inspire scientists to look for, and eventually find, the brain receptors and the endogenous chemical—serotonin—that activates them like a key in a lock, as a way to explain how such a small number of molecules could have such a profound effect on the mind. In this and other ways, Hofmann's discovery helped to launch modern brain science in the 1950s.

Now unfolds the world's first *bad* acid trip as Hofmann is plunged into what he is certain is irretrievable madness. He tells his lab assistant he needs to get home, and with the use of automobiles restricted during wartime, he somehow manages to pedal home by bicycle and lie down while his assistant summons the doctor. (Today LSD devotees celebrate "Bicycle Day" each year on April 19.) Hofmann describes how "familiar objects and pieces of furniture assumed grotesque, threatening forms. They were in continuous motion, animated as if driven by an inner restlessness." He experienced the disintegration of the outer world and the dissolution of his own ego. "A demon had invaded me, had taken possession of my body, mind, and soul. I jumped up and screamed, trying to free myself from him, but then sank down again and lay helpless on the sofa." Hofmann became convinced he was going to be rendered permanently insane or might actually be dying. "My ego was suspended somewhere in space and I saw my body lying dead on the sofa." When the doctor arrived and examined him, however, he found that all of Hofmann's vital signs—heartbeat, blood pressure, breathing—were perfectly normal. The only indication something was amiss were his pupils, which were dilated in the extreme.

Once the acute effects wore off, Hofmann felt the "afterglow" that frequently follows a psychedelic experience, the exact opposite of a hangover. When he walked out into his garden after a spring rain, "everything glistened and sparkled in a fresh light. The world was as if newly created." We've since learned that the experience of psychedelics is powerfully influenced by one's expectation; no other class of drugs are more suggestible in their effects. Because Hofmann's experiences with LSD are the only ones we have that are uncontaminated by previous accounts, it's interesting to note they exhibit neither the Eastern nor the Christian flavorings that would soon become conventions of the genre. However, his experience of familiar objects coming to life and the world "as if newly created"—the same rapturous Adamic moment that Aldous Huxley would describe so vividly a decade later in *The Doors of Perception*—would become commonplaces of the psychedelic experience.

Hofmann came back from his trip convinced, first, that LSD had somehow found him rather than the other way around and, second, that LSD would someday be of great value to medicine and especially psychiatry, possibly by offering researchers a model of schizophrenia. It never occurred to him that his "problem child," as he eventually would regard LSD, would also become a "pleasure drug" and a drug of abuse.

Yet Hofmann also came to regard the youth culture's adoption of LSD in the 1960s as an understandable response to the emptiness of what he described as a materialist, industrialized, and spiritually impoverished society that had lost its connection to nature. This master of chemistry—perhaps the most materialist of all disciplines—emerged from his experience with LSD-25 convinced the molecule offered civilization not only a potential therapeutic but also a spiritual balm—by opening a crack "in the edifice of materialist rationality." (In the words of his friend and translator, Jonathan Ott.)

Like so many who followed after him, the brilliant chemist became something of a mystic, preaching a gospel of spiritual renewal and reconnection with nature. Presented with a bouquet of roses that 2006 day in Basel, the scientist told the assembled that "the feel-ing of co-creatureliness with all things alive should enter our con-sciousness more fully and counterbalance the materialistic and nonsensical technological developments in order to enable us to re-turn to the roses, to the flowers, to nature, where we belong." The audience erupted in applause.

A skeptical witness to the event would not be entirely wrong to regard the little man on the stage as the founder of a new religion and the audience as his congregation. But if this is a religion, it's one with a significant difference. Typically, only the founder of a religion and perhaps a few early acolytes can lay claim to the kind of author-ity that flows from a direct experience of the sacred. For everyone coming after, there is the comparatively thin gruel of the stories, the symbolism of the sacrament, and faith. History attenuates the origi-nal power of it all, which now must be mediated by the priests. But the extraordinary promise on offer in the Church of Psychedelics is that anyone at any time may gain access to the primary religious experience by means of the sacrament, which happens to be a psy-choactive molecule. Faith is rendered superfluous.

Running alongside the celebration's spiritual undercurrent, how-ever, there also, perhaps somewhat incongruously, came science. During the weekend symposium following the observation of Hofmann's birthday, researchers from a variety of disciplines—including neuroscience, psychiatry, pharmacology, and conscious-ness studies, as well as the arts—explored the impact of Hofmann's invention on society and culture and its potential for expanding our understanding of consciousness and treating several intractable men-tal disorders. A handful of research projects, studying the effects of psychedelics on humans, had been approved or were under way in Switzerland and the United States, and scientists at the sympo-sium voiced their hope that the long hiatus in psychedelic research might finally be coming to an end. Irrational exuberance seems to be an occupational hazard among people working in this area, but in 2006 there was good reason to think the weather might actually be turning.

o   o   o

THE SECOND WATERSHED EVENT of 2006 came only five weeks later when the U.S. Supreme Court, in a unanimous decision written by the new chief justice, John G. Roberts Jr., ruled that the UDV, a tiny religious sect that uses a hallucinogenic tea called ayahuasca as its sacrament, could import the drink to the United States, even though it contains the schedule I substance dimethyltryptamine, or DMT. The ruling was based on the Religious Freedom Restoration Act of 1993, which had sought to reinstitute the right (under the First Amendment's religious freedom clause) of Native Americans to use peyote in their ceremonies, as they have done for generations. The 1993 law says that only if the government has a "compelling interest" can it interfere with one's practice of religion. In the UDV case, the Bush administration had argued that only Native Americans, because of their "unique relationship" to the government, had the right to use psychedelics as part of their worship, and even in their case this right could be abridged by the state.

The Court soundly rejected the government's argument, interpreting the 1993 law to mean that, absent a compelling state interest, the federal government cannot prohibit a recognized religious group from using psychedelic substances in their observances. Evidently, this includes relatively new and tiny religious groups specifically organized around a psychedelic sacrament, or "plant medicine," as the *ayahuasqueros* call their tea. The UDV is a Christian spiritist sect founded in 1961 in Brazil by José Gabriel da Costa, a rubber tapper inspired by revelations he experienced after receiving ayahuasca from an Amazonian shaman two years before. The church claims 17,000 members in six countries, but at the time of the ruling there were only 130 American members of the UDV. (The initials stand for União do Vegetal, or Union of the Plants, because ayahuasca is made by brewing together two Amazonian plant species, *Banisteriopsis caapi* and *Psychotria viridis*.)

The Court's decision inspired something of a religious awakening around ayahuasca in America. Today there are close to 525 American members of the church, with communities in nine locations. To supply them, the UDV has begun growing the plants needed to make the tea in Hawaii and shipping it to groups on the mainland without federal interference. But the number of Americans participating in ayahuasca ceremonies outside the UDV has also mushroomed in the years since, and any given night there are probably dozens if not hundreds of ceremonies taking place somewhere in America (with concentrations in the San Francisco Bay Area and Brooklyn). Federal prosecutions for possession or importation of ayahuasca appear to have stopped, at least for the time being.

With its 2006 decision, the Supreme Court seems to have opened up a religious path—narrow, perhaps, but based on the Bill of Rights—to the federal recognition of psychedelic drugs, at least when they're being used as a sacrament by a group deemed a religion by the government. It remains to be seen how wide or well trod that path will become, but it does make you wonder what the government, and the Court, will do when an American José Gabriel da Costa steps forward and attempts to turn his or her own psychedelic revelations into a new religion intent on using a psychoactive chemical as its sacrament. The jurisprudence of "cognitive liberty," as some in the psychedelic community call it, is still scant and limited (to religion), but now it had been affirmed, opening a new crack in the edifice of the drug war.

o o o

OF THE THREE 2006 EVENTS that helped bring psychedelics out
of their decades-long slumber, by far the most far-reaching in its impact was the publication that summer of the paper in *Psychopharmacology* described in the prologue—the one Bob Jesse e-mailed me at
the time but that I didn't bother to open. This event, too, had a distinctly spiritual cast, even though the experiment it reported was the
work of a rigorous and highly regarded scientist: Roland Griffiths. It
just so happens that Griffiths, a most unlikely psychedelic researcher,
was inspired to investigate the power of psilocybin to occasion a
"mystical-type" experience by a mystical experience of his own.

Griffiths's landmark paper, "Psilocybin Can Occasion Mystical-
Type Experiences Having Substantial and Sustained Personal Meaning and Spiritual Significance," was the first rigorously designed,
double-blind, placebo-controlled clinical study in more than four
decades—if not ever—to examine the psychological effects of a psychedelic. It received a small torrent of press coverage, most of it so
enthusiastic as to make you wonder if the moral panic around psychedelics that took hold in the late 1960s might finally have run its
course. No doubt the positive tenor of the coverage owed much to
the fact that, at Griffiths's urging, the journal had invited several of
the world's most prominent drug researchers—some of them decorated soldiers in the drug war—to comment on the study, giving the
journalists covering the study plenty of ideological cover.

All of the commentators treated the publication as a major event.
Herbert D. Kleber, a former deputy to William Bennett, George H. W.
Bush's drug czar, and later director of the Division on Substance
Abuse at Columbia University, applauded the paper for its method-
ological rigor and acknowledged there might be "major therapeutic

possibilities" in psychedelic research "merit[ing] NIH support." Charles "Bob" Schuster, who had served two Republican presidents as director of the National Institute on Drug Abuse (NIDA), noted that the term "psychedelic" implies a mind-expanding experience and expressed his "hope that this landmark paper will also be 'field expanding.'" He suggested that this "fascinating" class of drugs, and the spiritual experience they occasion, might prove useful in treating addiction.

Griffiths's paper and its reception served to reinforce an important distinction between the so-called classic psychedelics—psilocybin, LSD, DMT, and mescaline—and the more common drugs of abuse, with their demonstrated toxicity and potential for addiction. The American drug research establishment, such as it is, had signaled in the pages of one of its leading journals that these psychedelic drugs deserved to be treated very differently and had demonstrated, in the words of one commentator, "that, when used appropriately, these compounds can produce remarkable, possibly beneficial, effects that certainly deserve further study."

The story of how this paper came to be sheds an interesting light on the fraught relationship between science and that other realm of human inquiry that science has historically disdained and generally wants nothing to do with: spirituality. For in designing this, the first modern study of psilocybin, Griffiths had decided to focus not on a potential therapeutic application of the drug—the path taken by other researchers hoping to rehabilitate other banned substances, like MDMA—but rather on the spiritual effects of the experience on so-called healthy normals. What good was *that*?

In an editorial accompanying Griffiths's paper, the University of Chicago psychiatrist and drug abuse expert Harriet de Wit tried to address this tension, pointing out that the quest for experiences that "free oneself of the bounds of everyday perception and thought in a search for universal truths and enlightenment" is an abiding element of our humanity that has nevertheless "enjoyed little credibility in the mainstream scientific world." The time had come, she suggested, for science "to recognize these extraordinary subjective experiences . . . even if they sometimes involve claims about ultimate realities that lie outside the purview of science."

o   o   o

ROLAND GRIFFITHS might be the last scientist one would ever imagine getting mixed up with psychedelics, which surely helps explain his success in returning psychedelic research to scientific respectability. Six feet tall and rail thin, Griffiths, in his seventies, holds himself bolt upright; the only undisciplined thing about him is a thatch of white hair so dense it appears to have held his comb to a draw. At least until you get him talking about the ultimate questions, which light him up, he comes across as the ultimate straight arrow: sober, earnest, and methodical.

Born in 1946, Griffiths grew up in El Cerrito, California, in the Bay Area, and went to Occidental College for his undergraduate education (majoring in psychology) and then on to the University of Minnesota to study psychopharmacology. At Minnesota in the late 1960s, he came under the influence of B. F. Skinner, the radical behaviorist who helped shift the focus of psychology from the exploration of inner states and subjective experience to the study of outward behavior and how it is conditioned. Behaviorism has little interest in plumbing the depths of the human psyche, but the approach proved very useful in studying behaviors like drug use and dependence, which became Griffiths's specialty. Psychedelic drugs played no role in either his formal or his informal education. By the time Griffiths got to graduate school, Timothy Leary's notorious psychedelic research project at Harvard had already collapsed in scandal, and "it was clear from my mentors that these were compounds that had no future."

In 1972, right out of graduate school, Griffiths was hired at Johns Hopkins, where he has worked ever since, making his mark as a researcher studying the mechanisms of dependence in a variety of legal and illegal drugs, including the opiates, the so-called sedative hypnotics (like Valium), nicotine, alcohol, and caffeine. Working under grants from the National Institute on Drug Abuse, Griffiths helped pioneer the sorts of experiments in which an animal, often a baboon or a rat, is presented with a lever allowing it to self-administer various drugs intravenously, a powerful tool for researchers studying reinforcement, dependence, preferences (*lunch or more cocaine?*), and withdrawal. The fifty-five papers he published exploring the addictive properties of caffeine transformed the field, helping us to see coffee less as a food than as a drug, and led to the listing of "caffeine withdrawal" syndrome in the most recent edition of the *Diagnostic and Statistical Manual of Mental Disorders*, or *DSM 5*. By the time Griffiths turned fifty, in 1994, he was a scientist at the top of his game and his field.

But that year Griffiths's career took an unexpected turn, the result of two serendipitous introductions. The first came when a friend introduced him to Siddha Yoga. Despite his behaviorist orientation as a scientist, Griffiths had always been interested in what philosophers call phenomenology—the subjective experience of consciousness. He had tried meditation as a graduate student but found that "he couldn't sit still without going stark-raving mad. Three minutes felt like three hours." But when he tried it again in 1994, "something opened up for me." He started meditating regularly, going on retreats, and working his way through a variety of Eastern spiritual traditions. He found himself drawn "deeper and deeper into this mystery."

Somewhere along the way, Griffiths had what he modestly describes as "a funny kind of awakening"—a mystical experience. I was
surprised when Griffiths mentioned this during our first meeting in
his office, so I hadn't followed up, but even after I had gotten to
know him a little better, Griffiths was still reluctant to say much
more about exactly what happened and, as someone who had never
had such an experience, I had trouble gaining any traction with the
idea whatsoever. All he would tell me is that the experience, which
took place in his meditation practice, acquainted him with "something way, way beyond a material worldview that I can't really talk to
my colleagues about, because it involves metaphors or assumptions
that I'm really uncomfortable with as a scientist."

In time, what he was learning about "the mystery of consciousness and existence" in his meditation practice came to seem more
compelling to him than his science. He began to feel somewhat
alienated: "None of the people I was close to had any interest in en-
tertaining those questions, which fell into the general category of
the spiritual, and religious people I just didn't get.

"Here I am, a full professor, publishing like crazy, running off to
important meetings, and thinking I was a fraud." He began to lose
interest in the research that had organized his whole adult life. "I
could study a new sedative hypnotic, learn something new about
brain receptors, be on another FDA [Food and Drug Administra-
tion] panel, go to another conference, but so what? I was more emotionally and intellectually curious about where this other path might
lead. My drug research began to seem vacuous. I was going through
the motions at work, much more interested in going home in the
evening to meditate." The only way he could motivate himself to
continue writing grants was to think of it as a "service project" for
his graduate students and postdocs.

In the case of his caffeine research, Griffiths had been able to take his curiosity about a dimension of his own experience—why did he feel compelled to drink coffee every day?—and turn it into a productive line of scientific inquiry. But he could see no way to do that with his deepening curiosity about the dimensions of consciousness that meditation had opened up to him. "It never occurred to me there was any way to study it scientifically." Stymied and bored, Griffiths began to entertain thoughts of quitting science and going off to an ashram in India.

It was around this time that Bob Schuster, an old friend and colleague who had recently retired as head of the National Institute on Drug Abuse, phoned Griffiths to suggest he talk to a young man he had recently met at Esalen named Bob Jesse. Jesse had organized a small gathering of researchers, therapists, and religious scholars at the legendary Big Sur retreat center to discuss the spiritual and therapeutic potential of psychedelic drugs and how they might be rehabilitated. Jesse himself was neither a medical professional nor a scientist; he was a computer engineer, a vice president of business development at Oracle, who had made it his mission to revive the science of psychedelics—but as a tool not so much of medicine as of spiritual development.

Griffiths had told Schuster a little about his spiritual practice and confided in him his growing discontent with conventional drug research.

"You should talk to this guy," Schuster told him. "They have some interesting ideas about working with entheogens," he said. "You might have something in common."

o o o

WHEN THE HISTORY of second-wave psychedelic research is written, Bob Jesse will be seen as one of a pair of scientific outsiders in America—amateurs, really, and brilliant eccentrics—who worked tirelessly, often behind the scenes, to get it off the ground. Both found their vocation in the wake of transformative psychedelic experiences that convinced them these substances had the potential to heal not only individuals but humankind as a whole and that the best path to their rehabilitation was by way of credible scientific research. In many cases, these untrained researchers dreamed up the experiments first and then found (and funded) the scientists to conduct them. Often you will find their names on the papers, usually in the last position.

Of the two, Rick Doblin has been at it longer and is by far the more well known. Doblin founded the Multidisciplinary Association for Psychedelic Studies (MAPS) all the way back in the dark days of 1986—the year after MDMA was made illegal and a time when most wiser heads were convinced that restarting research into psychedelics was a cause beyond hopeless.

Doblin, born in 1953, is a great shaggy dog with a bone; he has been lobbying to change the government's mind about psychedelics since shortly after graduating from New College, in Florida, in 1987. After experimenting with LSD as an undergraduate, and later with MDMA, Doblin decided his calling in life was to become a psychedelic therapist. But after the banning of MDMA in 1985, that dream became unachievable without a change in federal laws and regulations, so he decided he'd better first get a doctorate in public policy at Harvard's Kennedy School. There, he mastered the intricacies of the FDA's drug approval process, and in his dissertation plotted the laborious path to official acceptance that psilocybin and MDMA are now following.

Doblin is disarmingly, perhaps helplessly, candid, happy to talk openly to a reporter about his formative psychedelic experiences as well as political strategy and tactics. Like Timothy Leary, Doblin is the happiest of warriors, never not smiling and exhibiting a degree of enthusiasm for the work you wouldn't expect from a man who has been knocking his head against the same wall for his entire adult life. Doblin works out of a somewhat Dickensian office tucked into the attic of his rambling colonial in Belmont, Massachusetts, at a desk stacked to the ceiling with precarious piles of manuscripts, journal articles, photographs, and memorabilia reaching back more than forty years. Some of the memorabilia commemorates the time early in his career when Doblin decided the best way to end sectarian strife would be to mail a group of the world's spiritual leaders tablets of MDMA, a drug famous for its ability to break down barriers between people and kindle empathy. Around the same time, he arranged to have a thousand doses of MDMA sent to people in the Soviet military who were working on arms control negotiations with President Reagan.

For Doblin, winning FDA approval for the medical use of psychedelics—which he believes is now in view, for both MDMA and psilocybin—is a means to a more ambitious and still more controversial end: the incorporation of psychedelics into American society and culture, not just medicine. This of course is the same winning strategy followed by the campaign to decriminalize marijuana, in which promoting the medical uses of cannabis changed the drug's image, leading to a more general public acceptance.

Not surprisingly, this sort of talk rankles more cautious heads in the community (Bob Jesse among them), but Rick Doblin is not one to soft-pedal his agenda or to even *think* about taking an interview off the record. This gets him a lot of press; how much it helps the cause is debatable. But there is no question that especially in the last several years Doblin has succeeded in getting important research approved and funded, especially in the case of MDMA, which has long been MAPS's main focus. MAPS has sponsored several small clinical trials that have demonstrated MDMA's value in treating post-traumatic stress disorder, or PTSD. (Doblin defines psychedelics generously, so as to include MDMA and even cannabis, even though their mechanisms of action in the brain are very different from that of the classic psychedelics.) But beyond helping those suffering with PTSD and other indications—MAPS is sponsoring a clinical study at Harbor-UCLA Medical Center that involves treating social anxiety in autistic adults with MDMA-assisted psychotherapy— Doblin believes fervently in the power of psychedelics to improve humankind by disclosing a spiritual dimension of consciousness we all share, regardless of our religious beliefs or lack thereof. "Mysticism," he likes to say, "is the antidote to fundamentalism."

o   o   o

COMPARED WITH RICK DOBLIN, Bob Jesse is a monk. There is nothing shaggy or uncareful about him. Taut, press shy, and disposed to choose his words with a pair of tweezers, Jesse, now in his fifties, prefers to do his work out of public view, and preferably from the one-room cabin where he lives by himself in the rugged hills north of San Francisco, off the grid except for a fast Internet connection.

"Bob Jesse is like the puppeteer," Katherine MacLean told me. MacLean is a psychologist who worked in Roland Griffiths's lab from 2009 until 2013. "He's the visionary guy working behind the scenes."

Following Jesse's meticulous directions, I drove north from the Bay Area, eventually winding up at the end of a narrow dirt road in a county he asked me not to name. I parked at a trailhead and made my way past the "No Trespassing" signs, following a path up a hill that brought me to his picturesque mountaintop camp. I felt as if I were going to visit the wizard. The shipshape little cabin is tight for two, so Jesse has set out among the fir trees and boulders some comfortable sofas, chairs, and tables. He's also built an outdoor kitchen and, on a shelf of rock commanding a spectacular view of the mountains, an outdoor shower, giving the camp the feeling of a house turned inside out.

We spent the better part of an early spring day outdoors in his living room, sipping herbal tea and discussing his notably quieter campaign to restore psychedelics to respectability—a master plan in which Roland Griffiths plays a central role. "I'm a little camera shy," he began, "so please, no pictures or recordings of any kind."

Jesse is a slender, compact fellow with a squarish head of closely cropped gray hair and rimless rectangular glasses that are unostentatiously stylish. Jesse seldom smiles and has some of the stiffness I associate with engineers, though occasionally he'll surprise you with a flash of emotion he will immediately then caption: "You may have noticed that thinking about that subject made my eyes get a little watery. Let me explain why . . ." Not only does he choose his own words with great care, but he insists that you do too, so, for example, when I carelessly deployed the term "recreational use," he stopped me in mid-sentence. "Maybe we need to reexamine that term. Typically, it is used to trivialize an experience. But why? In its literal meaning, the word 'recreation' implies something decidedly nontrivial. There is much more to be said, but let's bookmark this topic for another time. Please go on." My notes show that Jesse took our first conversation on and off the record half a dozen times.

Jesse grew up outside Baltimore and went to Johns Hopkins, where he studied computer science and electrical engineering. For several years in his twenties, he worked for Bell Labs, commuting weekly from Baltimore to New Jersey. During this period, he came out of the closet and persuaded management to recognize the company's first gay and lesbian employee group. (At the time, AT&T, the parent company, employed some 300,000 people.) Later, he persuaded AT&T management to fly a rainbow flag over headquarters during Gay Pride Week and send a delegation to march in the parade. This achievement formed Bob Jesse's political education, impressing on him the value of working behind the scenes without making a lot of noise or demanding credit.

Jesse moved to Oracle, and the Bay Area, in 1990, becoming employee number 8766—not one of the first, but early enough to have acquired a chunk of stock in the company. It wasn't long before Oracle fielded its own contingent in San Francisco's Gay Pride Parade, and after Jesse's gentle prodding of senior management Oracle became one of the first Fortune 500 companies to offer benefits to the same-sex partners of its employees.

Jesse's curiosity about psychedelics was first piqued during a drug education unit in his junior high school science class. This particular class of drugs was neither physically nor psychologically addictive, he was told (correctly); his teacher went on to describe the drugs' effects, including shifts in consciousness and visual perception that Jesse found intriguing. "I could sense there was even more here than they were telling us," he recalled. "So I made a mental note." But he would not be ready to see for himself what psychedelics were all about until much later. Why? He answered in the third person: "A closeted gay kid might be afraid of what might come out if he let his guard down."

In his twenties, while working at Bell Labs, Jesse fell in with a group of friends in Baltimore who decided, in a most deliberate way, to experiment with psychedelics. Someone would always remain "close to ground level" in case anyone needed help or the doorbell rang, and doses escalated gradually. It was during one of these Saturday afternoon experiments, in an apartment in Baltimore, that Jesse, twenty-five years old and having ingested a high dose of LSD, had a powerful "non-dual experience" that would prove transformative. I asked him to describe it, and after some hemming and hawing—"I hope you'll bracket what is sensitive"—he gingerly proceeded to tell the story.

"I was lying on my back underneath a ficus tree," he recalls. "I knew it was going to be a strong experience. And the point came where the little I still was just started slipping away. I lost all awareness of being on the floor in an apartment in Baltimore; I couldn't tell if my eyes were opened or closed. What opened up before me was, for lack of a better word, a space, but not our ordinary concept of space, just the pure awareness of a realm without form and void of content. And into that realm came a celestial entity, which was the emergence of the physical world. It was like the big bang, but without the boom or the blinding light. It was the birth of the physical universe. In one sense it was dramatic—maybe the most important thing that ever occurred in the history of the world—yet it just sort of happened."

I asked him where he was in all this.

"I was a diffusely located observer. I was coextensive with this emergence." Here I let him know he was losing me. Long pause. "I'm hesitating because the words are an awkward fit; words seem too constraining." Ineffability is of course a hallmark of the mystical experience. "The awareness transcends any particular sensory modality," he explained, unhelpfully. Was it scary? "There was no terror, only fascination and awe." Pause. "Um, maybe a little fear."

From here on, Jesse watched (or whatever you call it) the birth of . . . everything, in the unfolding of an epic sequence beginning with the appearance of cosmic dust leading to the creation of the stars and then the solar systems, followed by the emergence of life and from there the arrival of "what we call humans," then the acquisition of language and the unfolding of awareness, "all the way up to one's self, here in this room, surrounded by my friends. I had come all the way back to right where I was. How much clock time had elapsed? I had no idea.

"What stands out most for me is the quality of the awareness I experienced, something entirely distinct from what I've come to regard as Bob. How does this expanded awareness fit into the scope of things? To the extent I regard the experience as veridical—and about that I'm still not sure—it tells me that consciousness is primary to the physical universe. In fact, it precedes it." Did he now believe consciousness exists outside the brain? He's not certain. "But to go from being very sure that the opposite is true"—that consciousness is the product of our gray matter—"to be unsure is an immense shift." I asked him if he agreed with something I'd read the Dalai Lama had said, that the idea that brains create consciousness—an idea accepted without question by most scientists—"is a metaphysical assumption, not a scientific fact."

"Bingo," Jesse said. "And for someone with my orientation"—agnostic, enamored of science—"that changes everything."

o   o   o

HERE'S WHAT I DON'T GET about an experience like Bob Jesse's: Why in the world would you ever credit it at all? I didn't understand why you wouldn't simply file it under "interesting dream" or "drug-induced fantasy." But along with the feeling of ineffability, the conviction that some profound objective truth has been disclosed to you is a hallmark of the mystical experience, regardless of whether it has been occasioned by a drug, meditation, fasting, flagellation, or sensory deprivation. William James gave a name to this conviction: the noetic quality. People feel they have been let in on a deep secret of the universe, and they cannot be shaken from that conviction. As James wrote, "Dreams cannot stand this test." No doubt this is why some of the people who have such an experience go on to found religions, changing the course of history or, in a great many more cases, the course of their own lives. "No doubt" is the key.

I can think of a couple of ways to account for such a phenomenon, neither entirely satisfying. The most straightforward and yet hardest to accept explanation is that it's simply true: the altered state of consciousness has opened the person up to a truth that the rest of us, imprisoned in ordinary waking consciousness, simply cannot see. Science has trouble with this interpretation, however, because, whatever the perception is, it can't be verified by its customary tools. It's an anecdotal report, in effect, and so has no value. Science has little interest in, and tolerance for, the testimony of the individual; in this it is, curiously, much like an organized religion, which has a big problem crediting direct revelation too. But it's worth pointing out that there are cases where science has no choice but to rely on individual testimony—as in the study of subjective consciousness, which is inaccessible to our scientific tools and so can only be described by the person experiencing it. Here phenomenology is the all-important data. However, this is not the case when ascertaining truths about the world *outside* our heads.

The problem with crediting mystical experiences is precisely that they often seem to erase the distinction between inside and outside, in the way that Bob Jesse's "diffuse awareness" seemed to be his but also to exist outside him. This points to the second possible explanation for the noetic sense: when our sense of a subjective "I" disintegrates, as it often does in a high-dose psychedelic experience (as well as in meditation by experienced meditators), it becomes impossible to distinguish between what is subjectively and objectively true. What's left to do the doubting if not your I?

o o o

IN THE YEARS following that first powerful psychedelic journey, Bob Jesse had a series of other experiences that shifted the course of his life. Living in San Francisco in the early 1990s, he got involved in the rave scene and discovered that the "collective effervescence" of the best all-night dance parties, with or without psychedelic "materials," could also dissolve the "subject-object duality" and open up new spiritual vistas. He began to explore various spiritual traditions, from Buddhism to Quakerism to meditation, and found his priorities in life gradually shifting. "It began to occur to me that spending time in this area might actually be far more important and far more fulfilling than what I had been doing" as a computer engineer.

While on a sabbatical from Oracle (he would leave for good in 1995), Jesse set up a nonprofit called the Council on Spiritual Practices (CSP), with the aim of "making direct experience of the sacred more available to more people." The website downplays the organization's interest in promoting entheogens—Bob Jesse's preferred term for psychedelics—but does describe its mission in suggestive terms: "to identify and develop approaches to primary religious experience that can be used safely and effectively." The website (csp.org) offers an excellent bibliography of psychedelic research and regular updates on the work under way at Johns Hopkins. CSP would also play a role in supporting the UDV lawsuit that resulted in the 2006 Supreme Court decision.

The Council on Spiritual Practices grew out of Jesse's systematic exploration of the psychedelic literature and the psychedelic community in the Bay Area soon after he moved to San Francisco. In his highly deliberate, slightly obsessive, and scrupulously polite way, Jesse contacted the region's numerous "psychedelic elders"—the rich cast of characters who had been deeply involved in research and therapy in the years before most of the drugs were banned in 1970, with the passing of the Controlled Substances Act, and the classification of LSD and psilocybin as schedule I substances with a high potential for abuse and no recognized medical use. There was James Fadiman, the Stanford-trained psychologist who had done pioneering research on psychedelics and problem solving at the International Foundation for Advanced Study in Menlo Park, until the FDA halted the group's work in 1966. (In the early 1960s, there was at least as much psychedelic research going on around Stanford as there was at Harvard; it just didn't have a character of the wattage of a Timothy Leary out talking about it.) Then there was Fadiman's colleague at the institute Myron Stolaroff, a prominent Silicon Valley electrical engineer who worked as a senior executive at Ampex, the magnetic recording equipment maker, until an LSD trip inspired him to give up engineering (much like Bob Jesse) for a career as a psychedelic researcher and therapist. Jesse also found his way into the inner circle of Sasha and Ann Shulgin, legendary Bay Area figures who held weekly dinners for a community of therapists, scientists, and others interested in psychedelics. (Sasha Shulgin, who died in 2014, was a brilliant chemist who held a DEA license allowing him to synthesize novel psychedelic compounds, which he did in prodigious numbers. He also was the first to synthesize MDMA since it had been patented by Merck in 1912 and forgotten. Recognizing its psychoactive properties, he introduced the so-called empathogen to the Bay Area's psychotherapy community. Only later, did it become the club drug known as Ecstasy.) Jesse also befriended Huston Smith, the scholar of comparative religion, whose mind had been opened to the spiritual potential of psychedelics when, as an instructor/lecturer at MIT in 1962, he served as a volunteer in the Good Friday Experiment, from which he came away convinced that a mystical experience occasioned by a drug was no different from any other kind.

By way of these "elders" and his own reading, Jesse began unearthing the rich body of first-wave psychedelic research, much of which had been lost to science. He learned that there had been more than a thousand scientific papers on psychedelic drug therapy before 1965, involving more than forty thousand research subjects. Beginning in the 1950s and continuing into the early 1970s, psychedelic compounds had been used to treat a variety of conditions—including alcoholism, depression, obsessive-compulsive disorder, and anxiety at the end of life—frequently with impressive results. But few of the studies were well controlled by modern standards, and some of them were compromised by the enthusiasm of the researchers involved.

Of even keener interest to Bob Jesse was the early research exploring the potential of psychedelics to contribute to what, in a striking phrase, he calls "the betterment of well people." There had been studies in "healthy normals" of artistic and scientific creativity and spirituality. The most famous of these was the Good Friday, or Marsh Chapel, Experiment, conducted in 1962 by Walter Pahnke, a psychiatrist and minister working on a PhD dissertation at Harvard under Timothy Leary. In this double-blind experiment, twenty divinity students received a capsule of white powder during a Good Friday service at Marsh Chapel on the Boston University campus, ten of them containing psilocybin, ten an "active placebo"—in this case niacin, which creates a tingling sensation. Eight of the ten students receiving psilocybin reported a powerful mystical experience, while only one in the control group did. (Telling them apart was not difficult, rendering the double blind a somewhat hollow conceit: those on the placebo sat sedately in their pews while the others lay down or wandered about the chapel, muttering things like "God is everywhere" and "Oh, the Glory!") Pahnke concluded that the experiences of those who received the psilocybin were "indistinguishable from, if not identical with," the classic mystical experiences reported in the literature. Huston Smith agreed. "Until the Good Friday Experiment," he told an interviewer in 1996, "I had had no direct personal encounter with God."

In 1986, Rick Doblin conducted a follow-up study of the Good Friday Experiment in which he tracked down and interviewed all but one of the divinity students who received psilocybin at Marsh Chapel.

Most reported that the experience had reshaped their lives and work in profound and enduring ways. However, Doblin found serious flaws in Pahnke's published account: Pahnke had failed to mention that several subjects had struggled with acute anxiety during their experience. One had to be restrained and given an injection of Thorazine, a powerful antipsychotic, after he fled from the chapel and headed down Commonwealth Avenue, convinced he had been chosen to announce the news of the coming of the Messiah.

In this and a second review of another Timothy Leary–supervised experiment, of recidivism at Concord State Prison, Doblin had raised troubling questions about the quality of the research done in the Harvard Psilocybin Project, suggesting that the enthusiasm of the experimenters had tainted the reported results. If this research were going to be revived and taken seriously, Jesse concluded, it would have to be done with considerably more rigor and objectivity. And yet the results of the Good Friday Experiment were highly sugges-tive and, as Bob Jesse and Roland Griffiths would soon decide, well worth trying to reproduce.

○   ○   ○

BOB JESSE SPENT the early 1990s excavating the knowledge about psychedelics that had been lost when formal research was halted and informal research went underground. In this, he was a little like those Renaissance scholars who rediscovered the lost world of classi-cal thought in a handful of manuscripts squirreled away in monaster-ies. However, in this case, considerably less time had elapsed, so the knowledge remained in the brains of people still alive, like James Fadiman and Myron Stolaroff and Willis Harman (another Bay Area engineer turned psychedelic researcher), who merely had to be asked for it, and in scientific papers in libraries and databases, which merely had to be searched. But if there is a modern analogy to the medieval monastery where the world of classical thought was saved from oblivion, a place where the guttering flame of psychedelic knowledge was assiduously fanned during its own dark age, that place would have to be Esalen, the legendary retreat center in Big Sur, California.

Perched on a cliff overlooking the Pacific as if barely clinging to the continent, the Esalen Institute was founded in 1962 and ever since has been a center of gravity for the so-called human potential movement in America, serving as the unofficial capital of the New Age. A great many therapeutic and spiritual modalities were developed and taught here over the years, including the therapeutic and spiritual potential of psychedelics. Beginning in 1973, Stanislav Grof, the Czech émigré psychiatrist who is one of the pioneers of LSD-assisted psychotherapy, served as scholar in residence at Esalen, but he had conducted workshops there for years before. Grof, who has guided thousands of LSD sessions, once predicted that psychedelics "would be for psychiatry what the microscope is for biology or the telescope is for astronomy. These tools make it possible to study important processes that under normal circumstances are not available for direct observation." Hundreds came to Esalen to peer through that microscope, often in workshops Grof led for psychotherapists who wanted to incorporate psychedelics in their practices. Many if not most of the therapists and guides now doing this work underground learned their craft at the feet of Stan Grof in the Big House at Esalen.

Whether such work continued at Esalen after LSD was made illegal is uncertain, but it wouldn't be surprising: the place is perched so far out over the edge of the continent as to feel beyond the reach of federal law enforcement. But at least officially, such workshops ended when LSD became illegal. Grof began teaching instead something called Holotropic Breathwork, a technique for inducing a psychedelic state of consciousness without drugs, by means of deep, rapid, and rhythmic breathing, usually accompanied by loud drumming. Yet Esalen's role in the history of psychedelics did not end with their prohibition. It became the place where people hoping to bring these molecules back into the culture, whether as an adjunct to therapy or a means of spiritual development, met to plot their campaigns.

In January 1994, Bob Jesse managed to get himself invited to one such meeting at Esalen. While helping out with the dishes after a Friday night dinner at the Shulgins', Jesse learned that a group of therapists and scientists would be gathering in Big Sur to discuss the prospects for reviving psychedelic research. There were signs that the door Washington, D.C., had slammed shut on research in the late 1960s might be opening, if only a crack: Curtis Wright, a new administrator at the FDA (and, as it happens, a former student of Roland Griffiths's at Hopkins), had signaled that research protocols for psychedelics would be treated like any other—judged on their merits. Testing this new receptivity, a psychiatrist at the University of New Mexico named Rick Strassman had sought and received approval to study the physiological effects of DMT, a powerful psychedelic compound found in many plants. This small trial marked the first federally sanctioned experiment with a psychedelic compound since the 1970s—in retrospect, a watershed event.

Around the same time, Rick Doblin and Charles Grob, a psychiatrist at UCLA, had succeeded in persuading the government to approve the first human trial of MDMA. (Grob is one of the first psychiatrists to advocate for the return of psychedelics to psychotherapy; he later conducted the first modern trial of psilocybin for cancer patients.) The year before the Esalen gathering (which Grob and Doblin both attended), David Nichols, a Purdue University chemist and pharmacologist, launched the Heffter Research Institute (named for the German chemist who first identified the mescaline compound in 1897) with the then improbable ambition of funding serious psychedelic science. (Heffter has since helped fund many of the modern trials of psilocybin.) So there were scattered hopeful signs in the early 1990s that conditions were ripening for a revival of psychedelic research. The tiny community that had sustained such a dream through the dark ages began, tentatively, quietly, to organize.

Even though Jesse was new to this community, and neither a scientist nor a therapist, he asked if he could attend the Esalen meeting and offered to make himself useful, refilling water glasses if that's what it took. Most of the gathering was taken up with discussions of the potential medical applications of psychedelics, as well as the need for basic research on the neuroscience. Jesse was struck by the fact that so little attention was paid to the spiritual potential of these compounds. He left the meeting convinced that "okay, there is room to maneuver here. I was hoping one of these people would pick up the ball and run with it, but they were busy with the other ball. So I made a decision to seek a leave of absence from Oracle." Within a year, Jesse would launch the Council on Spiritual Practices, and within two the council would convene its own meeting at Esalen, in January 1996, with the aim of opening a second front in the campaign to resurrect psychedelics.

Fittingly, the gathering took place in the Maslow Room at Esalen, named for the psychologist whose writings on the hierarchy of human needs underscored the importance of "peak experiences" in self-actualization. Most of the fifteen in attendance were "psychedelic elders," therapists and researchers like James Fadiman and Willis Harman, Mark Kleiman, then a drug-policy expert at the Kennedy School (and Rick Doblin's thesis tutor there), and religious figures like Huston Smith, Brother David Steindl-Rast, and Jeffrey Bronfman, the head of the UDV church in America (and heir to the Seagram's liquor fortune). But Jesse wisely decided to invite an outsider as well: Charles "Bob" Schuster, who had served both Ronald Reagan and George H. W. Bush as director of the National Institute on Drug Abuse. Jesse didn't know Schuster well at all; they had once spoken briefly at a conference. But Jesse came away from the encounter thinking Schuster just might be receptive to an invitation.

Exactly why Bob Schuster—a leading figure in the academic establishment undergirding the drug war—would be open to the idea of coming to Esalen to discuss the spiritual potential of psychedelics was a mystery, at least until I had the opportunity to speak to his widow, Chris-Ellyn Johanson. Johanson, who is also a drug researcher, painted a picture of a man of exceptionally broad interests and deep curiosity.

"Bob was open-minded to a fault," she told me, with a laugh. "He would talk to anyone." Like many people in the NIDA community, Schuster well understood that psychedelics fit awkwardly into the profile of a drug of abuse; animals, given the choice, will not self-administer a psychedelic more than once, and the classic psychedelics exhibit remarkably little toxicity. I asked Johanson if Schuster had ever taken a psychedelic himself; Roland Griffiths had told me he thought it was possible. ("Bob was a jazz musician," Griffiths told me, "so I wouldn't be at all surprised.") But Johanson said no. "He was definitely curious about them," she told me, "but I think he was too afraid. We were martini people." I asked if he was a spiritual man. "Not really, though I think he would have liked to have been."

Jesse, not quite sure what Schuster would make of the meeting, arranged to have Jim Fadiman bunk with him, instructing Fadiman, a psychologist, to check him out. "Early the next morning Jim found me and said, 'Bob, mission accomplished. You have found a gem of a human being.'"

Schuster thoroughly enjoyed his time at Esalen, according to his wife. He took part in a drumming circle Jesse had helped arrange—you don't leave Esalen without doing some such thing—and was amazed to discover how easily he could slip into a trance. But Schuster also made some key contributions to the group's deliberations. He warned Jesse off working with MDMA, which he believed was toxic to the brain and had by then acquired an unsavory reputation as a club drug. He also suggested that psilocybin was a much better candidate for research than LSD, largely for political reasons: because so many fewer people had heard of it, psilocybin carried none of the political and cultural baggage of LSD.

By the end of the meeting, the Esalen group had settled on a short list of objectives, some of them modest—to draft a code of ethics for spiritual guides—and others more ambitious: "to get aboveboard, unimpeachable research done, at an institution with investigators beyond reproach," and, ideally, "do this without any pretext of clinical treatment."

"We weren't sure that was possible," Jesse told me, but he and his colleagues believed "it would be a big mistake if medicalization is all that happens." Why a mistake? Because Bob Jesse was ultimately less interested in people's mental problems than with their spiritual well-being—in using entheogens for the betterment of well people.

Shortly after the Esalen meeting, Schuster made what would turn out to be his most important contribution: telling Bob Jesse about his old friend Roland Griffiths, whom he described as exactly "the investigator beyond reproach" Jesse was looking for and "a scientist of the first order."

"Everything Roland's done he's devoted himself to completely," Jesse recalls Schuster saying, "including his meditation practice. We think it's changed him." Griffiths had shared with Schuster his growing dissatisfaction with science and his deepening interest in the kind of "ultimate questions" coming up in his meditation practice. Schuster then made the call to Griffiths telling him about the interesting young man he'd just met at Esalen, explaining that they shared an interest in spirituality, and suggesting they should meet. After an exchange of e-mails, Jesse flew to Baltimore to have lunch with Griffiths in the cafeteria on the Bayview medical campus, inaugurating a series of conversations and meetings that would eventually lead to their collaboration on the 2006 study of psilocybin and mystical experience at Johns Hopkins.

o   o   o

BUT THERE WAS STILL one missing piece of the puzzle and the scientific team. Most of the drug trials Griffiths had run in the past involved baboons and other nonhuman primates; he had much less clinical experience working with humans and realized he needed a skilled therapist to join the project—a "master clinician," as he put it. As it happened, Bob Jesse had met a psychologist at a psychedelic conference a few years before who not only filled the bill but lived in Baltimore. Still more fortuitous, this psychologist, whose name was Bill Richards, probably has more experience guiding psychedelic journeys in the 1960s and 1970s than anyone alive, with the possible exception of Stan Grof (with whom he had once worked). In fact, Bill Richards administered the very last legal dose of psilocybin to an American, at the Maryland Psychiatric Research Center at Spring Grove State Hospital in the spring of 1977. In the decades since, he had been practicing more conventional psychotherapy out of his home in a leafy Baltimore neighborhood called Windsor Hills, biding his time and waiting patiently for the world to come around so that he might work with psychedelics once again.

"In the big picture," he told me the first time we met in his home office, "these drugs have been around at least five thousand years, and many times they have surfaced and have been repressed, so this was another cycle. But the mushroom still grows, and eventually this work would come around again. Or so I hoped." When he got the

call from Bob Jesse in 1998, and met Roland Griffiths shortly there-
after, he couldn't quite believe his good fortune. "It was thrilling."

Bill Richards, a preternaturally cheerful man in his seventies, is a
bridge between the two eras of psychedelic therapy. Walter Pahnke
was the best man at his wedding; he worked closely with Stan Grof
at Spring Grove and visited Timothy Leary in Millbrook, New York,
where Leary landed after his exile from Harvard. Though Richards
left the Midwest half a century ago, he's retained the speech patterns
of rural Michigan, where he was born in 1940. Richards today sports
a white goatee, laughs with an infectious cackle, and ends many of
his sentences with a cheerful, up-spoken "y'know?"

Richards, who holds graduate degrees in both psychology and divinity, had his first psychedelic experience while a divinity student at
Yale in 1963. He was spending the year studying in Germany, at the
University of Göttingen, and found himself drawn to the Department of Psychiatry, where he learned about a research project involving a drug called psilocybin.

"I had no idea what that was, but two friends of mine had participated and had had interesting experiences." One of them, whose
father had been killed in the war, had regressed to childhood to find
himself sitting on his father's lap. The other had hallucinations of SS
men marching in the street. "I had never had a decent hallucination,"
Richards said with a chuckle, "and I was trying to get some insight
into my childhood. In those days, I viewed my own mind as a psychological laboratory, so I decided to volunteer.

"This was before the importance of set and setting was understood. I was brought to a basement room, given an injection, and left
alone." A recipe for a bad trip, surely, but Richards had precisely the
opposite experience. "I felt immersed in this incredibly detailed imagery that looked like Islamic architecture, with Arabic script, about
which I knew nothing. And then I somehow *became* these exquisitely

intricate patterns, losing my usual identity. And all I can say is that the eternal brilliance of mystical consciousness manifested itself. My awareness was flooded with love, beauty, and peace beyond anything I ever had known or imagined to be possible. 'Awe,' 'glory,' and 'gratitude' were the only words that remained relevant."

Descriptions of such experiences always sound a little thin, at least when compared with the emotional impact people are trying to convey; for a life-transforming event, the words can seem paltry. When I mentioned this to Richards, he smiled. "You have to imagine a caveman transported into the middle of Manhattan. He sees buses, cell phones, skyscrapers, airplanes. Then zap him back to his cave. What does he say about the experience? 'It was big, it was impressive, it was loud.' He doesn't have the vocabulary for 'skyscraper,' 'elevator,' 'cell phone.' Maybe he has an intuitive sense there was some sort of significance or order to the scene. But there are words we need that don't yet exist. We've got five crayons when we need fifty thousand different shades."

In the middle of his journey, one of the psychiatric residents stopped by the room to look in on Richards, asking him to sit up so he could test his reflexes. As the resident tapped his patellar tendon with his little rubber hammer, Richards remembers feeling "compassion for the infancy of science. The researchers had no idea what really was happening in my inner experiential world, of its unspeakable beauty or of its potential importance for all of us." A few days after the experience, Richards returned to the lab and asked, "What was that drug you gave me? How is it spelled?

"And the rest of my life is footnotes!"

Yet after several subsequent psilocybin sessions failed to produce another mystical experience, Richards started to wonder if perhaps he had exaggerated that first trip. Some time later, Walter Pahnke arrived at the university, fresh from his graduate work with Timothy

Leary at Harvard, and the two became friends. (It was Richards who gave Pahnke his first psychedelic trip while the two were in Germany; he had apparently never taken LSD or psilocybin at Harvard, thinking it might compromise the objectivity of the Good Friday Experiment.) Pahnke suggested Richards try one more time, but in a room with soft lighting, plants, and music and using a higher dose. Once again, Richards had "an incredibly profound experience. I realized I had not exaggerated the first trip but in fact had forgotten 80 percent of it.

"I have never doubted the validity of these experiences," Richards told me. "This was the realm of mystical consciousness that Shankara was talking about, that Plotinus was writing about, that Saint John of the Cross and Meister Eckhart were writing about. It's also what Abraham Maslow was talking about with his 'peak experiences,' though Abe could get there without the drugs." Richards would go on to study psychology under Maslow at Brandeis University. "Abe was a natural Jewish mystic. He could just lie down in the backyard and have a mystical experience. Psychedelics are for those of us who aren't so innately gifted."

Richards emerged from those first psychedelic explorations in possession of three unshakable convictions. The first is that the experience of the sacred reported both by the great mystics and by people on high-dose psychedelic journeys is the same experience and is "real"—that is, not just a figment of the imagination.

"You go deep enough or far out enough in consciousness and you will bump into the sacred. It's not something we generate; it's something out there waiting to be discovered. And this reliably happens to nonbelievers as well as believers." Second, that, whether occasioned by drugs or other means, these experiences of mystical consciousness are in all likelihood the primal basis of religion. (Partly for this reason Richards believes that psychedelics should be part of a divinity student's education.) And third, that consciousness is a property of the universe, not brains. On this question, he holds with Henri Bergson, the French philosopher, who conceived of the human mind as a kind of radio receiver, able to tune in to frequencies of energy and information that exist outside it. "If you wanted to find the blonde who delivered the news last night," Richards offered by way of an analogy, "you wouldn't look for her in the TV set." The television set is, like the human brain, necessary but not sufficient.

After Richards finished with his graduate studies in the late 1960s, he found work as a research fellow at the Spring Grove State Hospital outside Baltimore, where a most improbable counterfactual history of psychedelic research was quietly unfolding, far from the noise and glare surrounding Timothy Leary. Indeed, this is a case where the force of the Leary narrative has bent the received history out of shape, such that many of us assume there was no serious psychedelic research before Leary arrived at Harvard and no serious research after he was fired. But until Bill Richards administered psilocybin to his last volunteer in 1977, Spring Grove was actively (and without much controversy) conducting an ambitious program of psychedelic research—much of it under grants from the National Institute of Mental Health—with schizophrenics, alcoholics and other addicts, cancer patients struggling with anxiety, religious and mental health professionals, and patients with severe personality disorders. Several hundred patients and volunteers received psychedelic therapy at Spring Grove between the early 1960s and the mid-1970s. In many cases, the researchers were getting very good results in well-designed studies that were being regularly published in peer-reviewed journals such as *JAMA* and the *Archives of General Psychiatry*. (Roland Griffiths is of the opinion that much of this research is "suspect," but Richards told me, "These studies weren't as bad as people like Roland might imply.") It is remarkable just how much of the work being done today, at Hopkins and NYU and other places, was prefigured at Spring Grove; indeed, it is hard to find a contemporary experiment with psychedelics that wasn't already done in Maryland in the 1960s or 1970s.

At least at the beginning, the Spring Grove psychedelic work enjoyed lots of public support. In 1965, CBS News broadcast an admiring hour-long "special report" on the hospital's work with alcoholics, called *LSD: The Spring Grove Experiment.* The response to the program was so positive that the Maryland state legislature established a multimillion-dollar research facility on the campus of the Spring Grove State Hospital, called the Maryland Psychiatric Research Center. Stan Grof, Walter Pahnke, and Bill Richards were hired to help run it, along with several dozen other therapists, psychiatrists, pharmacologists, and support staff. Equally hard to believe today is the fact that, as Richards told me, "whenever we hired someone, they would receive a couple of LSD sessions as part of their training to do the work. We had authorization! How else could you be sensitive to what was going on in the mind of the patient? I wish we could do that at Hopkins."

The fact that such an ambitious research program continued at Spring Grove well into the 1970s suggests the story of the suppression of psychedelic research is a little more complicated than the conventional narrative would indicate. While it is true that some research projects—such as Jim Fadiman's creativity trials in Palo Alto—received orders from Washington to stop, other projects on long-term grants were allowed to continue until the money ran out, as it eventually did. Rather than shut down all research, as many in the psychedelic community believe happened, the government simply made it more difficult to get approvals, and funding gradually dried up. As time went on, researchers found that on top of all the bureaucratic and financial hurdles they also had to deal with "the snicker test": How would your colleagues react when you told them you were running experiments with LSD? By the mid-1970s, psychedelics had become something of a scientific embarrassment—not because they were a failure, but because they had become identified with the counterculture and with disgraced scientists such as Timothy Leary.

But there was nothing embarrassing about psychedelic research at Spring Grove in the late 1960s and early 1970s. Then, and there, it looked like the future. "We thought this was the most incredible frontier in psychiatry," Richards recalls. "We would all sit around the conference table talking about how we were going to train the hundreds if not thousands of therapists that would be needed to do this work. (And look, we're having the same conversation again today!) There were international conferences on psychedelic research, and we had colleagues throughout Europe doing similar work. The field was taking off. But in the end the societal forces were stronger than we were."

In 1971, Richard Nixon declared Timothy Leary, a washed-up psychology professor, "the most dangerous man in America." Psychedelics were nourishing the counterculture, and the counterculture was sapping the willingness of America's young to fight. The Nixon administration sought to blunt the counterculture by attacking its neurochemical infrastructure.

Was the suppression of psychedelic research inevitable? Many of the researchers I interviewed feel that it might have been avoided had the drugs not leaped the laboratory walls—a contingency that, fairly or not, most of them blame squarely on the "antics," "misbehavior," and "evangelism" of Timothy Leary.

Stanislav Grof believes that psychedelics loosed "the Dionysian element" on 1960s America, posing a threat to the country's puritan values that was bound to be repulsed. (He told me he also thinks the same thing could happen again.) Roland Griffiths points out that ours is not the first culture to feel threatened by psychedelics: the reason R. Gordon Wasson had to rediscover magic mushrooms in Mexico was that the Spanish had suppressed them so effectively, deeming them dangerous instruments of paganism.

"That says something important about how reluctant cultures are to expose themselves to the changes these kinds of compounds can occasion," he told me the first time we met. "There is so much authority that comes out of the primary mystical experience that it can be threatening to existing hierarchical structures."

o   o   o

BY THE MID-1970S, the LSD work at Spring Grove, much of which was state funded, had become a political hot potato in Annapolis. In 1975, the Rockefeller Commission investigating the CIA disclosed that the agency had also been running LSD experiments in Maryland, at Fort Detrick, as part of a mind-control project called MK-Ultra. (An internal memo the commission released concisely set forth the agency's objective: "Can we get control of an individual to the point where he will do our bidding against his will and even against fundamental laws of nature, such as self-preservation?") It was revealed that the CIA was dosing both government employees and civilians without their knowledge; at least one person had died. The news that Maryland taxpayers were also supporting research with LSD promptly blew up into a scandal, and pressure to close down psychedelic research at Spring Grove became irresistible.

"Pretty soon it was just me and two secretaries," Richards recalls. "And then it was over."

Today Roland Griffiths, who would pick up the thread of research that was dropped when the work at Spring Grove ended, marvels at the fact that the first wave of psychedelic research, promising as it was, would end for reasons having nothing to do with science. "We ended up demonizing these compounds. Can you think of another area of science thought to be so dangerous and taboo that all research gets shut down for decades? It's unprecedented in modern science." So too, perhaps, is the sheer amount of scientific knowledge that was simply erased.

In 1998, Griffiths, Jesse, and Richards began designing a pilot study loosely based on the Good Friday Experiment. "It wasn't a psychotherapy study," Richards points out. "It was a study designed to determine whether psilocybin can elicit a transcendental experience. That we were able to obtain permission to give it to healthy normals is a tribute to Roland's long history of commanding respect both at Hopkins and in Washington." In 1999, the protocol was approved, but only after wending its way through five layers of review at Hopkins as well as the FDA and the DEA. (Many of Griffiths's Hopkins colleagues were skeptical of the proposal, worried psychedelic research might jeopardize federal funding; one told me there were "people in the Department of Psychiatry and the broader institution who questioned the work, because this class of compounds carries a lot of baggage from the '60s.")

"We had faith that the people on all these committees would be good scientists," Richards told me. "And with luck maybe a few of them had tried mushrooms in college!" Roland Griffiths became the principal investigator of the trial, Bill Richards became the clinical director, and Bob Jesse continued to work behind the scenes.

"I can vividly remember the first session I ran after that long twenty-two-year hiatus," Richards recalled. He and I were together in the session room at Hopkins; I was sitting on the couch where the volunteers lie down during their journeys, and Richards was in the chair where he has now sat and guided more than a hundred psilocybin journeys since 1999. The room feels more like a den or living room than a room in a laboratory, with a plush sofa, vaguely spiritual paintings on the walls, a sculpture of the Buddha on a side table, and shelves holding a giant stone mushroom and various other nondenominational spiritual artifacts, as well as the small chalice in which the volunteers receive their pills.

"This guy is lying on the couch right there where you are, with tears streaming down his face, and I'm thinking, how absolutely beautiful and meaningful this experience is. How *sacred*. How can this ever have been illegal? It's as if we made entering Gothic cathedrals illegal, or museums, or sunsets!

"I honestly never knew if this would happen again in my lifetime. And look at where we are now: the work at Hopkins has been going on now for fifteen years—five years longer than Spring Grove."

o   o   o

In 1999, an odd but intriguing advertisement began appearing in weeklies in the Baltimore and Washington, D.C., area, under the headline "Interested in the Spiritual Life?"

> University research with entheogens (roughly, God-evoking substances such as peyote and sacred mushrooms) has returned. The field of study includes pharmacology, psychology, creativity enhancement, and spirituality. To explore the possibility of participating in confidential entheogen research projects, call 1-888-585-8870, toll free. www.csp.org.

Not long after, Bill Richards and Mary Cosimano, a social worker and school guidance counselor Richards recruited to help him guide psychedelic sessions, administered the first legal dose of psilocybin to an American in twenty-two years. In the years since, the Hopkins team has conducted more than three hundred psilocybin sessions, working in a variety of populations, including healthy normals, long-term and novice meditators, cancer patients, smokers seeking to break their habit, and religious professionals. I was curious to get the volunteer's-eye view of the experience from all these types, but especially from that first cohort of healthy normals, partly because they were participants in a study that would turn out to be historically important and partly because I figured they would be the most like, well, me. What is it like to have a legally sanctioned, professionally guided, optimally comfortable high-dose psilocybin experience?

Yet the volunteers in the first experiments were not *exactly* like me, because at the time I doubt I would have read past "Interested in the Spiritual Life?" There were no stone-cold atheists in the original group, and interviews with nearly a dozen of them suggested many if not most of them came into the study with spiritual leanings to one degree or another. There was an energy healer, a man who'd done the whole Iron John trip, a former Franciscan friar, and an herbalist. There was also a physicist with an interest in Zen and a philosophy professor with an interest in theology. Roland Griffiths acknowledged, "We were interested in a spiritual effect and were biasing the condition initially [in that direction]."

That said, Griffiths went to great lengths in the design of the study to control for "expectancy effects." In part this owed to Griffiths's skepticism that a drug could occasion the same kind of mystical experience he had had in his meditation: "This is all truth to Bill and hypothesis to me. So we needed to control for Bill's biases." All of the volunteers were "hallucinogen naive," so had no idea what psilocybin felt like, and neither they nor their monitors knew in any given session whether they were getting psilocybin or a placebo, and whether that placebo was a sugar pill or any one of half a dozen different psychoactive drugs. In fact the placebo was Ritalin, and as it turned out, the monitors guessed wrong nearly a quarter of the time as to what was in the pill a volunteer had received.

Even years after their experiences in the trials, the volunteers I spoke to recalled them in vivid detail and at considerable length; the interviews lasted hours. These people had big stories to tell; in several cases, these were the most meaningful experiences of their lives, and they clearly relished the opportunity to relive them for me in great detail, whether in person, by Skype, or on the telephone. The volunteers were also required to write a report of their experiences soon after they occurred, and all of the ones I interviewed were happy to share these reports, which made for strange and fascinating reading.

Many of the volunteers I spoke to reported initial episodes of intense fear and anxiety before surrendering themselves to the experience—as the sitters encourage them to do. The sitters work from a set of "flight instructions" prepared by Bill Richards, based on the hundreds of psychedelic journeys he has guided. The guides go over the instructions with the volunteers during the eight hours of preparation all of them receive before commencing their journeys.

The flight instructions advise guides to use mantras like "Trust the trajectory" and "TLO—Trust, Let Go, Be Open." Some guides like to quote John Lennon: "Turn off your mind, relax and float downstream."

Volunteers are told they may experience the "death/transcendence of your ego or everyday self," but this is "always followed by Rebirth/ Return to the normative world of space & time. Safest way to return to normal is to entrust self unconditionally to the emerging experiences." Guides are instructed to remind volunteers they'll never be left alone and not to worry about the body while journeying because the guides are there to keep an eye on it. If you feel as if you are "dying, melting, dissolving, exploding, going crazy etc.—go ahead."

Volunteers are quizzed: "If you see a door, what do you do? If you see a staircase, what do you do?" "Open it" and "climb up it" are of course the right answers.

This careful preparation means that a certain expectancy effect is probably unavoidable. After all, the researchers are preparing people for a major experience, involving death and rebirth and holding the potential for transformation. "It would be irresponsible *not* to warn volunteers these things could happen," Griffiths pointed out when I asked if his volunteers were being "primed" for a certain kind of experience. One volunteer—the physicist—told me that the "mystical experience questionnaire" he filled out after every session also planted expectations. "I long to see some of the stuff hinted at in the questionnaire," he wrote after an underwhelming session—perhaps on the placebo. "Seeing everything as alive and connected, meeting the void, or some embodiment of deities and things like that." In this and so many other ways, it seems, the Hopkins psilocybin experience is the artifact not only of this powerful molecule but also of the preparation and expectations of the volunteer, the skills and world-views of the sitters, Bill Richards's flight instructions, the decor of the room, the inward focus encouraged by the eyeshades and the music (and the music itself, much of which to my ears sounds notably religious), and, though they might not be pleased to hear it, the minds of the designers of the experiments.

The sheer suggestibility of psychedelics is one of their defining characteristics, so in one sense it is no wonder that so many of the first cohort of volunteers at Hopkins had powerful mystical experiences: the experiment was designed by three men intensely interested in mystical states of consciousness. (And it is likewise no wonder that the European researchers I interviewed all failed to see as many instances of mystical experience in their subjects as the Americans did in theirs.) And yet, for all the priming going on, the fact remains that the people who received a placebo simply didn't have the kinds of experiences that volunteer after volunteer described to me as the most meaningful or significant in their lives.

Soon after a volunteer takes her pill from the little chalice, but before she feels any effects, Roland Griffiths will usually drop by the session room to wish her bon voyage. Griffiths often uses a particular metaphor that made an impression on many of the volunteers I spoke to. "Think of yourself as an astronaut being blasted into outer space," Richard Boothby recalled him saying. Boothby is a philosophy professor who was in his early fifties when he volunteered at Hopkins. "You're going way out there to take it all in and engage with whatever you find there, but you can be confident that we'll be here keeping an eye on things. Think of us as ground control. We've got you covered."

For the astronaut being blasted into space, the shudder of liftoff and strain of escaping Earth's gravitational field can be wrenching—even terrifying. Several volunteers describe trying to hold on for dear life as they felt their sense of self rapidly disintegrating. Brian Turner, who at the time of his journey was a forty-four-year-old physicist working for a military contractor (with a security clearance), put it this way:

> I could feel my body dissolving, beginning with my feet, until it all disappeared but the left side of my jaw. It was really unpleasant; I could count only a few teeth left and the bottom part of my jaw. I knew that if that went away I would be gone. Then I remembered what they told me, that whenever you encounter anything scary, go toward it. So instead of being afraid of dying I got curious about what was going on. I was no longer trying to avoid dying. Instead of recoiling from the experience, I began to interrogate it. And with

that, the whole situation dissolved into this pleasant floaty
feeling, and I became the music for a while.

Soon after, he found himself "in a large cave where all my past
relationships were hanging down as icicles: the person who sat next
to me in second grade, high school friends, my first girlfriend, all of
them were there, encased in ice. It was very cool. I thought about
each of them in turn, remembering everything about our relationship. It was a review—something about the trajectory of my life. All
these people had made me what I had become."

Amy Charnay, a nutritionist and herbalist in her thirties, came to
Hopkins after a crisis. An avid runner, she had been studying forest
ecology when she fell from a tree and shattered her ankle, ending
both her running and her forestry careers. In the early moments of
her journey, Amy was overcome by waves of guilt and fear.

"The visual I had was from the 1800s and I was up on this stage.
Two people next to me were slipping a noose around my neck while
a crowd of people watched, cheering for my death. I felt drenched
with guilt, just terrified. I was in a hell realm. And I remember Bill
asking, 'What's going on?'

"'I'm experiencing a lot of guilt.' Bill replied, 'That's a very common human experience,' and with that, the whole image of being
hanged pixilated and then just disappeared, to be replaced by this
tremendous sensation of freedom and interconnectedness. This was
huge for me. I saw that if I can name and admit a feeling, confess it
to someone, it would let go. A little older and wiser, now I can do this
for myself."

Some time later, Charnay found herself flying around the world
and through time perched on the back of a bird. "I was aware enough
to know my body was on the couch, but I was leaving my body and
experiencing these things firsthand. I found myself in a drumming

circle with an indigenous tribe somewhere, and I was being healed but was also being the healer. This was very profound for me. Not having that traditional lineage [of a healer], I had always felt like I was a phony doing plant medicine, but this made me see I was connected to the plants and to people who use plants, whether for rituals or psychedelics or salad!"

During a subsequent session, Charnay reconnected with a boyfriend from her youth who had died in a car accident at nineteen. "All of a sudden there is a piece of Phil living in my left shoulder. I've never had an experience like that, but it was so real. I don't know why he's yellow and lives in my left shoulder—what does that even mean?—but I don't care. He's back with me." Such reconnections with the dead are not uncommon. Richard Boothby, whose twenty-three-year-old son had committed suicide a year earlier after years of drug addiction, told me, "Oliver was more present to me now than he had ever been before."

The supreme importance of surrendering to the experience, however frightening or bizarre, is stressed in the preparatory sessions and figures largely in many people's journeys, and beyond. Boothby, the philosopher, took the advice to heart and found that he could use the idea as a kind of tool to shape the experience in real time. He wrote:

> Early on I began to perceive that the effects of the drug respond strikingly to my own subjective determination. If, in response to the swelling intensity of the whole experience, I began to tense up with anxiety, the whole scene appears to tighten in some way. But if I then consciously remind myself to relax, to let myself go into the experience, the effect is dramatic. The space in which I seem to find myself, already enormous, suddenly yawns open even further and the shapes that undulate before my eyes appear

to explode with new and even more extravagant patterns. Over and over again I had the overwhelming sense of infinity being multiplied by another infinity. I joked to my wife as she drove me home that I felt as if I had been repeatedly sucked into the asshole of God.

Boothby had what sounds very much like a classic mystical experience, though he may be the first in the long line of Western mystics to enter the divine realm through that particular aperture.

At the depths of this delirium I conceived that I was either dying or, most bizarrely, I was already dead. All points of secure attachment to a trustworthy sense of reality had fallen away. Why not think that I am dead? And if this is dying, I thought, then so be it. How can I say no to this?

At this point, at the greatest depth of the experience, I felt all my organizing categories of opposition—dreaming and wakefulness, life and death, inside and outside, self and other—collapse into each other . . . Reality appeared to fold in on itself, to implode in a kind of ecstatic catastrophe of logic. Yet in the midst of this hallucinatory hurricane I was having a weird experience of ultra-sublimity. And I remember repeating to myself again and again, "Nothing matters, nothing matters any more. I see the point! Nothing matters at all."

And then it was over.

During the last few hours, reality began slowly, effortlessly, to stitch itself back together. In sync with some particularly wowing choral music, I had an incredibly moving sense of triumphant reawakening, as if a new day were dawning after a long and harrowing night.

o   o   o

AT THE SAME TIME I was interviewing Richard Boothby and his fellow volunteers, I was reading William James's account of mystical consciousness in *The Varieties of Religious Experience* in the hope of orienting myself. And indeed much of what James had to say helped me get my bearings amid the torrent of words and images I was collecting. James prefaced his discussion of mystical states of consciousness by admitting that "my own constitution shuts me out from their enjoyment almost entirely." *Almost* entirely: what James knows about mystical states was gleaned not just from his reading but also from his own experiments with drugs, including nitrous oxide.

Rather than attempt to define something as difficult to grab hold of as a mystical experience, James offers four "marks" by which we may recognize one. The first and, to his mind, "handiest" is ineffability: "The subject of it immediately says that it defies expression, that no adequate report of its contents can be given in words." With the possible exception of Boothby, all the volunteers I spoke to at one point or another despaired of conveying the full force of what they had experienced, gamely though they tried. "You had to be there" was a regular refrain.

The noetic quality is James's second mark: "Mystical states seem to those who experience them to be also states of knowledge . . . They are illuminations, revelations full of significance and importance . . . and as a rule they carry with them a curious sense of authority."

For every volunteer I've interviewed, the experience yielded many more answers than questions, and—curiously for what is after all a drug experience—these answers had about them a remarkable sturdiness and durability. John Hayes, a psychotherapist in his fifties who was one of the first volunteers at Hopkins,

> felt like mysteries were being unveiled and yet it all felt familiar and more like I was being reminded of things I had already known. I had a sense of initiation into dimensions of existence most people never know exist, including the distinct sense that death was illusory, in the sense that it is a door we walk through into another plane of existence, that we're sprung from an eternity to which we will return.

Which is true enough, I suppose, but to someone having a mystical experience, such an insight acquires the force of revealed truth.

So many of the specific insights gleaned during the psychedelic journey exist on a knife-edge poised between profundity and utter banality. Boothby, an intellectual with a highly developed sense of irony, struggled to put words to the deep truths about the essence of our humanity revealed to him during one of his psilocybin journeys.

> I have at times been almost embarrassed by them, as if they give voice to a cosmic vision of the triumph of love that one associates derisively with the platitudes of Hallmark cards. All the same, the basic insights afforded to me during the session still seem for the most part compelling.

What was the philosophy professor's compelling insight?

> "Love conquers all."

James touches on the banality of these mystical insights: "that deepened sense of the significance of a maxim or formula which occasionally sweeps over one. 'I've heard that said all my life,' we exclaim, 'but I never realized its full meaning until now.'" The mystical journey seems to offer a graduate education in the obvious. Yet people come out of the experience understanding these platitudes in a new way; what was merely known is now felt, takes on the authority of a deeply rooted conviction. And, more often than not, that conviction concerns the supreme importance of love.

Karin Sokel, a life coach and energy healer in her fifties, described an experience "that changed everything and opened me profoundly." At the climax of her journey, she had an encounter with a god who called himself "I Am." In its presence, she recalled, "every one of my chakras was exploding. And then there was this light, it was the pure light of love and divinity, and it was with me and no words were needed. I was in the presence of this absolute pure divine love and I was merging with it, in this explosion of energy . . . Just talking about it my fingers are getting electric. It sort of penetrated me. The core of our being, I now knew, is love. At the peak of the experience, I was literally holding the face of Osama bin Laden, looking into his eyes, feeling pure love from him and giving it to him. The core is not evil, it is love. I had the same experience with Hitler, and then someone from North Korea. So I think we are divine. This is not intellectual, this is a core knowingness."

I asked Sokel what made her so sure this wasn't a dream or drug-induced fantasy—a suggestion that proved no match for her noetic sense. "This was no dream. This was as real as you and I having this conversation. I wouldn't have understood it either if I hadn't had the direct experience. Now it is hardwired in my brain so I can connect to it and do often."

This last point James alludes to in his discussion of the third mark of mystical consciousness, which is "transiency." For although the mystical state cannot be sustained for long, its traces persist and recur, "and from one recurrence to another it is susceptible of continuous development in what is felt as inner richness and importance."

The fourth and last mark in James's typology is the essential "passivity" of the mystical experience. "The mystic feels as if his own will were in abeyance, and indeed sometimes as if he were grasped and held by a superior power." This sense of having temporarily surrendered to a superior force often leaves the person feeling as if he or she has been permanently transformed.

For most of the Hopkins volunteers I interviewed, their psilocybin journeys had taken place ten or fifteen years earlier, and yet their effects were still keenly felt, in some cases on a daily basis. "Psilocybin awakened my loving compassion and gratitude in a way I had never experienced before," a psychologist who asked not to be named told me when I asked her about lasting effects. "Trust, Letting go, Openness, and Being were the touchstones of the experience for me. Now I know these things instead of just believing." She had turned Bill Richards's flight instructions into a manual for living.

Richard Boothby did much the same thing, converting his insight about letting go into a kind of ethic:

> During my session this art of relaxation itself became the basis of an immense revelation, as it suddenly appeared to me that something in the spirit of this relaxation, something in the achievement of a perfect, trusting and loving openness of spirit, is the very essence and purpose of life. Our task in life consists precisely in a form of letting go of fear and expectations, an attempt to purely give oneself to the impact of the present.

John Hayes, the psychotherapist, emerged with "his sense of the concrete destabilized," replaced by a conviction "that there's a reality beneath the reality of ordinary perceptions. It informed my cosmology—that there is a world beyond this one." Hayes particularly recommends the experience to people in middle age for whom, as Carl Jung suggested, experience of the numinous can help them negotiate the second half of their lives. Hayes added, "I would not recommend it to young people."

Charnay's journey at Hopkins solidified her commitment to herbal medicine (she now works for a supplement maker in Northern California); it also confirmed her in a decision to divorce her husband. "Everything was now so clear to me. I came out of the session, and my husband was late to pick me up. I realized, this is the theme with us. We're just really different people. I just got my ass kicked today, and I needed him to be on time." She broke the news to him in the car going home and has not looked back.

To listen to these people describe the changes in their lives inspired by their psilocybin journeys is to wonder if the Hopkins session room isn't a kind of "human transformation factory," as Mary Cosimano, the guide who has probably spent more time there than anyone else, described it to me. "From now on," one volunteer told me, "I think of my life as before and after psilocybin." Soon after his psilocybin experience, Brian Turner, the physicist, quit his job with the military contractor and moved to Colorado to study Zen. He had had a meditation practice before psilocybin, but "now I had the motivation, because I had tasted the destination"; he was willing to do the hard work of Zen now that he had gotten a preview of the new modes of consciousness it could make available to him.

Turner is now an ordained Zen monk, yet he is also still a physicist, working for a company that makes helium neon lasers. I asked him if he felt any tension between his science and his spiritual practice. "I don't feel there's a contradiction. Yet what happened at Hopkins has influenced my physics. I realize there are just some domains that science will not penetrate. Science can bring you to the big bang, but it can't take you beyond it. You need a different kind of apparatus to peer into that."

These anecdotal reports of personal transformation found strong support in a follow-up study done on the first groups of healthy normals studied at Hopkins. Katherine MacLean, a psychologist on the Hopkins team, crunched the survey data produced by fifty-two volunteers, including follow-up interviews with friends and family members they had designated, and discovered that in many cases the psilocybin experience had led to lasting changes in their personalities. Specifically, those volunteers who had "complete mystical experiences" (as determined by their scores on the Pahnke-Richards Mystical Experience Questionnaire) showed, in addition to lasting improvements in well-being, long-term increases in the personality trait of "openness to experience." One of the five traits psychologists use to assess personality (the other four are conscientiousness, extroversion, agreeableness, and neuroticism), openness encompasses aesthetic appreciation and sensitivity, fantasy and imagination, as well as tolerance of others' viewpoints and values; it also predicts creativity in both the arts and the sciences, as well as, presumably, a willingness to entertain ideas at odds with those of current science. Such pronounced and lasting changes in the personalities of adults are rare.

Yet not all these shifts in the direction of greater openness were confined to the volunteers in the Hopkins experiments; the sitters, too, speak of having been changed by the experience of witnessing these journeys, sometimes in surprising ways. Katherine MacLean, who guided dozens of sessions during her time at Hopkins, told me, "I started out on the atheist side, but I began seeing things every day in my work that were at odds with this belief. My world became more and more mysterious as I sat with people on psilocybin."

During my last interview with Richard Boothby, toward the end of a leisurely Sunday brunch at the modern art museum in Baltimore, he looked at me with an expression that mixed an almost evangelical fervor about the "treasures" he had glimpsed at Hopkins with a measure of pity for his still-hallucinogen-naive interlocutor.

"I don't blame you for being envious."

o   o   o

MY ENCOUNTERS with the Hopkins volunteers had indeed left me feeling somewhat envious, but also with a great many more questions than answers. How are we to evaluate the "insights" these people bring back from their psychedelic journeys? What sort of authority should we grant them? Where in the world does the material that makes up these waking dreams or, as one volunteer put it, "intrapsychic movies," come from? The unconscious? From the suggestions of their guides and the setting of the experiment? Or, as many of the volunteers believe, from somewhere "out there" or "beyond"? What do these mystical states of consciousness ultimately *mean* for our understanding of either the human mind or the universe?

For his part, Roland Griffiths's own encounters with the volunteers in the 2006 study reignited his passion for science, but they also left him with a deeper respect for all that science does not know—for what he is content to call "the mysteries."

"For me the data [from those first sessions] were . . . I don't want to use the word mind-blowing, but it was unprecedented the kinds of things we were seeing there, in terms of the deep meaning and lasting spiritual significance of these effects. I've given lots of drugs to lots of people, and what you get are drug experiences. What's unique about the psychedelics is the meaning that comes out of the experience."

Yet how real is that meaning? Griffiths himself is agnostic, but strikingly open-minded, even about his volunteers' firsthand reports of a "beyond," however they define it. "I'm willing to hold the possibility these experiences may or may not be true," he told me. "The exciting part is to use the tools we have to explore and pick apart this mystery."

Not all of his colleagues share his open-mindedness. During one of our meetings, over breakfast on the sunporch of his modest ranch house in suburban Baltimore, Griffiths mentioned a colleague at Hopkins, a prominent psychiatrist named Paul McHugh, who dismisses the psychedelic experience as nothing more than a form of "toxic delirium." He encouraged me to google McHugh.

"Doctors encounter this strange and colorful state of mind in patients suffering from advanced hepatic, renal, or pulmonary disease, in which toxic products accumulate in the body and do to the brain and mind just what LSD does," McHugh had written in a review of a book about the Harvard Psilocybin Project in *Commentary*. "The vividness of color perception, the merging of physical sensations, the hallucinations, the disorientation and loss of a sense of time, the delusional joys and terrors that come and go evoking unpredictable feelings and behaviors—are sadly familiar symptoms doctors are called to treat in hospitals every day."

Griffiths admits it is possible that what he's seeing is some form of temporary psychosis, and he plans to test for delirium in an upcoming experiment, but he seriously doubts that diagnosis accurately describes what is going on with his volunteers. "Patients suffering from delirium find it really unpleasant," he points out, "and they certainly don't report months later, 'Wow, that was one of the greatest and most meaningful experiences of my life.'"

William James grappled with these questions of veracity in his discussion on mystical states of consciousness. He concluded that the import of these experiences is, and should be, "authoritative over the individuals to whom they come" but that there is no reason the rest of us must "accept their revelations uncritically." And yet he believed that the very possibility people can experience these states of consciousness should bear on our understanding of the mind and world: "The existence of mystical states absolutely overthrows the pretension of non-mystical states to be the sole and ultimate dictators of what we may believe." These alternate forms of consciousness "might, in spite of all the perplexity, be indispensable stages in our approach to the final fullness of the truth." He detected in such experiences, in which the mind "ascend[s] to a more enveloping point of view," hints of a grand metaphysical "reconciliation": "It is as if the opposites of the world, whose contradictoriness and conflict make all our difficulties and troubles, were melted into unity." This ultimate unity, he suspected, was no mere delusion.

o o o

ROLAND GRIFFITHS today sounds like a scientist deeply committed—or rather recommitted—to his research. "I described to you how when I first got into meditation, I felt disconnected from my work life and considered dropping it entirely. I would say I'm now reengaged in a way that's more integrated than it has ever been. I'm more interested in the final questions and existential truths and with the sense of well-being, compassion, and love that come from these practices. Now I'm bringing these gifts to the laboratory. And it feels great."

The idea that we can now approach mystical states of consciousness with the tools of science is what gets Roland Griffiths out of bed in the morning. "As a scientific phenomenon, if you can create a condition in which 70 percent of people will say they have had one of the most meaningful experiences of their lives . . . well, as a scientist that's just incredible." For him the import of the 2006 result is that it proved "we can now do prospective studies" of mystical states of consciousness "because we can occasion them with a high degree of probability. That's the way science gains real traction." He believes the psilocybin work has opened a whole new frontier of human consciousness to scientific exploration. "I describe myself as a kid in a candy shop."

The gamble Roland Griffiths took with his career in 1998, when he decided to devote himself to the investigation of psychedelics and mystical experience, has already paid off. A month before our breakfast, Griffiths had received the Eddy Award from the College on Problems of Drug Dependence, perhaps the most prestigious lifetime achievement prize in the field. The nominators all cited Griffiths's psychedelic work as one of his signal contributions. The scope of that work has expanded significantly since the 2006 paper; when I last visited Hopkins, in 2015, some twenty people were working on various studies involving psychedelics. Not since Spring Grove has there been such strong institutional support for the study of psychedelics, and never before has an institution of Hopkins's reputation devoted so many resources to what is, after all, the study of mystical states of consciousness.

The Hopkins lab remains keenly interested in exploring spirituality and the "betterment of well people"—there are trials under way giving psilocybin to long-term meditators and religious professionals—but the transformative effect of the mystical experience has obvious therapeutic implications that the lab has been investigating. Completed studies suggest that psilocybin—or rather the mystical state of consciousness that psilocybin occasions—may be useful in treating both addiction (a pilot study in smoking cessation achieved an 80 percent success rate, which is unprecedented) and the existential distress that often debilitates people facing a terminal diagnosis. When we last met, Griffiths was about to submit an article reporting striking results in the lab's trial using psilocybin to treat the anxiety and depression of cancer patients; the study found one of the largest treatment effects ever demonstrated for a psychiatric intervention. The majority of volunteers who had a mystical experience reported that their fear of death had either greatly diminished or completely disappeared.

Once again, hard questions arise about the meaning and authority of such experiences, especially ones that appear to convince people that consciousness is not confined to brains and might somehow survive our deaths. Yet even to questions of this kind Griffiths brings an open and curious mind. "The phenomenology of these experiences is so profoundly reorganizing and profoundly compelling that I'm willing to hold there's a mystery here we can't understand."

Griffiths has clearly traveled a long way from the strict behaviorism that once informed his scientific worldview; the experience of alternate states of consciousness, both his own and those of his volunteers, has opened him to possibilities about which few scientists will dare speak openly.

"So what happens after you die? All I need is one percent [of uncertainty]. I can't think of anything more interesting than what I may or may not discover at the time I die. That's the most interesting question going." For that reason, he fervently hopes he isn't hit by a bus but rather has enough time to "savor" the experience without the distraction of pain. "Western materialism says the switch gets turned off and that's it. But there are so many other descriptions. It could be a beginning! Wouldn't that be amazing?"

This is when Griffiths turned the tables and started asking me about my own spiritual outlook, questions for which I was completely unprepared.

"How sure are you there is nothing after death?" he asked. I demurred, but he persisted. "What do you think the chances are there is something beyond death? In percentages."

"Oh, I don't know," I stammered. "Two or three percent?" To this day I have no idea where that estimate came from, but Griffiths seized on it. "That's a lot!" So I turned the table back again, put the same question to him.

"I don't know if I want to answer it," he said with a laugh, glancing at my tape recorder. "It depends on which hat I'm wearing."

Roland Griffiths had more than one hat! I only had one, I realized, and that made me feel a little jealous.

Compared with many scientists—or for that matter many spiritual types—Roland Griffiths possesses a large measure of what Keats, referring to Shakespeare, described as "negative capability," the ability to exist amid uncertainties, mysteries, and doubt without reaching for absolutes, whether those of science or spirituality. "It makes no more sense to say I'm 100 percent convinced of a material worldview than to say I'm 100 percent convinced of the literal version of the Bible."

At our last meeting, a dinner at a bistro in his Baltimore neighborhood, I tried to engage Griffiths in a discussion of the ostensible conflict between science and spirituality. I asked him if he agreed with E. O. Wilson, who has written that all of us must ultimately choose: either the path of science or the path of spirituality. But Griffiths doesn't see the two ways of knowing as mutually exclusive and has little patience for absolutists on either side of the supposed divide. Rather, he hopes the two ways can inform each other and correct each other's defects, and in that exchange help us to pose and then, possibly, answer the big questions we face. I then read to him a letter from Huston Smith, the scholar of comparative religion who in 1962 had volunteered in Walter Pahnke's Good Friday Experiment. It was written to Bob Jesse shortly after the publication of Griffiths's landmark 2006 paper; Jesse had shared it with me.

"The Johns Hopkins experiment shows—proves—that under controlled, experimental conditions, psilocybin can occasion genuine mystical experiences. It uses science, which modernity trusts, to undermine modernity's secularism. In doing so, it offers hope of nothing less than a re-sacralization of the natural and social world, a spiritual revival that is our best defense against not only soullessness, but against religious fanaticism. And it does so in the very teeth of the unscientific prejudices built into our current drug laws."

As I read Smith's letter aloud, a smile bloomed across Griffiths's face; he was clearly moved but had little to add except to say, "That's beautiful."

CHAPTER TWO

NATURAL HISTORY
Bemushroomed

AT THE END of my first meeting with Roland Griffiths, the session in his Johns Hopkins office where he engaged me on the topics of his own mystical experience, my assessment of the odds of an afterlife, and the potential of psilocybin to change people's lives, the scientist stood up from his desk, unfolding his lanky frame, and reached into the pocket of his trousers to take out a small medallion.

"A little gift for you," he explained. "But first, you must answer a question.

"At this moment," Griffiths began, locking me in firm eye contact, "are you aware that you are aware?" Perplexed, I thought for a long, self-conscious moment and then replied in the affirmative. This must have been the correct answer, because Griffiths handed me the coin. On one side was a quartet of tall, slender, curving *Psilocybe cubensis*, one of the more common species of magic mushroom. On the back was a quotation from William Blake that, it occurred to me later, neatly aligned the way of the scientist with that of the mystic: "The true method of knowledge is experiment."

It seems that the previous summer Roland Griffiths had gone for the first time to Burning Man (*had I heard of it?*), and when he learned that no money is exchanged in the temporary city, only gifts, he had the mushroom medallions minted so he would have something suitable to give away or trade. Now, he gives the coins to volunteers in the research program as a parting gift. Griffiths had surprised me once again. Or twice. First, that the scientist had attended the arts-and-psychedelics festival in the Nevada desert. And, second, that he had seen fit in choosing his gift to honor the psilocybin mushroom itself.

On one level, a mushroom medallion made perfect sense: the molecule that Griffiths and his colleagues have been working with for the last fifteen years does, after all, come from a fungus. Both the mushroom and its psychoactive compound were unknown to science until the 1950s, when the psilocybin mushroom was discovered in southern Mexico, where Mazatec Indians had been using "the flesh of the gods," in secret, for healing and divination since before the Spanish conquest. Yet, apart from the decorative ceramic mushroom on the shelf in the session room, there are few if any reminders of "magic mushrooms" in the lab. No one I spoke to at Hopkins ever mentioned the rather astonishing fact that the life-changing experiences their volunteers were reporting owed to the action of a chemical compound found in nature—in a mushroom.

In the laboratory context, it can be easy to lose sight of this astonishment. All of the scientists doing psychedelic research today work exclusively with a synthetic version of the psilocybin molecule. (The mushroom's psychoactive compound was first identified, synthesized, and named in the late 1950s by Albert Hofmann, the Swiss chemist who discovered LSD.) So the volunteers ingest a little white pill made in a lab, rather than a handful of gnarly and acrid-tasting mushrooms. Their journeys unfold in a landscape of medical suites populated, figuratively speaking, by men and women in white coats.

I suppose this is the usual distancing effect of modern science at work, but here it is compounded by a specific desire to distance psilocybin from its tangled roots (or I should say, mycelia) in the worlds of 1960s counterculture, Native American shamanism, and, perhaps, nature itself. For it is there—in nature—that we bump up against the mystery of a little brown mushroom with the power to change the consciousness of the animals that eat it. LSD too, it is easy to forget, was derived from a fungus, *Claviceps purpurea*, or ergot. Somehow, for some reason, these remarkable mushrooms produce, in addition to spores, meanings in human minds.

In the course of my days spent hanging around the Hopkins lab and hours spent interviewing people about their psilocybin journeys, I became increasingly curious to explore this other territory—that is, the natural history of these mushrooms and their strange powers. Where did these mushrooms grow, and how? Why did they evolve the ability to produce a chemical compound so closely related to serotonin, the neurotransmitter, that it can slip across the blood-brain barrier and temporarily take charge of the mammalian brain? Was it a defense chemical, intended to poison mushroom eaters? That would seem to be the most straightforward explanation, yet it is undermined by the fact the fungus produces the hallucinogen almost exclusively in its "fruiting body"—that part of the organism it is *happiest* to have eaten. Was there perhaps some benefit to the mushroom in being able to change the minds of the animals that eat it?*

---

* Technically, a mushroom is the "fruiting body" of a fungus—its reproductive organ. Think of mushrooms as the apples on a tree that grows entirely underground. Most of the fungal organism exists belowground, in the form of mycelia—the typically white cobwebby single-cell-wide filaments that extend through the soil. But because it is hard to observe and study these delicate subterranean structures— they can't be unearthed without breaking—we tend to focus on the mushrooms we can see, even though they are just the tip of a kind of fungal iceberg.

There were also the more philosophical questions posed by the existence of a fungus that could not only change consciousness but occasion a profound mystical experience in humans. This fact can be interpreted in two completely different ways. On the first interpretation, the mind-altering power of psilocybin argues for a firmly materialist understanding of consciousness and spirituality, because the changes observed in the mind can be traced directly to the presence of a chemical—psilocybin. What is more material than a chemical? One could reasonably conclude from the action of psychedelics that the gods are nothing more than chemically induced figments of the hominid imagination.

Yet, surprisingly, most of the people who have had these experiences don't see the matter that way at all. Even the most secular among them come away from their journeys convinced there exists something that transcends a material understanding of reality: some sort of a "Beyond." It's not that they deny a naturalistic basis for this revelation; they just interpret it differently.

If the experience of transcendence is mediated by molecules that flow through both our brains and the natural world of plants and fungi, then perhaps nature is not as mute as Science has told us, and "Spirit," however defined, exists *out there*—is immanent in nature, in other words, just as countless premodern cultures have believed. What to my (spiritually impoverished) mind seemed to constitute a good case for the *disenchantment* of the world becomes in the minds of the more psychedelically experienced irrefutable proof of its fundamental enchantment. Flesh of the gods, indeed.

So here was a curious paradox. The same phenomenon that pointed to a materialist explanation for spiritual and religious belief gave people an experience so powerful it convinced them of the existence of a nonmaterial reality—the very basis of religious belief.

I hoped that getting to know the psychoactive LBMs (mycologist

shorthand for "little brown mushrooms") at the bottom of this paradox might clarify the matter or, perhaps, somehow dissolve it. I was
already something of a mushroom hunter, secure in my ability to
identify a handful of edible woodland species (chanterelles, morels,
black trumpets, and porcini) with a high enough degree of confidence to eat what I found. However, I had been told by all my teachers that the world of LBMs was far more daunting in its complexity
and peril; many if not most of the species that can kill you are LBMs.
But perhaps with some expert guidance, I could add a *Psilocybe* or two
to my mushroom hunting repertoire and in the process begin to unpack the mystery of their existence and spooky powers.

o   o   o

THERE WAS NEVER any doubt who could best help me on this
quest, assuming he was willing. Paul Stamets, a mycologist from
Washington State who literally wrote the book on the genus *Psilocybe,* in the form of the authoritative 1996 field guide *Psilocybin Mushrooms of the World*. Stamets has himself "published"—that is, identified and described in a peer-reviewed journal—four new species of
Psilocybe, including *azurescens*, named for his son Azureus[†] and the
most potent species yet known. But while Stamets is one of the country's most respected mycologists, he works entirely outside the academy, has no graduate degree, funds most of his own research,[‡] and
holds views of the role of fungi in nature that are well outside the

---

* Pronounced sill-OSS-a-bee.

† Complicating matters, Stamets first named his son for the bluish color that *Psilocybes* turn, then named the bluest of *Psilocybes* after his son.

‡ Since 1984, Stamets has run a very successful company called Fungi Perfecti,
which sells medicinal mushroom supplements, spores, and growing kits for edible
mushrooms, as well as various other mushroom-related paraphernalia.

scientific mainstream and that, he will gladly tell you, owe to insights granted to him by the mushrooms themselves, in the course of both close study and regular ingestion.

I've known Stamets for years, though not very well and always from what I confess has been a somewhat skeptical distance. His extravagant claims for the powers of mushrooms and eyebrow-elevating boasts about his mushroom work with institutions like DARPA (the Pentagon's Defense Advanced Research Projects Agency) and NIH (the National Institutes of Health) are bound to set off a journalist's bullshit detector, rightly or—as often happens in his case—wrongly.

Over the years, we've found ourselves at some of the same conferences, so I've had several opportunities to hear his talks, which consist of a beguiling (often brilliant) mash-up of hard science and visionary speculation, with the line between the two often impossible to discern. His 2008 TED talk, which is representative, has been viewed online more than four million times.

Stamets, who was born in 1955 in Salem, Ohio, is a big hairy man with a beard and a bearish mien; I was not surprised to learn he once worked as a lumberjack in the Pacific Northwest. Onstage, he usually wears what appears to be a felt hat in the alpine style but which, as he'll explain, is in fact made in Transylvania from something called amadou, the spongy inner layer of the horse's hoof fungus (*Fomes fomentarius*), a polypore that grows on several species of dead or dying trees. Amadou is flammable and in ancient times was used to start and transport fires. Ötzi, the five-thousand-year-old "Ice Man" found mummified in an alpine glacier in 1991, was carrying a pouch in which he had a piece of amadou. Because of its antimicrobial properties, *Fomes fomentarius* was also used to dress wounds and preserve food. Stamets is so deep into the world of fungi there's frequently one perched on top of his head.

Fungi constitute the most poorly understood and underappreciated kingdom of life on earth. Though indispensable to the health of the planet (as recyclers of organic matter and builders of soil), they are the victims not only of our disregard but of a deep-seated ill will, a mycophobia that Stamets deems a form of "biological racism." Leaving aside their reputation for poisoning us, this is surprising in that we are closer, genetically speaking, to the fungal kingdom than to that of the plants. Like us, they live off the energy that plants harvest from the sun. Stamets has made it his life's work to right this wrong, by speaking out on their behalf and by demonstrating the potential of mushrooms to solve a great many of the world's problems. Indeed, the title of his most popular lecture, and the subtitle of his 2005 book, *Mycelium Running*, is "How Mushrooms Can Help Save the World." By the end of his presentation, this claim no longer sounds hyperbolic.

I can remember the first time I heard Stamets talk about "mycoremediation"—his term for the use of mushrooms to clean up pollution and industrial waste. One of the jobs of fungi in nature is to break down complex organic molecules; without them, the earth would long ago have become a vast, uninhabitable waste heap of dead but undecomposed plants and animals. So after the *Exxon Valdez* ran aground off the coast of Alaska in 1989, spilling millions of gallons of crude oil into Prince William Sound, Stamets revived a long-standing idea of putting fungi to work breaking down petrochemical waste. He showed a slide of a steaming heap of oily black sludge before inoculating it with the spores of oyster mushrooms, and then a second photograph of the same pile taken four weeks later, when it was reduced by a third and covered in a thick mantle of snowy white oyster mushrooms. It was a performance, and a feat of alchemy, I won't soon forget.

But Stamets's aspirations for the fungal kingdom go well beyond turning petrochemical sludge into arable soil. Indeed, in his view there is scarcely an ecological or medical problem that mushrooms can't help solve.

Cancer? Stamets's extract of turkey tail mushrooms (*Trametes versicolor*) has been shown to help cancer patients by stimulating their immune systems. (Stamets claims to have used it to help cure his mother's stage 4 breast cancer.)

Bioterrorism? After 9/11, the federal government's Bioshield program asked to screen hundreds of the rare mushroom strains in Stamets's collection and found several that showed strong activity against SARS, smallpox, herpes, and bird and swine flu. (If this strikes you as implausible, remember that penicillin is the product of a fungus.)

Colony collapse disorder (CCD)? After watching honeybees visiting a woodpile to nibble on mycelium, Stamets identified several species of fungus that bolster the bees' resistance to infection and CCD.

Insect infestation? A few years ago, Stamets won a patent for a "mycopesticide"—a mutant mycelium from a species of *Cordyceps* that, after being eaten by carpenter ants, colonizes their bodies and kills them, but not before chemically inducing the ant to climb to the highest point in its environment and then bursting a mushroom from the top of its head that releases its spores to the wind.

The second or third time I watched Stamets show a video of a *Cordyceps* doing its diabolical thing to an ant—commandeering its body, making it do its bidding, and then exploding a mushroom from its brain in order to disseminate its genes—it occurred to me that Stamets and that poor ant had rather a lot in common. Fungi haven't killed him, it's true, and he probably knows enough about their wiles to head off that fate. But it's also true that this man's life—his brain!—has been utterly taken over by fungi; he has dedicated himself to their cause, speaking for the mushrooms in the same way that Dr. Seuss's Lorax speaks for the trees. He disseminates fungal spores far and wide, helping them, whether by mail order or sheer dint of his enthusiasm, to vastly expand their range and spread their message.

o   o   o

I DON'T THINK I'm saying anything about Paul Stamets to which he would object. He writes in his book that mycelia—the vast, cob-webby whitish net of single-celled filaments, called hyphae, with which fungi weave their way through the soil—are intelligent, form-ing "a sentient membrane" and "the neurological network of nature." The title of his book *Mycelium Running* can be read in two ways. The mycelium is indeed always running through the ground, where it plays a critical role in forming soils, keeping plants and animals in good health, and knitting together the forest. But the mycelium are also, in Stamets's view, running the show—that of nature in general and, like a neural software program, the minds of certain creatures, including, he would be the first to tell you, Paul Stamets himself. "Mushrooms are bringing us a message from nature," he likes to say. "This is a call I'm hearing."

Yet even some of Stamets's airier notions turn out to have a scien-tific foundation beneath them. For years now, Stamets has been talking about the vast web of mycelia in the soil as "Earth's natu-ral Internet"—a redundant, complexly branched, self-repairing, and scalable communications network linking many species over tre-mendous distances. (The biggest organism on earth is not a whale or a tree but a mushroom—a honey fungus in Oregon that is 2.4 miles wide.) Stamets contends that these mycelial networks are in some sense "conscious": aware of their environment and able to respond to challenges accordingly. When I first heard these ideas, I thought they were, at best, fanciful metaphors. Yet in the years since, I've watched as a growing body of scientific research has emerged to suggest they are much more than metaphors. Experiments with slime molds have demonstrated these organisms can navigate mazes in search of food—sensing its location and then growing in that direction. The mycelia in a forest *do* link the trees in it, root to root, not only supplying them with nutrients, but serving as a medium that conveys information about environmental threats and allows trees to selectively send nutrients to other trees in the forest.* A forest is a far more complex, sociable, and intelligent entity than we knew, and it is fungi that organize the arboreal society.

Stamets's ideas and theories have turned out to be far more durable, and practicable, than I ever would have guessed. This was the other reason I became eager to spend some time with Stamets: I was curious to find out how his own experience with psilocybin had colored his thinking and lifework. Yet I wasn't at all certain he would be willing to talk on the record about psilocybin, much less take me 'shroom hunting, now that he had a successful business, had eight or nine patents to his name, and was collaborating with institutions like DARPA and NIH and the Lawrence Livermore National Laboratory. In the more recent interviews and lectures I could find online, he seldom talked about psilocybin and often omitted mention of the field guide from his list of publications. What's more, he had just received prestigious honors from the Mycological Society of Amer-

---

* Scientists at the University of British Columbia (UBC) injected fir trees with radioactive carbon isotopes, then followed the spread of the isotopes through the forest community using a variety of sensing methods, including a Geiger counter. Within a few days, stores of the radioactive carbon had been routed from tree to tree. Every tree in a plot thirty meters square was connected to the network; the oldest trees functioned as hubs, some with as many as forty-seven connections. The diagram of the forest network resembled a map of the Internet. In what is surely a tip of the hat to Stamets, a paper by one of the UBC scientists dubbed it the "wood-wide web."

ica and the American Association for the Advancement of Science (AAAS). Paul Stamets, it seemed, had gone legit. Bad timing for me.

o   o   o

THANKFULLY, I WAS WRONG. When I reached Stamets at his home in Kamilche, Washington, and told him what I was up to, he couldn't have been more forthcoming or cooperative. We talked for a long time about psilocybin mushrooms, and it soon became clear they remained a subject of keen interest to him. He knew all about the work going on at Hopkins—in fact had consulted with the Hopkins team when they were first looking for a source of psilocybin. My impression was that the revival of legitimate university research had made Stamets more comfortable reopening this particular chapter in his life. He mentioned he was in the process of updating the 1996 psilocybin field guide. The only discordant note in the conversation came when I casually dropped the slang expression for psilocybin when asking him about going hunting for 'shrooms.

"I really, really hate that word," he said, almost gravely, adopting the tone of a parent upbraiding a potty-mouthed child.

The word never crossed my lips again.

By the end of the call, Stamets had invited me up to his place in Washington State, on the Little Skookum Inlet at the base of the Olympic Peninsula. I asked him, gingerly, if I could come at a time when the *Psilocybes* were fruiting. "Most of them have already come and gone," he said. "But if you come right after Thanksgiving, and the weather's right, I can take you to the only place in the world where *Psilocybe azurescens* has been consistently found, at the mouth of the Columbia River." He mentioned the name of the park where he had found them in the past and told me to book a yurt there, adding, "Probably best not to use my name."

o o o

IN THE WEEKS BEFORE my trip to Washington State, I pored over Stamets's field guide, hoping to prepare myself for the hunt. It seems there are more than two hundred species of *Psilocybe*, distributed all over the world; it's not clear whether that's always been the case, or if the mushrooms have followed in the footsteps of the animals who have taken such a keen interest in them. (Humans have been using psilocybin mushrooms sacramentally for at least seven thousand years, according to Stamets. But animals sometimes ingest them too, for reasons that remain obscure.)

Psilocybes are saprophytes, living off dead plant matter and dung. They are denizens of disturbed land, popping up most often in the habitats created by ecological catastrophe, such as landslides, floods, storms, and volcanoes. They also prosper in the ecological catastrophes caused by our species: clear-cut forests, road cuts, the wakes of bulldozers, and agriculture. (Several species live in and fruit from the manure of ruminants.) Curiously, or perhaps *not* so curiously, the most potent species occur less often in the wild than in cities and towns; their predilection for habitats disturbed by us has allowed them to travel widely, "following streams of debris," including our own. In recent years, the practice of mulching with wood chips has vastly expanded the range of a handful of potent *Psilocybes* once confined to the Pacific Northwest. They now thrive in all those places we humans now "landscape": suburban gardens, nurseries, city parks, churchyards, highway rest stops, prisons, college campuses, even, as Stamets likes to point out, on the grounds of courthouses and police stations. "Psilocybe mushrooms and civilization continue to coevolve," Stamets writes.

So you would think these mushrooms would be fairly easy to find.

In fact after I published an article about psilocybin research, I was informed by a student that after the December rains *Psilocybes* can be found on the Berkeley campus, where I teach. "Look in the wood chips," he advised. Yet as soon as I began studying the photographs in Stamets's field guide, I began to despair of ever identifying any mushroom as a member of the genus, much less learning how to distinguish one species of *Psilocybe* from another.

To judge from the pictures, the genus is just a big bunch of little brown mushrooms, most of them utterly nondescript. By comparison, the edible species with which I was familiar were as distinct as tulips are from roses, poodles from Great Danes. Yes, all the *Psilocybes* have gills, but that isn't much help, because thousands of other mushrooms have gills, too. After that, you're trying to sort out a bewildering array of characteristics, not all of which are shared by the class. Some *Psilocybes* have a little nipple-like knob or protrusion on top—it's called an umbo, I learned; others don't. Some were "viscid"—slippery or slimy when wet, giving them a shiny appearance. Others were dull and matte gray; some, like *azurescens*, were a milky caramel color. Many but not all *Psilocybes* sport a "pellicle"—a condom-like layer of gelatinous material covering the cap that can be peeled off. My fungal vocabulary might be expanding, but my confidence was rapidly collapsing, much like the mushroom that, in the course of a single day, decomposes into an inky puddle.

By the time I got to chapter four, "The Dangers of Mistaken Identification," I was ready to throw in the towel. "Mistakes in mushroom identification can be lethal," Stamets begins, before displaying a photograph in which a *Psilocybe stuntzii* is seen growing cheek by jowl with a trio of indistinguishable *Galerina autumnalis*, an unremarkable little mushroom that, when eaten, "can result in an agonizing death."

But while Stamets urges extreme circumspection in amateurs hoping to identify *Psilocybes*, he also equips the mushroom hunter who hasn't been completely discouraged with something he calls "The Stametsian Rule": a three-pronged test that, he (sort of) assures us, can head off death and disaster.

"How do I know if a mushroom is a psilocybin producing species or not?"

"If a gilled mushroom has purplish brown to black spores, *and* the flesh bruises bluish, the mushroom in question is very likely a psilocybin-producing species." This is definitely a big help, though I wouldn't mind something more categorical than "very likely." He then offers a sobering caveat. "I know of no exceptions to this rule," he adds, "but that does not mean there are none!"

After committing to memory the Stametsian Rule, I began picking promising-looking gilled LBMs—in my neighbors' yards, on my walk to work, in the parking lot of the bank—and then roughing them up a bit to see if they would turn black and blue. The blue pigment is in fact evidence of oxidized psilocin, one of the two main psychoactive compounds in a *Psilocybe*. (The other is psilocybin, which breaks down into psilocin in the body.) To determine if the mushroom in question had purplish-brown or black spores, I began making spore prints. This involves cutting the cap off a mushroom and placing it, gill side down, on a piece of white paper. (Or black paper if you have reason to believe the mushroom has white spores.) Within hours, the mushroom cap releases its microscopic spores, which will form a pretty, shadowy pattern on the paper (reminiscent of a lipstick kiss) that you can then try to decide is purplish brown or black—or rust colored, in which case you might have a deadly *Galerina* on your hands.

Certain things are perhaps best learned in person, rather than from a book. I decided I should probably wait before making any irreversible decisions until I had spent some time in the company of my mycological Virgil.

o   o   o

AT THE TIME OF MY VISIT, Paul Stamets lived with his partner, Dusty Yao, and their two big dogs, Plato and Sophie, in a sprawling new house on the Little Skookum Inlet that is constructed inside and out of a small forest's worth of the most gorgeous clear Douglas fir and cedar. Like many species of fungi, Stamets has a passionate attachment to trees and wood. I arrived on a Friday; our reservation at the campsite wasn't until Sunday night, so we had the better part of the weekend to talk *Psilocybes*, eat (other kinds of) mushrooms, tour the Fungi Perfecti facilities, and ramble the surrounding woods and shoreline with the dogs before driving south to the Oregon border Sunday morning to hunt azzies.

This was the house that mushrooms built, Stamets explained, launching into its story before I had a chance to unpack my bag. It replaced a rickety old farmhouse on the site that, when Stamets moved in, was slowly succumbing to an infestation of carpenter ants. Stamets set about devising a mycological solution to the problem. He knew precisely which species of *Cordyceps* could wipe out the ant colony, but so did the ants: they scrupulously inspect every returning member for *Cordyceps* spores and promptly chew off the head of any ant bearing spores, dumping the body far away from the colony. Stamets outwitted the ants' defense by breeding a mutant *Cordyceps*-like fungus that postponed sporulation. He put some of its mycelium in his daughter's dollhouse bowl, left that on the floor of the kitchen, and during the night watched as a parade of ants carried the mycelium into the nest—having mistaken it for a safe food source. When the fungus eventually sporulated, it was already deep inside the colony and the ants were done for: the *Cordyceps* colonized their bodies and sent fruiting bodies bursting forth from their heads. It was too late to save the farmhouse, but with the proceeds from the sale of his patent on the fungus Stamets was able to erect this far grander monument to mycological ingenuity.

The house was spacious and comfortable; I had a whole upstairs wing of bedrooms to myself. The living room, where we spent most of a rainy December weekend, had a soaring cathedral ceiling, a big wood-burning fireplace, and, looming over the room from across the way, a seven-and-a-half-foot-tall skeleton of a cave bear. A painting of Albert Hofmann hangs over the fireplace. Overhead, beneath the peak, is a massive round stained glass depicting "The Universality of the Mycelial Archetype"—an intricate tracery of blue lines on a night sky, the lines representing at once mycelium, roots, neurons, the Internet, and dark matter.

Displayed on the walls heading upstairs from the living room are framed artworks, photographs, and keepsakes, including a diploma signifying the successful completion of one of the Merry Pranksters' Acid Tests, signed by Ken Kesey and Neal Cassady. There are several photographs of Dusty posing in old-growth forests with impressive specimens of fungi and a colorfully grotesque print by Alex Grey, the dean of American psychedelic artists. The print is Grey's interpretation of the so-called stoned ape theory, depicting an early, electrified-looking hominid clutching a *Psilocybe* while a cyclone of abstractions flies out of its mouth and forehead. The only reason I could make any sense of the image at all was that a few days earlier I had received an e-mail from Stamets referring to the theory in question: "I want to discuss the high likelihood that the Stoned Ape Theory, first presented by Roland Fischer and then popularized/restated by Terence McKenna, is probably true—[ingestion of psilocybin] causing a rapid development of the hominid brain for analytical thinking and societal bonding. Did you know that 23 primates (including humans) consume mushrooms and know how to distinguish 'good' from 'bad'?"

I did not.

But the brief, elliptical e-mail nicely prefigured the tenor of my weekend with Stamets as I struggled to absorb a torrent of mycological fact and speculation that, like a rushing river, is impossible to ford without being knocked sideways. The sheer brilliance of Stamets's mushroom's-eye view of the world can be dazzling, but after a while it can also make you feel claustrophobic, as only the true monomaniac or autodidact—and Stamets is both—can do. *Everything is connected* is ever the subtext with such people; in this case what connects everything you could possibly think of just happens to be fungal mycelia.

I was curious to find out how Stamets came by his mycocentric worldview and what role psilocybin mushrooms, in particular, might have contributed to it. Stamets grew up in an Ohio town outside Youngstown called Columbiana, the youngest of five children. His father's engineering company went belly-up when Paul was a boy, the family "going from riches to rags pretty quickly." Dad began to drink heavily, and Paul began looking up to his older brother John as a role model.

Five years his senior, John was an aspiring scientist—he would receive a scholarship to study neurophysiology—who kept "an exquisite laboratory in the basement," a realm that was Paul's idea of heaven, but to which John seldom granted his little brother admittance. "I thought all houses had laboratories, so whenever I went over to a friend's house, I would ask where the laboratory was. I didn't understand why they would always point me to the bathroom instead—the lavatory." Winning John's approval became a motive force in Paul's life, which perhaps explains the value Stamets places on mainstream scientific recognition of work. John had died, of a heart attack, six months before my visit and, as it happened, on the same day Paul received word of his AAAS honor. His death was a loss from which Paul hadn't yet recovered.

When Paul was fourteen, John told him about magic mushrooms, and when he went off to Yale, John left behind a book, *Altered States of Consciousness*, that made a tremendous impression on Paul. Edited by Charles T. Tart, a psychologist, the book is a doorstop of an anthology of scholarly writings about non-ordinary mental states, covering the spectrum from dreaming and hypnosis to meditation and psychedelics. But the reason the book made such a lasting impression on Stamets had less to do with its contents, provocative as these were, than with the reaction the book elicited in certain adults.

"My friend Ryan Snyder wanted to borrow it. His parents were really conservative. A week later, when I told him I wanted it back, he stalls and delays. Another week goes by, I ask him again, and he finally confesses what happened. 'My parents found it and they burned it.'

"*They burned my book?!?* That was a pivotal moment for me. I saw the Snyders as the enemy, trying to suppress the exploration of consciousness. But if this was such powerful information that they felt compelled to destroy it, then this was powerful information I now had to have. So I owe them a debt of gratitude."

Stamets went off to Kenyon College, where, as a freshman, he had "a profound psychedelic experience" that set his course in life. As long as he could remember, Stamets had been stymied by a debilitating stutter. "This was a huge issue for me. I was always looking down at the ground because I was afraid people would try to speak to me. In fact, one of the reasons I got so good at finding mushrooms was because I was always looking down."

One spring afternoon toward the end of his freshman year, walking alone along the wooded ridgeline above campus, Stamets ate a whole bag of mushrooms, perhaps ten grams, thinking that was a proper dose. (Four grams is a lot.) As the psilocybin was coming on, Stamets spied a particularly beautiful oak tree and decided he would climb it. "As I'm climbing the tree, I'm literally getting higher as I'm climbing higher." Just then the sky begins to darken, and a thunderstorm lights up the horizon. The wind surges as the storm approaches, and the tree begins to sway.

"I'm getting vertigo but I can't climb down, I'm too high, so I just wrapped my arms around the tree and held on, hugging it tightly. The tree became the *axis mundi*, rooting me to the earth. 'This is the tree of life,' I thought; it was expanding into the sky and connecting me to the universe. And then it hits me: I'm going to be struck by lightning! Every few seconds there's another strike, here, then there, all around me. On the verge of enlightenment, I'm going to be electrocuted. This is my destiny! The whole time, I'm being washed by warm rains. I am crying now, there is liquid everywhere, but I also feel one with the universe.

"And then I say to myself, what are my issues if I survive this? Paul, I said, you're not stupid, but stuttering is holding you back. You can't look women in the eyes. What should I do? *Stop stuttering now*—that became my mantra. *Stop stuttering now*, I said it over and over and over.

"The storm eventually passed. I climbed down from the tree and walked back to my room and went to sleep. That was the most important experience of my life to that point, and here's why: The next morning, I'm walking down the sidewalk, and here comes this girl I was attracted to. She's way beyond my reach. She's walking toward me, and she says, 'Good morning, Paul. How are you?' I look at her and say, 'I'm doing great.' I wasn't stuttering! And I have hardly ever stuttered since.

"And that's when I realized I wanted to look into these mushrooms."

o   o   o

IN A REMARKABLY SHORT SPAN of time, Stamets made himself into one of the country's leading experts on the genus *Psilocybe*. In 1978, at the age of twenty-three, he published his first book, *Psilocybe Mushrooms and Their Allies*—their allies understood to be us, the animal that had done the most to spread their genes and, as Stamets now saw as his calling, their planetary gospel.

Stamets got his mycological education not at Kenyon, which he left after one year, but at the Evergreen State College, which in the mid-1970s was a new experimental college in Olympia, Washington, where students could design their own course of independent study. A young professor named Michael Beug, who had a degree in environmental chemistry, agreed to take under his wing Stamets and two other equally promising mycologically obsessed students: Jeremy Bigwood and Jonathan Ott. Beug was not himself a mycologist by training, but the four of them mastered the subject together, with the help of an electron microscope and a DEA license that Beug had somehow secured. Thus armed, the four trained their attention on a genus that the rest of the field generally chose to pass over in uncomfortable silence.

Illegal since 1970, psilocybin mushrooms were at the time chiefly of interest to the counterculture, as a gentler, more natural alternative to LSD, but very little was known about their habitat, distribution, life cycle, or potency. It was believed that psychedelic mushrooms were native to southern Mexico, where R. Gordon Wasson had "discovered" them in 1955. By the 1970s, most of the psilocybin in circulation in America was being imported from Latin America or grown domestically from spores of Latin American species, mainly *cubensis*.

The Evergreen group chalked up several notable accomplishments: they identified and published three new psilocybin species, perfected methods for growing them indoors, and developed techniques for measuring levels of psilocin and psilocybin in mushrooms. But perhaps the group's most important contribution was to shift the focus of attention among people who cared about *Psilocybes* from southern Mexico to the Pacific Northwest. Stamets and his colleagues were finding new species of psilocybin mushrooms all around them and publishing their findings. "You could almost feel the earth's axis tilting to this corner of the world." Anywhere you went in the Pacific Northwest, Stamets recalls, you could see people tracing peculiar patterns through farm fields and lawns, bent over in what he calls "the psilocybin stoop."

During this period, the Pacific Northwest emerged as a new center of gravity in American psychedelic culture, with the Evergreen State College serving as its de facto intellectual hub and R&D facility. Beginning in 1976, Stamets and his Evergreen colleagues organized a series of now-legendary mushroom conferences, bringing together the leading lights of both the credentialed and the amateur wings of the psychedelic world, and during my first evening at his house Stamets dug out some VHS tapes of the last of these conferences, held in 1999. The footage had been shot by Les Blank, but as often happened with coverage of such psychedelic gatherings, no one could ever quite get it together to edit the raw footage, so raw it remains.

"Conference" might not do justice to what now appeared on Stamets's television. We watched as several of the attendees—I spotted Dr. Andrew Weil, best known for his books on holistic medicine; the psychedelic chemist Sasha Shulgin and his wife, Ann; and the New York Botanical Garden mycologist Gary Lincoff—arrived to great fanfare in a psychedelically painted school bus piloted by Ken Kesey. (The bus was called Farther, the successor to Further, the original Merry Prankster bus, evidently no longer roadworthy.) The proceedings looked more like a Dionysian revel than a conference, yet there were some serious talks. Jonathan Ott delivered a brilliant lecture on the history of "entheogens"—a term he helped coin. He traced their use all the way back to the Eleusinian mysteries of the Greeks, through the "pharmocratic inquisition," when the Spanish conquest suppressed the Mesoamerican mushroom cults, and forward to the "entheogenic reformation" that has been under way since R. Gordon Wasson's discovery that those cults had survived in Mexico. Along the way, Ott made an offhand reference to the "placebo sacraments" of the Catholic Eucharist.

Then came footage of a big costume ball with lingering close-ups of a giant punch bowl that had been spiked with dozens of different kinds of psychedelic mushrooms. Stamets pointed out several prominent mycologists and ethnobotanists among the revelers; many of them dressed as specific kinds of fungus—*Amanita muscaria*, button mushrooms, and so on. Stamets himself appeared dressed as a bear.

When one is screening raw footage of people in costume tripping on mushrooms and dancing sloppily to a reggae band, a little goes a long way, so after a few minutes we flicked off the TV. I asked Stamets about earlier iterations of the conference, some of which seemed to have a slightly more interesting ratio of intellectual substance to Dionysian revelry. In 1977, for instance, Stamets had the opportunity to play host to two of his heroes: Albert Hofmann and R. Gordon Wasson, whose 1957 article in *Life* magazine describing the first psilocybin journey ever taken by a Westerner—his own—helped launch the psychedelic revolution in America.

Stamets mentioned that he collected original copies of that issue of *Life*, which occasionally show up on eBay and at flea markets, and on my way upstairs to bed that night I stopped in his office so I could have a look at it. The issue was dated May 13, 1957, and Bert

Lahr was on the cover, mugging for the camera in a morning suit and a bowler hat. But the most prominent cover line was devoted to Wasson's notorious article: "The Discovery of Mushrooms That Cause Strange Visions." Stamets said I could have a copy, and I took it to bed.

o   o   o

FROM THE VANTAGE OF TODAY, it is hard to believe that psilocybin was introduced to the West by a vice president of J. P. Morgan in the pages of a mass-circulation magazine owned by Henry Luce; two more establishment characters it would be difficult to dream up. But in 1957, psychedelic drugs had not yet acquired any of the cultural and political stigmas that, a decade later, would weigh on our attitudes toward them. At the time, LSD was not well known outside the small community of medical professionals who regarded it as a potential miracle drug for psychiatric illness and alcohol addiction.

As it happened, the Time-Life founder and editor in chief, Henry Luce, along with his wife, Clare Boothe Luce, had personal knowledge of psychedelic drugs, and they shared the enthusiasm of the medical and cultural elites who had embraced them in the 1950s. In 1964, Luce told a gathering of his staff that he and his wife had been taking LSD "under doctor's supervision"; Clare Boothe Luce recalled that during her first trip in the 1950s she saw the world "through the eyes of a happy and gifted child." Before 1965, when a moral panic erupted over LSD, Time-Life publications were enthusiastic boosters of psychedelics, and Luce took a personal interest in directing his magazine's coverage of them.

So when R. Gordon Wasson approached *Life* magazine with his story, he could not have knocked on a more receptive door. *Life* gave him a generous contract that, in addition to the princely sum of eighty-five hundred dollars, granted him final approval on the editing of his article, as well as the wording of headlines and captions. It specified that Wasson's account include a "description of your own sensations and fantasies under the influence of the mushroom."

As I paged through the issue in bed that evening, the world of 1957 seemed like a faraway planet, even though I lived on it, albeit as a two-year-old. My parents subscribed to *Life*, so the issue probably sat in the big pile in our den for a stretch of my childhood. *Life* magazine was a mass medium in 1957, with a circulation of 5.7 million.

"Seeking the Magic Mushroom," in which "a New York banker goes to Mexico's mountains to participate in the age-old rituals of Indians who chew strange growths that produce visions," opened on a spread with a full-page color photograph of a Mazatec woman turning a mushroom over a smoky fire and goes on for no fewer than fifteen pages. The headline is the first known reference to "magic mushrooms," a phrase that, it turns out, was coined not by a stoned hippie but by a Time-Life headline writer.

"We chewed and swallowed these acrid mushrooms, saw visions, and emerged from the experience awestruck," Wasson tells us, somewhat breathlessly, in the first paragraph. "We had come from afar to attend a mushroom rite but had expected nothing so staggering as the virtuosity of the performing *curanderas* [healers] and the astonishing effects of the mushrooms. [The photographer] and I were the first white men in recorded history to eat the divine mushrooms, which for centuries had been a secret of certain Indian peoples living far from the great world in southern Mexico."

Wasson then proceeds to tell the improbable tale of how someone like him, "a banker by occupation," would end up eating magic mushrooms in the dirt-floored basement of a thatch-roofed, adobe-walled home in a Oaxacan town so remote it could only be reached by means of an eleven-hour trek through the mountains by mule.

The story begins in 1927, during Wasson's honeymoon in the

Catskills. During an afternoon stroll in the autumn woods, his bride, a Russian physician named Valentina, spotted a patch of wild mushrooms, before which "she knelt in poses of adoration." Wasson knew nothing of "those putrid, treacherous excrescences" and was alarmed when Valentina proposed to cook them for dinner. He refused to partake. "Not long married," Wasson wrote, "I thought to wake up the next morning a widower."

The couple became curious as to how two cultures could hold such diametrically opposed attitudes toward mushrooms. They soon embarked on a research project to understand the origins of both "mycophobia" and "mycophilia," terms that the Wassons introduced. They concluded that each Indo-European people is by cultural inheritance either mycophobic (for example, the Anglo-Saxons, Celts, and Scandinavians) or mycophilic (the Russians, Catalans, and Slavs) and proposed an explanation for the powerful feelings in both camps: "Was it not probable that, long ago, long before the beginnings of written history, our ancestors had worshipped a divine mushroom? This would explain the aura of the supernatural in which all fungi seem to be bathed."* The logical next question presented itself to the Wassons—"What kind of mushroom was once worshipped, and why?"—and with that question in hand they embarked on a thirty-year quest to find the divine mushroom. They hoped to obtain evidence for the audacious theory that Wasson had developed and that would occupy him until his death: that the religious impulse in humankind had been first kindled by the visions inspired by a psychoactive mushroom.

---

* The Wassons either dismissed or overlooked a somewhat simpler explanation: that powerful feelings and a cult of mystery could be expected to gather around a "plant" that, depending on knowledge and context, could either nourish and delight or lead to an agonizing death.

As a prominent financier, R. Gordon Wasson had the resources and the connections to enlist all manner of experts and scholars in his quest. One of these was the poet Robert Graves, who shared the Wassons' interest in the role of mushrooms in history and in the common origins of the world's myths and religions. In 1952, Graves sent Wasson a clipping from a pharmaceutical journal that made reference to a psychoactive mushroom used by sixteenth-century Mesoamerican Indians. The article was based on research done in Central America by Richard Evans Schultes, a Harvard ethnobotanist who studied the uses of psychoactive plants and fungi by indigenous cultures. Schultes was a revered professor whom students recall shooting blowguns in class and keeping a basket of peyote buttons outside his Harvard office; he trained a generation of American ethnobotanists, including Wade Davis, Mark Plotkin, Michael Balick, Tim Plowman, and Andrew Weil. Along with Wasson, Schultes is one of a handful of figures whose role in bringing psychedelics to the West has gone underappreciated; indeed, some of the first seeds of that movement have quite literally sat in the Harvard herbarium since the 1930s, more than a quarter century before Timothy Leary set foot on the campus. For it was Schultes who first identified *teonanácatl*—the sacred mushroom of the Aztecs and their descendants—as well as *ololiuqui*, the seeds of the morning glory, which the Aztecs also consumed sacramentally and which contain an alkaloid closely related to LSD.

Up to this point, the Wassons had been looking toward Asia for their divine mushroom; Schultes reoriented their quest, pointing them toward the Americas, where there were scattered reports, from missionaries and anthropologists, suggesting that an ancient mushroom cult might yet survive in the remote mountain villages of southern Mexico.

In 1953, Wasson made the first of ten trips to Mexico and Central

America, several of them to the village of Huautla de Jiménez, deep in the mountains of Oaxaca, where one of his informants—a missionary—had told him healers were using mushrooms. At first the locals were tight-lipped. Some told Wasson they had never heard of the mushrooms, or that they were no longer used, or that the practice survived only in some other, distant village.

Their reticence was not surprising. The sacramental use of psychoactive mushrooms had been kept secret from Westerners for four hundred years, since shortly after the Spanish conquest, when it was driven underground. The best account we have of the practice is that of the Spanish missionary priest Bernardino de Sahagún, who in the sixteenth century described the use of mushrooms in an Aztec religious observance:

> These they ate before dawn with honey, and they also drank cacao before dawn. The mushrooms they ate with honey when they began to get heated from them, they began to dance, and some sang, and some wept . . . Some cared not to sing, but would sit down in their rooms, and stayed there pensive-like. And some saw in a vision that they were dying, and they wept, and others saw in a vision that some wild beast was eating them, others saw in a vision that they were taking captives in war . . . others saw in a vision that they were to commit adultery and that their heads were to be bashed in therefor . . . Then when the drunkenness of the mushrooms had passed, they spoke one with another about the visions that they had seen.

The Spanish sought to crush the mushroom cults, viewing them, rightly, as a mortal threat to the authority of the church. One of the first priests Cortés brought to Mexico to Christianize the Aztecs declared that the mushrooms were the flesh of "the devil that they worshipped, and . . . with this bitter food they received their cruel god in communion." Indians were interrogated and tortured into confessing the practice, and mushroom stones—many of them foot-tall chiseled basalt sculptures of the sacred fungi, presumably used in religious ceremonies—were smashed. The Inquisition would bring dozens of charges against Native Americans for crimes involving both peyote and psilocybin, in what amounted to an early battle in the war on drugs—or, to be more precise, the war on certain plants and fungi. In 1620, the Roman Catholic Church declared that the use of plants for divination was "an act of superstition condemned as opposed to the purity and integrity of our Holy Catholic Faith."

It's not hard to see why the church would have reacted so violently to the sacramental use of mushrooms. The Nahuatl word for the mushrooms—flesh of the gods—must have sounded to Spanish ears like a direct challenge to the Christian Sacrament, which of course was also understood to be the flesh of the gods, or rather of the one God. Yet the mushroom sacrament enjoyed an undeniable advantage over the Christian version. It took an act of faith to believe that eating the bread and wine of the Eucharist gave the worshipper access to the divine, an access that had to be mediated by a priest and the church liturgy. Compare that with the Aztec sacrament, a psychoactive mushroom that granted anyone who ate it direct, unmediated access to the divine—to visions of another world, a realm of the gods. So who had the more powerful sacrament? As a Mazatec Indian told Wasson, the mushrooms "carry you there where god is."

The Roman Catholic Church might have been the first institution to fully recognize the threat to its authority posed by a psychedelic plant, but it certainly wouldn't be the last.

o   o   o

ON THE NIGHT OF JUNE 29–30, 1955, R. Gordon Wasson experienced the sacred mushrooms firsthand. On his third trip to Huautla,
he had persuaded María Sabina, a sixty-one-year-old Mazatec and a
respected *curandera* in the village, to let him and his photographer not
only observe but take part in a ceremony in which no outsider had ever
participated. The *velada*, as the ceremony was called, took place after
dark in the basement of the home of a local official Wasson had enlisted
in his cause, before a simple altar "adorned with Christian images." To
protect her identity, Wasson called Sabina "Eva Mendez," discerning
"a spirituality in her expression that struck us at once." After cleaning
the mushrooms and passing them through the purifying smoke of
incense, Sabina handed Wasson a cup containing six pairs of mushrooms; she called them "the little children." They tasted awful: "acrid
with a rancid odor that repeated itself." Even so, "I could not have
been happier: this was the culmination of six years of pursuit."

The visions that now arrived "were in vivid color, always harmo-
nious. They began with art motifs, angular such as might decorate
carpets or textiles or wallpaper . . . Then they evolved into palaces
with courts, arcades, gardens—resplendent palaces all laid over with
semiprecious stone. Then I saw a mythological beast drawing a regal
chariot." And so forth.

Wasson's original field notebooks are in the botanical library at
Harvard. In a neat but somewhat idiosyncratic hand, he kept meticulous track of the time that night, from arrival (8:15) to ingestion
(10:40) to the snuffing out of the last candle (10:45).

After that, the handwriting disintegrates. Some sentences now
appear upside down, and Wasson's descriptions of what he felt and
saw gradually break into fragments:

Nausea as vision distorted. Touching wall—made the world of visions seem to crumble. Light from above door and below—moon. Table took new forms—creatures, great processional vehicle, architectural patterns of radiant color. Nausea. No photos once the [illegible] seized us.

Architectural

Eyes out of focus—the candles we saw them double.

Oriental splendor—Alhambra—chariot

Table transformed

Contrast vision and reality—I touch wall.

"The visions were not blurred or uncertain," he writes. Indeed, "they seemed more real to me than anything I had ever seen with my own eyes." At this point, the reader begins to feel the literary hand of Aldous Huxley exerting a certain pressure on both Wasson's prose and his perceptions: "I felt that I was now seeing plain, whereas ordinary vision gives us an imperfect view." Wasson's own doors of perception had been flung wide open: "I was seeing the archetypes, the Platonic ideas, that underlie the imperfect images of everyday life." To read Wasson is to feel as if you were witnessing the still-fresh and malleable conventions of the psychedelic narrative gradually solidifying before your eyes. Whether Aldous Huxley invented these tropes, or was merely their stenographer, is hard to say, but they would inform the genre, as well as the experience, from here on. "For the first time the word ecstasy took on real meaning," Wasson recalls. "For the first time it did not mean someone else's state of mind."

Wasson concluded from his experience that his working hypothesis about the roots of the religious experience in psychoactive fungi had been vindicated. "In man's evolutionary past . . . there must have come a moment in time when he discovered the secret of the hallucinatory mushrooms. Their effect on him, as I see it, could only have been profound, a detonator to new ideas. For the mushrooms revealed to him worlds beyond the horizons known to him, in space and time, even worlds on a different plane of being, a heaven and perhaps a hell . . . One is emboldened to the point of asking whether they may not have planted in primitive man the very idea of a God."

Whatever one thinks about this idea, it's worth pointing out that Wasson came to Huautla with it already firmly planted and he was willing to subtly twist various elements of his experience there in order to confirm it. As much as he wants us to see María Sabina as a religious figure, and her ceremony as a form of what he calls "Holy communion," she saw herself quite differently. The mushroom might well have served as a sacrament five hundred years earlier, but by 1955 many Mazatecs had become devout Catholics, and they now used mushrooms not for worship but for healing and divination—to locate missing people and important items. Wasson knew this perfectly well, which is why he employed the ruse he did to gain access to a ceremony: he told María Sabina he was worried about his son back home and wanted information about his whereabouts and well-being. (Spookily enough, he received what he discovered on his return to New York to be accurate information on both counts.) Wasson was distorting a complex indigenous practice in order to fit a preconceived theory and conflating the historical significance of that practice with its contemporary meaning. As Sabina told an interviewer some years later, "Before Wasson nobody took the mushrooms only to find God. They were always taken for the sick to get well." As one of Wasson's harsher critics, the English writer Andy Letcher, acidly put it, "To find God, Sabina—like all good Catholics—went to Mass."

o   o   o

WASSON'S ARTICLE IN *LIFE* was read by millions of people (including a psychology professor on his way to Harvard named Timothy Leary). Wasson's story reached tens of millions more when he shared it on the popular CBS news program *Person to Person*, and in the months to follow several other magazines, including *True: The Man's Magazine*, ran first-person accounts of magic mushroom journeys ("The Vegetable That Drives Men Mad"), journeys for which Wasson supplied the mushrooms. (He had brought back a supply and would conduct ceremonies in his Manhattan apartment.) An exhibition on magic mushrooms soon followed at the American Museum of Natural History in New York.

Shortly after the article in *Life* was published, Wasson arranged to have some specimens of the Mexican mushrooms sent to Albert Hofmann in Switzerland for analysis. In 1958, Hofmann isolated and named the two psychoactive compounds, psilocybin and psilocin, and developed the synthetic version of psilocybin used in the current research. Hofmann also experimented with the mushrooms himself. "Thirty minutes after my taking the mushrooms," he wrote, "the exterior world began to undergo a strange transformation. Everything assumed a Mexican character." In 1962, Hofmann joined Wasson on one of his return trips to Huautla, during which the chemist gave María Sabina psilocybin in pill form. She took two of the pills and declared they did indeed contain the spirit of the mushroom.*

It didn't take long for thousands of other people—including,

---

* On another return trip, Wasson was joined by James Moore, who had introduced himself as a chemist for a pharmaceutical company. But Moore was really a CIA agent eager to obtain psilocybin for the agency's own psychedelic research program, MK-Ultra.

eventually, celebrities such as Bob Dylan, John Lennon, and Mick Jagger—to find their way to Huautla and to María Sabina's door.* For María Sabina and her village, the attention was ruinous. Wasson would later hold himself responsible for "unleash[ing] on lovely Huautla a torrent of commercial exploitation of the vilest kind," as he wrote in a plaintive 1970 *New York Times* op-ed. Huautla had become first a beatnik, then a hippie mecca, and the sacred mushrooms, once a closely guarded secret, were now being sold openly on the street. María Sabina's neighbors blamed her for what was happening to their village; her home was burned down, and she was briefly jailed. Nearing the end of her life, she had nothing but regret for having shared the divine mushrooms with R. Gordon Wasson and, in turn, the world. "From the moment the foreigners arrived," she told a visitor, "the saint children lost their purity. They lost their force; the foreigners spoiled them. From now on they won't be any good."

o   o   o

WHEN THE NEXT MORNING I came downstairs, Paul Stamets was in the living room, arranging his collection of mushroom stones on the coffee table. I had read about these artifacts but had never seen or held one, and they were impressive objects: roughly carved chunks of basalt in a variety of sizes and shapes. Some were simple and looked like gigantic mushrooms; others had a tripod or four-footed base, and still others had a figure carved into the stipe (or stem). Thousands of these stones were smashed by the Spanish, but two hundred are known to survive, and Stamets owns sixteen of them. Most of the surviving stones have been found in the Guatemalan

---

* Wasson was halfhearted in his desire to protect María Sabina's identity. The same week that the *Life* article appeared, he self-published a book, *Mushrooms, Russia, and History*, in which he retold her story but neglected to disguise her name.

highlands, often when farmers are plowing their fields; some have been dated to at least 1000 B.C.

As Stamets carried the heavy stones, one by one, from their cabinet to the coffee table, where he arranged them with great care, he looked like an altar boy, handling them with the sobriety appropriate to irreplaceable sacred objects. It occurred to me Paul Stamets is R. Gordon Wasson's rightful heir. (Wasson, too, collected mushroom stones, some of which I saw at Harvard.) He shares his radically mycocentric cosmology and sees evidence wherever he looks for the centrality of psychoactive mushrooms in culture, religion, and nature. Stamets's laptop is crammed with images of *Psilocybes* taken not only from nature (he's a superb photographer) but also from cave paintings, North African petroglyphs, medieval church architecture, and Islamic designs, some of which recall the forms of mushrooms or, with their fractal geometric patternings, mushroom experiences. I confess that try as I might, I often failed to find the mushrooms lurking in the pictures. No doubt the mushrooms themselves could help.

This brings us to Terence McKenna's stoned ape theory, the epitome of all mycocentric speculation, which Stamets had wanted to make sure we discussed. Though reading is no substitute for hearing McKenna expound his thesis (you can find him on YouTube), he summarizes it in *Food of the Gods* (1992): *Psilocybes* gave our hominid ancestors "access to realms of supernatural power," "catalyzed the emergence of human self-reflection," and "brought us out of the animal mind and into the world of articulated speech and imagination." This last hypothesis about the invention of language turns on the concept of synesthesia, the conflation of the senses that psychedelics are known to induce: under the influence of psilocybin, numbers can take on colors, colors attach to sounds, and so on. Language, he contends, represents a special case of synesthesia, in which otherwise meaningless sounds become linked to concepts. Hence, the stoned ape: by giving us the gifts of language and self-reflection psilocybin mushrooms made us who we are, transforming our primate ancestors into *Homo sapiens*.

The stoned ape theory is not really susceptible to proof or disproof. The consumption of mushrooms by early hominids would be unlikely to leave any trace in the fossil record, because the mushrooms are soft tissue and can be eaten fresh, requiring no special tools or processing methods that might have survived. McKenna never really explains how the consumption of psychoactive mushrooms could have influenced biological evolution—that is, selected for changes at the level of the genome. It would have been easier for him to make an argument for psychoactive fungi's influence on *cultural* evolution—such as the one Wasson made—but evidently the fungi had more ambitious plans for the mind of Terence McKenna, and Terence McKenna was more than happy to oblige.

Stamets became good friends with McKenna during the last few years of his life, and ever since McKenna's death (at age fifty-three, from brain cancer), he has been carrying the stoned ape's torch, recounting McKenna's theory in many of his talks. Stamets acknowledges the challenges of ever proving it to anyone's satisfaction yet deems it "more likely than not" that psilocybin "was pivotal in human evolution." What is it about these mushrooms, I wondered, and the experience they sponsor in the minds of men, that fires this kind of intellectual extravagance and conviction?

The stories of myco-evangelists like McKenna read like conversion narratives, in which certain people who have felt the power of these mushrooms firsthand emerge from the experience convinced that these fungi are prime movers—gods, of a sort—that can explain everything. Their prophetic mission in life becomes clear: bring this news to the world!

Now consider all this from the mushroom's point of view: what might have started as a biochemical accident has turned into an ingenious strategy for enlarging the species' range and number, by winning the fervent devotion of an animal as ingenious and well traveled (and well spoken!) as *Homo sapiens*. In McKenna's vision, it is the mushroom itself that helped form precisely the kind of mind— endowed with the tools of language and fired by imagination—that could best advance its interests. How diabolically brilliant! No wonder Paul Stamets is convinced of their intelligence.

o   o   o

THE NEXT MORNING, before we packed up the cars for our trip south, Stamets had another gift he wanted to give me. We were in his office, looking at some images on his computer, when he pulled off the shelf a small pile of amadou hats. "See if one of these fits you." Most of the mushroom hats were too big for me, but I found one that sat comfortably on my head and thanked him for the gift. The hat was surprisingly soft and almost weightless, but I felt a little silly with a mushroom on my head, so I carefully packed it in my luggage.

Early Sunday morning we drove west toward the Pacific coast and then south to the Columbia River, stopping for lunch and camping provisions in the resort town of Long Beach. This being the first week of December, the town was pretty well buttoned up and sleepy. Stamets requested that I not publish the exact location where we went hunting for *Psilocybe azurescens*. But what I can say is that there are three public parks bordering the wide-open mouth of the Columbia— Fort Stevens, Cape Disappointment, and the Lewis and Clark National Historical Park—and we stayed at one of them. Stamets, who has been coming here to hunt azzies for years, was mildly paranoid about being recognized by a ranger, so he stayed in the car while I checked in at the office and picked up a map giving directions to our yurt.

As soon as we unloaded and stowed our gear, we laced up our boots and headed out to look for mushrooms. Which really just meant walking around with eyes cast downward, tracing desultory patterns through the scrub along the sand dunes and in the grassy areas adjoining the yurts. We adopted the posture of the psilocybin stoop, except that we raised our heads every time we heard a car coming. Foraging mushrooms is prohibited in most state parks, and being in possession of psilocybin mushrooms is both a state and a federal crime.

The weather was overcast in the high forties—balmy for this far north on the Pacific coast in December, when it can be cold, wet, and stormy. We pretty much had the whole park to ourselves. It was a stunning, desolate landscape, with pine trees pruned low and angular by the winds coming off the ocean, endless dead-flat sandy beaches with plenty of driftwood, and giant storm-tossed timbers washed up and jack-strawed here and there along the beach. These logs had somehow slipped out from under the thumb of the lumber industry, floating down the Columbia from the old-growth forests hundreds of miles upriver and washing up here.

Stamets suspects that *Psilocybe azurescens* might originally have ridden out of the forest in the flesh of those logs and found its way here to the mouth of the Columbia—thus far the only place the species has ever been found. Some mycelium will actually insinuate itself into the grain of trees, taking up residence and forming a symbiotic relationship with the tree. Stamets believes the mycelium functions as a kind of immune system for its arboreal host, secreting antibacterial, antiviral, and insecticidal compounds that protect the trees from diseases and pests, in exchange for nourishment and habitat.

As we walked in widening spirals and figure eights over the grassy dunes, Stamets kept up a steady mycological patter; one nice thing about hunting mushrooms is that you don't have to worry about scaring them away with the sound of your voice. Every now and then he paused to show me a mushroom. Little brown mushrooms are notoriously difficult to identify, but Stamets almost always had its Latin binomial and a few interesting facts about it at his fingertips. At one point, he handed me a *Russula*, explaining it was good to eat. I only nibbled at the ruddy cap before I had to spit it out, it was so fiery. Evidently, offering newbies this particular *Russula* is an old mycologist hazing ritual.

I saw plenty of LBMs that might or might not be psilocybin and was constantly interrupting Stamets for another ID, and every time he had to prick my bubble of hope that I had at last found the precious quarry. After an hour or two of fruitless searching, Stamets wondered aloud if maybe we had come too late for the azzies.

And then all of a sudden, in an excited stage whisper, he called out, "Got one!" I raced over, asking him to leave the mushroom in place so I could see where and how it grew. This would, I hoped, allow me to "get my eyes on," as mushroom hunters like to say. Once we register on our retinas the visual pattern of the object we're searching for, it's much more likely to pop out of the visual field. (In fact the technical name for this phenomenon is "the pop-out effect.")

It was a handsome little mushroom, with a smooth, slightly glossy caramel-colored cap. Stamets let me pick it; it had a surprisingly tenacious grip, and when it came out of the ground, it brought with it some leaf litter, soil, and a little knot of bright white mycelium. "Bruise the stipe a bit," Stamets suggested. I did, and within minutes a blue tinge appeared where I'd rubbed it. "That's the psilocin." I never expected to actually *see* the chemical I had read so much about.

The mushroom had been growing a stone's throw from our yurt, right on the edge of a parking spot. Stamets says that like many psilocybin species "azzies are organisms of the ecological edge. Look at where we are: at the edge of the continent, the edge of an ecosystem, the edge of civilization, and of course these mushrooms bring us to the edge of consciousness." At this point, Stamets, who when it comes to mushrooms is one serious dude, made the first joke I had ever heard him make: "You know one of the best indicator species for *Psilocybe azurescens* are Winnebagos." We're obviously not the first people to hunt for azzies in this park, and anyone who picks a mushroom trails an invisible cloud of its spore behind him; this, he believes, is the origin of the idea of fairy dust. At the end of many of those trails is apt to be a campsite, a car, or a Winnebago.

We found seven azzies that afternoon, though by we I mean Stamets; I only found one, and even then I wasn't at all certain it was a *Psilocybe* until Stamets gave me a smile and a thumbs-up. I could swear it looked exactly like half a dozen other species I was finding. Stamets patiently tutored me in mushroom morphology, and by the following day my luck had improved, and I found four little caramel beauties on my own. Not much of a haul, but then Stamets had said that even just one of these mushrooms could underwrite a major psychic expedition.

That evening, we carefully laid out our seven mushrooms on a paper towel and photographed them before putting them in front of the yurt's space heater to dry. Within hours, the hot air had transformed a mushroom that was unimpressive to begin with into a tiny, shriveled gray-blue scrap it would be easy to overlook. The idea that something so unprepossessing could have such consequence was hard to credit.

I had been looking forward to trying an azzie, but before the evening was over, Stamets had dampened my enthusiasm. "I find *azurescens* almost *too* strong," he told me when we were standing around the fire pit outside our yurt, having a beer. After nightfall, we had driven out onto the beach to hunt for razor clams by headlight; now we were sautéing them with onions over the fire.

"And azzies have one potential side effect that some people find troubling."

Yes?

"Temporary paralysis," he said matter-of-factly. He explained that some people on azzies find they can't move their muscles for a period of time. That might be tolerable if you're in a safe place, he suggested, "but what if you're outdoors and the weather turns cold and wet? You could die of hypothermia." Not much of an advertisement for *azurescens*, especially coming from the man who discovered the species and named it. I was suddenly in much less of a hurry to try one.

o   o   o

THE QUESTION I KEPT returning to that weekend is this: Why in the world would a fungus go to the trouble of producing a chemical compound that has such a radical effect on the minds of the animals that eat it? What, if anything, did this peculiar chemical do for the mushroom? One could construct a quasi-mystical explanation for this phenomenon, as Stamets and McKenna have done: both suggest that neurochemistry is the language in which nature communicates with us, and it's trying to tell us something important by way of psilocybin. But this strikes me as more of a poetic conceit than a scientific theory.

The best answer I've managed to find arrived a few weeks later courtesy of Paul Stamets's professor at Evergreen State, Michael Beug, the chemist. When I reached him by phone at his home in the Columbia River Gorge, 160 miles upriver of our campsite, Beug said he was retired from teaching and hadn't spent much time thinking about *Psilocybes* recently, but he was intrigued by my question.

I asked him if there is reason to believe that psilocybin is a defense chemical for the mushroom. Defense against pests and diseases is the most common function of the so-called secondary metabolites produced in plants. Curiously, many plant toxins don't directly kill pests, but often act as psychostimulants as well as poisons, which is why we use many of them as drugs to alter consciousness. Why wouldn't plants just kill their predators outright? Perhaps because that would quickly select for resistance, whereas messing with its neurotransmitter networks can distract the predator or, better still, lead it to engage in risky behaviors likely to shorten its life. Think of an inebriated insect behaving in a way that attracts the attention of a hungry bird.

But Beug pointed out that if psilocybin were a defense chemical, "my former student Paul Stamets would have jumped on it long ago and found a use for it as an antifungal, antibacterial, or insecticide." In fact Beug has tested fungi for psilocybin and psilocin levels and found that they occur only in minute quantities in the mycelium—the part of the organism most likely to be well defended. "Instead the chemicals are in the fruiting bodies—sometimes at over two percent by dry weight!"—a stupendous quantity, and in a part of the organism it is not a priority to defend.

Even if psilocybin in mushrooms began as "an accident of a metabolic pathway," the fact that it wasn't discarded during the course of the species' evolution suggests it must have offered some benefit. "My best guess," Beug says, "is that the mushrooms that produced the most psilocybin got selectively eaten and so their spores got more widely disseminated."

Eaten by whom, or what? And why? Beug says that many animals are known to eat psilocybin mushrooms, including horses, cattle, and dogs. Some, like cows, appear unaffected, but many animals appear to enjoy an occasional change in consciousness too. Beug is in charge of gathering mushroom-poisoning reports for the North American Mycological Association and over the years has seen accounts of horses tripping in their paddocks and dogs that "zero in on *Psilocybes* and appear to be hallucinating." Several primate species (aside from our own) are also known to enjoy psychedelic mushrooms. Presumably animals with a taste for altered states of consciousness have helped spread psilocybin far and wide. "The strains of a species that produced more rather than less psilocybin and psilocin would tend to be favored and so gradually become more widespread."

Eaten in small doses, psychedelic mushrooms might well increase fitness in animals, by increasing sensory acuity and possibly focus as well. A 2015 review article in the *Journal of Ethnopharmacology* reported that several tribes around the world feed psychoactive plants to their dogs in order to improve their hunting ability.*

At higher doses, however, one would think that animals tripping on psychedelic mushrooms would be at a distinct disadvantage for survival, and no doubt many of them are. But for a select few, the effects *may* offer some adaptive value, not only for themselves, but also possibly for the group and even the species.

Here we venture out onto highly speculative, slightly squishy ground, guided by an Italian ethnobotanist named Giorgio Samorini. In a book called *Animals and Psychedelics: The Natural World and*

---

* The authors concluded that "hallucinogenic plants alter perception in hunting dogs by diminishing extraneous signals and by enhancing sensory perception (most likely olfaction) that is directly involved in the detection and capture of game." Bradley C. Bennett and Rocío Alarcón, "Hunting and Hallucinogens: The Use Psychoactive and Other Plants to Improve the Hunting Ability of Dogs," *Journal of Ethnopharmacology* 171 (2015): 171–83.

the Instinct to Alter Consciousness, Samorini hypothesizes that during times of rapid environmental change or crisis it may avail the survival of a group when a few of its members abandon their accustomed conditioned responses and experiment with some radically new and different behaviors. Much like genetic mutations, most of these novelties will prove disastrous and be discarded by natural selection. But the laws of probability suggest that a few of the novel behaviors might end up being useful, helping the individual, the group, and possibly the species to adapt to rapid changes in their environment.

Samorini calls this a "depatterning factor." There are times in the evolution of a species when the old patterns no longer avail, and the radical, potentially innovative perceptions and behaviors that psychedelics sometimes inspire may offer the best chance for adaptation. Think of it as a neurochemically induced source of variation in a population.

It is difficult to read about Samorini's lovely theory without thinking about our own species and the challenging circumstances in which we find ourselves today. *Homo sapiens* might have arrived at one of those periods of crisis that calls for some mental and behavioral depatterning. Could *that* be why nature has sent us these psychedelic molecules now?

o   o   o

SUCH A NOTION would not strike Paul Stamets as the least bit far-fetched. As we stood around the fire pit, the warm light flickering across our faces while our dinner sizzled in its pan, Stamets talked about what mushrooms have taught him about nature. He was expansive, eloquent, grandiose, and, at times, in acute danger of slipping the surly bonds of plausibility. We had had a few beers, and while we hadn't touched our tiny stash of azzies, we had smoked a little pot. Stamets dilated on the idea of psilocybin as a chemical messenger sent from Earth, and how we had been elected, by virtue of the gift of consciousness and language, to hear its call and act before it's too late.

"Plants and mushrooms have intelligence, and they want us to take care of the environment, and so they communicate that to us in a way we can understand." Why us? "We humans are the most populous bipedal organisms walking around, so some plants and fungi are especially interested in enlisting our support. I think they have a consciousness and are constantly trying to direct our evolution by speaking out to us biochemically. We just need to be better listeners."

These were riffs I'd heard Stamets deliver in countless talks and interviews. "Mushrooms have taught me the interconnectedness of all life-forms and the molecular matrix that we share," he explains in another one. "I no longer feel that I am in this envelope of a human life called Paul Stamets. I am part of the stream of molecules that are flowing through nature. I am given a voice, given consciousness for a time, but I feel that I am part of this continuum of stardust into which I am born and to which I will return at the end of this life." Stamets sounded very much like the volunteers I met at Hopkins who had had full-blown mystical experiences, people whose sense of themselves as individuals had been subsumed into a larger whole—a form of "unitive consciousness," which, in Stamets's case, had folded him into the web of nature, as its not so humble servant.

"I think *Psilocybes* have given me new insights that may allow me to help steer and speed fungal evolution so that we can find solutions to our problems." Especially in a time of ecological crisis, he suggests, we can't afford to wait for evolution, unfolding at its normal pace, to put forth these solutions in time. Let the depatterning begin.

As Stamets held forth, and forth, I couldn't help but picture in my mind Alex Grey's wacked painting of the stoned ape, with the tornadoes of thought flying out of his hairy head. So much of what Stamets has to say treads a perilously narrow ledge, perched between the autodidact's soaring speculative flights and the stoned crank's late night riffings that eventually send everyone in earshot off to bed. But just when I was beginning to grow impatient with his meanders, and could hear the call of my sleeping bag from inside the yurt, he, or I, turned a corner, and his mycological prophecies suddenly appeared to me in a more generous light.

The day before, Stamets had given me a tour of the labs and grow rooms at Fungi Perfecti, the company he founded right out of college. Tucked into the evergreen forest a short walk from his house, the Fungi Perfecti complex consists of a series of long white metal buildings that look like Quonset huts or small hangars. Outside are piles of wood chips, discarded fungi, and growing media. Some of the buildings house the grow rooms where he raises medicinal and edible species; others contain his research facility, with clean rooms and laminar flow chambers in which Stamets reproduces fungi from tissue culture and conducts his experiments. On the office walls hang several of his patents, framed. Amid the torrent of words, what I observed in these buildings was a salutary reminder that while Stamets is surely a big talker, he is not *just* a talker. He is a big doer too, a successful researcher and entrepreneur who is using fungi to make original contributions across a remarkably wide range of fields, from medicine and environmental restoration to agriculture and forestry and even national defense. Stamets is in fact a scientist, albeit of a special kind.

Exactly what kind of scientist I didn't completely understand until a few weeks later, when I happened to read a wonderful biography of Alexander von Humboldt, the great early nineteenth-century German scientist (and colleague of Goethe's) who revolutionized our understanding of the natural world. Humboldt believed it is only with our feelings, our senses, and our imaginations—that is, with the faculties of human subjectivity—that we can ever penetrate nature's secrets. "Nature everywhere speaks to man in a voice" that is "familiar to his soul." There is an order and beauty organizing the system of nature—a system that Humboldt, after briefly considering the name "Gaia," chose to call "Cosmos"—but it would never have revealed itself to us if not for the human imagination, which is itself of course a product of nature, of the very system it allows us to comprehend. The modern conceit of the scientist attempting to observe nature with perfect objectivity, as if from a vantage located outside it, would have been anathema to Humboldt. "I myself am identical with nature."

If Stamets is a scientist, as I believe he is, it is in the Humboldtian mold, making him something of a throwback. I don't mean to suggest his contribution is on the same order as Humboldt's. But he too is an amateur in the best sense, self-taught, uncredentialed, and blithe about trespassing disciplinary borders. He too is an accomplished naturalist and inventor, with several new species and patents to his credit. He too hears nature's voice, and it is his imagination—wild as it often is—that allows him to see systems where others have not, such as what is going on beneath our feet in a forest. I'm thinking, for example, of the "earth's Internet," "the neurological network of nature," and the "forest's immune system"—three Romantic-sounding metaphors that it would be foolish to bet against.

What strikes me about both Stamets and many of the so-called Romantic scientists (like Humboldt and Goethe, Joseph Banks, Erasmus Darwin, and I would include Thoreau) is how very much more alive nature seems in their hands than it would soon become in the cooler hands of the professionals. These more specialized scientists (a word that wasn't coined until 1834) gradually moved science indoors and increasingly gazed at nature through devices that allowed them to observe it at scales invisible to the human eye. These moves subtly changed the object of study—indeed, made it *more* of an object.

Instead of seeing nature as a collection of discrete objects, the Romantic scientists—and I include Stamets in their number—saw a densely tangled web of subjects, each acting on the other in the great dance that would come to be called coevolution. "Everything," Humboldt said, "is interaction and reciprocal." They could see this dance of subjectivities because they cultivated the plant's-eye view, the animal's-eye view, the microbe's-eye view, and the fungus's-eye view—perspectives that depend as much on imagination as observation.

I suspect that imaginative leap has become harder for us moderns to make. Our science and technology encourage us in precisely the opposite direction, toward the objectification of nature and of all species other than our own. Surely we need to acknowledge the practical power of this perspective, which has given us so much, but we should at the same time acknowledge its costs, material as well as spiritual. Yet that older, more enchanted way of seeing may still pay dividends, as it does (to cite just one small example) when it allows Paul Stamets to figure out that the reason honeybees like to visit woodpiles is to medicate themselves, by nibbling on a saprophytic mycelium that produces just the right antimicrobial compound that the hive needs to survive, a gift the fungus is trading for . . . what? Something yet to be imagined.

Coda

You are probably wondering what ever happened to the azzies Stamets and I found that weekend. Many months later, in the middle of a summer week spent in the house in New England where we used to live, a place freighted with memories, I ate them, with Judith. I crumbled two little mushrooms in each of two glasses and poured hot water over them to make a tea; Stamets had recommended that I "cook" the mushrooms to destroy the compounds that can upset the stomach. Judith and I each drank half a cup, ingesting both the liquid and the crumbles of mushroom. I suggested we take a walk on the dirt road near our house while we waited for the psilocybin to come on.

However, after only about twenty minutes or so, Judith reported she was "feeling things," none of them pleasant. She didn't want to be walking anymore, she said, but now we were at least a mile from home. She told me her mind and her body seemed to be drifting apart and then that her mind had flown out of her head and up into the trees, like a bird or insect.

"I need to get home and feel safe," she said, now with some urgency. I tried to reassure her as we abruptly turned around and picked up our pace. It was hot and the air was thick with humidity. She said, "I really don't want to run into anybody." I assured her we wouldn't. I still felt more or less myself, but it may be that Judith's distress was keeping me from feeling the mushrooms; somebody had to be ready to act normally if a neighbor happened to drive by and roll down his window for a chat, a prospect that was quickly taking on the proportions of nightmare. In fact shortly before we got back to home base—so it now felt to both of us—we spotted a neighbor's pickup truck bearing down on us and, like guilty children, we ducked into the woods until it passed.

Judith made a beeline for the couch in the living room, where she lay down with the shades drawn, while I went into the kitchen to polish off my cup of mushroom tea, because I wasn't yet feeling very much. I was a little worried about her, but once she reached her base on the living room couch, her mood lightened and she said she was fine.

I couldn't understand her desire to be indoors. I went out and sat on the screened porch for a while, listening to the sounds in the garden, which suddenly grew very loud, as if the volume had been turned way up. The air was stock-still, but the desultory sounds of flying insects and the digital buzz of hummingbirds rose to form a cacophony I had never heard before. It began to grate on my nerves, until I decided I would be better off regarding the sound as beautiful, and then all at once it was. I lifted an arm, then a foot, and noted with relief that I wasn't paralyzed, though I also didn't feel like moving a muscle.

Whenever I closed my eyes, random images erupted as if the insides of my lids were a screen. My notes record: *Fractal patterns, tunnels plunging through foliage, ropy vines forming grids.* But when I started to feel panic rise at the lack of control I had over my visual field, I discovered that all I needed to do to restore a sense of semi-normalcy was to open my eyes. To open or close my eyes was like changing the channel. I thought, "I am learning how to manage this experience."

Much happened, or seemed to happen, during the course of that August afternoon, but I want to focus here on just one element of the experience, because it bears on the questions of nature and our place in it that psilocybin seems to provoke, at least for me. I decided I wanted to walk out to my writing house, a little structure I had built myself twenty-five years ago, in what is now another life, and which holds a great many memories. I had written two and a half books in the little room (including one about building it), sitting before a broad window that looked back over a pond and the garden to our house.

However, I was still vaguely worried about Judith, so before wandering too far from the house, I went inside to check on her. She was stretched out on the couch, with a cool damp cloth over her eyes. She was fine. "I'm having these very interesting visuals," she said, something having to do with the stains on the coffee table coming to life, swirling and transforming and rising from the surface in ways she found compelling. She made it clear she wanted to be left alone to sink more deeply into the images—she is a painter. The phrase "parallel play" popped into my mind, and so it would be for the rest of the afternoon.

I stepped outside, feeling unsteady on my feet, legs a little rubbery. The garden was thrumming with activity, dragonflies tracing complicated patterns in the air, the seed heads of plume poppies rattling like snakes as I brushed by, the phlox perfuming the air with its sweet, heavy scent, and the air itself so palpably dense it had to be forded. The word and sense of "poignance" flooded over me during the walk through the garden, and it would return later. Maybe because we no longer live here, and this garden, where we spent so many summers as a couple and then a family, and which at this moment seemed so acutely *present*, was in fact now part of an irretrievable past. It was as if a precious memory had not just been recalled but had actually come back to life, in a reincarnation both beautiful and cruel. Also heartrending was the fleetingness of this moment in time, the ripeness of a New England garden in late August on the verge of turning the corner of the season. Before dawn one cloudless night very soon and without warning, the thrum and bloom and perfume would end all at once, with the arrival of the killing frost. I felt wide open emotionally, undefended.

When at last I arrived at the writing house, I stretched out on the daybed, something I hardly ever took the time to do in all the years when I was working here so industriously. The bookshelves had been emptied, and the place felt abandoned, a little sad. From where I lay, I could see over my toes to the window screen and, past that, to the grid of an arbor that was now densely woven with the twining vines of what had become a venerable old climbing hydrangea, a *petiolaris*. I had planted the hydrangea decades ago, in hopes of creating just this sort of intricately tangled prospect. Backlit by the late afternoon sunlight streaming in, its neat round leaves completely filled the window, which meant you gazed out at the world through the fresh green scrim they formed. It seemed to me these were the most beautiful leaves I had ever seen. It was as if they were emitting their own soft green glow. And it felt like a kind of privilege to gaze out at the world through their eyes, as it were, as the leaves drank up the last draughts of sunlight, transforming those photons into new matter. *A plant's-eye view of the world*—it *was* that, and for real! But the leaves were also looking back at me, fixing me with this utterly benign gaze. I could feel their curiosity and what I was certain was an attitude of utter benevolence toward me and my kind. (Do I need to say that I know how crazy this sounds? I do!)

I felt as though I were communing directly with a plant for the first time and that certain ideas I had long thought about and written about—having to do with the subjectivity of other species and the way they act upon us in ways we're too self-regarding to appreciate—had taken on the flesh of feeling and reality. I looked through the negative spaces formed by the hydrangea leaves to fix my gaze on the swamp maple in the middle of the meadow beyond, and it too was now more alive than I'd ever known a tree to be, infused with some kind of spirit—this one, too, benevolent. The idea that there had ever been a disagreement between matter and spirit seemed risible, and I felt as though whatever it is that usually divides me from the world out there had begun to fall away. Not completely: the battlements of ego had not fallen; this was not what the researchers would deem a "complete" mystical experience, because I retained the sense of an observing I. But the doors and windows of perception had opened wide, and they were admitting more of the world and its myriad nonhuman personalities than ever before.

Buoyed by this development, I sat up now and looked out over my desk, through the big window that faced back to the house. When I sited the building, I carefully framed the main view between two very old and venerable trees, a stolidly vertical ash on the right and an elegantly angled and intricately branched white oak on the left. The ash has seen better days; storms have shorn several important limbs from it, wrecking its symmetry and leaving some ragged stumps. The oak was somewhat healthier, in full leaf now with its upturned limbs reaching into the sky like the limbs of a dancer. But the main trunk, which had always leaned precariously to one side, now concerned me: a section of it had rotted out at ground level, and for the first time it was possible to look clear through it and see daylight. How was it possibly still standing?

As I gazed at the two trees I had gazed at so many times before from my desk, it suddenly dawned on me that these trees were— *obviously!*—my parents: the stolid ash my father, the elegant oak my mother. I don't know exactly what I mean by that, except that thinking about those trees became identical to thinking about my parents. They were completely, indelibly, present in those trees. And so I thought about all they had given me, and about all that time had done to them, and what was going to become of this prospect, this place (*this me!*), when they finally fell, as eventually they would. That parents die is not exactly the stuff of epiphany, but the prospect, no longer distant or abstract, pierced me more deeply than it ever had, and I was disarmed yet again by the pervasive sense of poignancy that trailed me all that afternoon. Yet I must have still had *some* wits about me, because I made a note to call the arborist tomorrow; maybe something could be done to reduce the weight on the leaning side of the oak, in order to prevent it from falling, if only for a while longer.

My walk back to the house was, I think, the peak of the experience and comes back to me now in the colors and tones of a dream. There was, again, the sense of pushing my body through a mass of air that had been sweetened by phlox and was teeming, almost frenetic, with activity. The dragonflies, big as birds, were now out in force, touching down just long enough to kiss the phlox blossoms and then lift off, before madly crisscrossing the garden path. These were more dragonflies than I had ever seen in one place, so many in fact that I wasn't completely sure if they were real. (Judith later confirmed the sighting when I got her to come outside.) And as they executed their flight patterns, they left behind them contrails that persisted in the air, or so at least it appeared. Dusk now approaching, the air traffic in the garden had built to a riotous crescendo: the pollinators making their last rounds of the day, the plants still signifying to them with their flowers: *me, me, me!* In one way I knew this scene well—the garden coming briefly back to life after the heat of a summer day has relented—but never had I felt so integral to it. I was no longer the alienated human observer, gazing at the garden from a distance, whether literal or figural, but rather felt part and parcel of all that was transpiring here. So the flowers were addressing me as much as the pollinators, and perhaps because the very air that afternoon was such a felt presence, one's usual sense of oneself as a subject observing objects in space—objects that have been thrown into relief and rendered discrete by the apparent void that surrounds them— gave way to a sense of being deep inside and fully implicated in this scene, one more being in relation to the myriad other beings and to the whole.

"Everything is interaction and reciprocal," wrote Humboldt, and that felt very much the case, and so, for the first time I can remember, did this: "I myself am identical with nature."

o   o   o

I HONESTLY DON'T KNOW what to make of this experience. In a certain light at certain moments, I feel as though I had had some kind of spiritual experience. I had felt the personhood of other beings in a way I hadn't before; whatever it is that keeps us from feeling our full implication in nature had been temporarily in abeyance. There had also been, I felt, an opening of the heart, toward my parents, yes, and toward Judith, but also, weirdly, toward some of the plants and trees and birds and even the damn bugs on our property. Some of this openness has persisted. I think back on it now as an experience of wonder and immanence.

The fact that this transformation of my familiar world into something I can only describe as numinous was occasioned by the eating of a little brown mushroom that Stamets and I had found growing on the edge of a parking lot in a state park on the Pacific coast—well, that fact can be viewed in one of two ways: either as an additional wonder or as support for a more prosaic and materialist interpretation of what happened to me that August afternoon. According to one interpretation, I had had "a drug experience," plain and simple. It was a kind of waking dream, interesting and pleasurable but signifying nothing. The psilocin in that mushroom unlocked the 5-hydroxytryptamine 2-A receptors in my brain, causing them to fire wildly and set off a cascade of disordered mental events that, among other things, permitted some thoughts and feelings, presumably from my subconscious (and, perhaps, my reading too), to get cross-wired with my visual cortex as it was processing images of the trees and plants and insects in my field of vision.

Not quite a hallucination, "projection" is probably the psychological term for this phenomenon: when we mix our emotions with certain objects that then reflect those feelings back to us so that they appear to glisten with meaning. T. S. Eliot called these things and situations the "objective correlatives" of human emotion. Emerson had a similar phenomenon in mind when he said that "Nature always wears the colors of the spirit," suggesting it is our minds that dress her in such significance.

I'm struck by the fact there was nothing supernatural about my heightened perceptions that afternoon, nothing that I needed an idea of magic or a divinity to explain. No, all it took was another perceptual slant on the same old reality, a lens or mode of consciousness that invented nothing but merely (*merely!*) italicized the prose of ordinary experience, disclosing the wonder that is *always* there in a garden or wood, hidden in plain sight—another form of consciousness "parted from [us]," as William James put it, "by the filmiest of screens." Nature does in fact teem with subjectivities—call them spirits if you like—other than our own; it is only the human ego, with its imagined monopoly on subjectivity, that keeps us from recognizing them all, our kith and kin. In this sense, I guess Paul Stamets is right to think the mushrooms are bringing us messages from nature, or at least helping us to open up and read them.

Before this afternoon, I had always assumed access to a spiritual dimension hinged on one's acceptance of the supernatural—of God, of a Beyond—but now I'm not so sure. The Beyond, whatever it consists of, might not be nearly as far away or inaccessible as we think. Huston Smith, the scholar of religion, once described a spiritually "realized being" as simply a person with "an acute sense of the astonishing mystery of everything." Faith need not figure. Maybe to be in a garden and feel awe, or wonder, in the presence of an astonishing mystery, is nothing more than a recovery of a misplaced perspective, perhaps the child's-eye view; maybe we regain it by means of a neurochemical change that disables the filters (of convention, of ego) that prevent us in ordinary hours from seeing what is, like those lovely leaves, staring us in the face. I don't know. But if those dried-up little scraps of fungus taught me anything, it is that there are other, stranger forms of consciousness available to us, and, whatever they mean, their very existence, to quote William James again, "forbid[s] a premature closing of our accounts with reality."

Open-minded. And bemushroomed. That was me, now, ready to reopen my own accounts with reality.

CHAPTER THREE

HISTORY
The First Wave

When the federal authorities came down hard on Timothy Leary in the mid-1960s, hitting him with a thirty-year sentence for attempting to bring a small amount of marijuana over the border at Laredo, Texas, in 1966,* the embattled former psychology professor turned to Marshall McLuhan for some advice. The country was in the throes of a moral panic about LSD, inspired in no small part by Leary's own promotion of psychedelic drugs as a means of personal and cultural transformation and by his recommendation to America's youth that they "turn on, tune in, drop out." Dated and goofy as those words sound to our ears, there was a moment when they were treated as a credible threat to the social order, an invitation to America's children not only to take mind-altering drugs but to

---

* Because possession of LSD wouldn't be a federal crime until 1968, the government often had to rely on marijuana prosecutions when moving against people in the counterculture.

reject the path laid out for them by their parents and their government—including the path taking young men to Vietnam. Also in 1966, Leary was called before a committee of the U.S. Senate to defend his notorious slogan, which he gamely if not very persuasively attempted to do. In the midst of the national storm raging around him—a storm, it should be said, he quite enjoyed—Leary met with Marshall McLuhan over lunch at the Plaza hotel in New York, the LSD guru betting that the media guru might have some tips on how best to handle the public and the press.

"Dreary Senate hearing and courtrooms are not the platforms for your message, Tim," McLuhan advised, in a conversation that Leary recounts in *Flashbacks*, one of his many autobiographies. (Leary would write another one every time legal fees and alimony payments threatened to empty his bank account.) "To dispel fear you must use your public image. You are the basic product endorser." The product by this point was of course LSD. "Whenever you are photographed, smile. Wave reassuringly. Radiate courage. Never complain or appear angry. It's okay if you come off as flamboyant and eccentric. You're a professor after all. But a confident attitude is the best advertisement. You must be known for your smile."

Leary took McLuhan's advice to heart. In virtually all of the many thousands of photographs taken of him from that lunch date forward, Leary made sure to present the gift of his most winning grin to the camera. It didn't matter if he was coming into or out of a courthouse, addressing a throng of youthful admirers in his love beads and white robes, being jostled into a squad car freshly handcuffed, or perched on the edge of John and Yoko's bed in a Montreal hotel room, Timothy Leary always managed to summon a bright smile and a cheerful wave for the camera.

So, ever smiling, the charismatic figure of Timothy Leary looms large over the history of psychedelics in America. Yet it doesn't take many hours in the library before you begin to wonder if maybe Timothy Leary looms a little *too* large in that history, or at least in our popular understanding of it. I was hardly alone in assuming that the Harvard Psilocybin Project—launched by Leary in the fall of 1960, immediately after his first life-changing experience with psilocybin in Mexico—represented the beginning of serious academic research into these substances or that Leary's dismissal from Harvard in 1963 marked the end of that research. But in fact neither proposition is even remotely true.

Leary played an important role in the modern history of psychedelics, but it's not at all the pioneering role he wrote for himself. His success in shaping the popular narrative of psychedelics in the 1960s obscures as much as it reveals, creating a kind of reality distortion field that makes it difficult to see everything that came either before or after his big moment onstage.

In a truer telling of the history, the Harvard Psilocybin Project would appear more like the beginning of the end of what had been a remarkably fertile and promising period of research that unfolded during the previous decade far from Cambridge, in places as far flung as Saskatchewan, Vancouver, California, and England, and, everywhere, with a lot less sound and fury or countercultural baggage. The larger-than-life figure of Leary has also obscured from view the role of a dedicated but little-known group of scientists, therapists, and passionate amateurs who, long before Leary had ever tried psilocybin or LSD, developed the theoretical framework to make sense of these unusual chemicals and devised the therapeutic protocols to put them to use healing people. Many of these researchers eventually watched in dismay as Leary (and his "antics," as they inevitably referred to his various stunts and pronouncements) ignited what would become a public bonfire of all their hard-won knowledge and experience.

In telling the modern history of psychedelics, I want to put aside the Leary saga, at least until the crack-up where it properly belongs, to see if we can't recover some of that knowledge and the experience that produced it without passing it through the light-bending prism of the "Psychedelic Sixties." In doing so, I'm following in the steps of several of the current generation of psychedelic researchers, who, beginning in the late 1990s, set out to excavate the intellectual ruins of this first flowering of research into LSD and psilocybin and were astounded by what they found.

Stephen Ross is one such researcher. A psychiatrist specializing in addiction at Bellevue, he directed an NYU trial using psilocybin to treat the existential distress of cancer patients, to which I will return later; since then, he has turned to the treatment of alcoholics with psychedelics, what had been perhaps the single most promising area of clinical research in the 1950s. When several years ago an NYU colleague mentioned to Ross that LSD had once been used to treat thousands of alcoholics in Canada and the United States (and that Bill Wilson, the founder of Alcoholics Anonymous, had sought to introduce LSD therapy into AA in the 1950s), Ross, who was in his thirties at the time, did some research and was "flabbergasted" by all that he—as an expert on the treatment of alcoholism—did not know and hadn't been told. His own field had a secret history.

"I felt a little like an archaeologist, unearthing a completely buried body of knowledge. Beginning in the early fifties, psychedelics had been used to treat a whole host of conditions," including addiction, depression, obsessive-compulsive disorder, schizophrenia, autism, and end-of-life anxiety. "There had been forty thousand research participants and more than a thousand clinical papers! The American Psychiatric Association had whole meetings centered around LSD, this new wonder drug." In fact, there were six international scientific meetings devoted to psychedelics between 1950 and 1965.

"Some of the best minds in psychiatry had seriously studied these compounds in therapeutic models, with government funding." But after the culture and the psychiatric establishment turned against psychedelics in the mid-1960s, an entire body of knowledge was effectively erased from the field, as if all that research and clinical experience had never happened. "By the time I got to medical school in the 1990s, no one even talked about it."

o o o

WHEN LSD BURST onto the psychiatric scene in 1950, the drug's effects on patients (and researchers, who routinely tried the drug on themselves) were so novel and strange that scientists struggled for the better part of a decade to figure out what these extraordinary experiences were or meant. How, exactly, did this new mind-altering drug fit into the existing paradigms for understanding the mind and the prevailing modes of psychiatry and psychotherapy? A lively debate over these questions went on for more than a decade. What wasn't known at the time is that beginning in 1953, the CIA was conducting its own (classified) research into psychedelics and was struggling with similar issues of interpretation and application: Was LSD best regarded as a potential truth serum, or a mind-control agent, or a chemical weapon?

The world's very first LSD trip, and the only one undertaken with no prior expectations, was the one Albert Hofmann took in 1943. While it left him uncertain whether he had experienced madness or transcendence, Hofmann immediately sensed the potential importance of this compound for neurology and psychiatry. So Sandoz, the pharmaceutical company for which he worked at the time of his discovery, did something unusual: in effect, it crowd-sourced a worldwide research effort to figure out what in the world Delysid— its brand name for LSD-25—might be good for. Hoping someone somewhere would hit upon a commercial application for its spookily powerful new compound, Sandoz offered to supply, free of charge, however much LSD any researcher requested. The company defined the term "researcher" liberally enough to include any therapist who promised to write up his or her clinical observations. This policy remained more or less unchanged from 1949 to 1966 and was in large part responsible for setting off the first wave of psychedelic research—the one that crashed in 1966, when Sandoz, alarmed at the controversy that had erupted around its experimental drug, abruptly withdrew Delysid from circulation.

So what was learned during that fertile and freewheeling period of investigation? A straightforward question, and yet the answer is complicated by the very nature of these drugs, which is anything but straightforward. As the literary theorists would say, the psychedelic experience is highly "constructed." If you are told you will have a spiritual experience, chances are pretty good that you will, and, likewise, if you are told the drug may drive you temporarily insane, or acquaint you with the collective unconscious, or help you access "cosmic consciousness," or revisit the trauma of your birth, you stand a good chance of having exactly that kind of experience.

Psychologists call these self-fulfilling prophecies "expectancy effects," and they turn out to be especially powerful in the case of psychedelics. So, for example, if you have ever read Aldous Huxley's *Doors of Perception*, which was published in 1954, your own psychedelic experience has probably been influenced by the author's mysticism and, specifically, the mysticism of the East to which Huxley was inclined. Indeed, even if you have *never* read Huxley, his construction of the experience has probably influenced your own, for that Eastern flavoring—think of the Beatles song "Tomorrow Never Knows"—would come to characterize the LSD experience from 1954 on. (Leary would pick up this psychedelic orientalism from

Huxley and then greatly amplify it when he and his Harvard colleagues wrote a bestselling manual for psychedelic experience based on the *Tibetan Book of the Dead*.) Further complicating the story and adding another feedback loop, Huxley was inspired to try psychedelics and write about the experience by a scientist who gave him mescaline in the explicit hope that a great writer's descriptions and metaphors would help him and his colleagues make sense of an experience they were struggling to interpret. So did Aldous Huxley "make sense" of the modern psychedelic experience, or did he in some sense invent it?

This hall of epistemological mirrors was just one of the many challenges facing the researchers who wanted to bring LSD into the field of psychiatry and psychotherapy: psychedelic therapy could look more like shamanism or faith healing than medicine. Another challenge was the irrational exuberance that seemed to infect any researchers who got involved with LSD, an enthusiasm that might have improved the results of their experiments at the same time it fueled the skepticism of colleagues who remained psychedelic virgins. Yet a third challenge was how to fit psychedelics into the existing structures of science and psychiatry, if indeed that was possible. How do you do a controlled experiment with a psychedelic? How do you effectively blind your patients and clinicians or control for the powerful expectancy effect? When "set" and "setting" play such a big role in the patient's experience, how can you hope to isolate a single variable or design a therapeutic application?

Part I: The Promise

The drugs weren't called "psychedelics" at the beginning; that term wasn't introduced until 1957. In the same way that Sandoz couldn't

figure out what it had on its hands with LSD, the researchers experimenting with the drug couldn't figure out what to call it. Over the
course of the 1950s, this class of drugs underwent a succession of
name changes as our understanding of the chemicals and their action
evolved, each new name reflecting the shifting interpretation—or
was it a construction?—of what these strange and powerful molecules meant and did.

The first name was perhaps the most awkward: beginning around
1950, shortly after LSD was made available to researchers, the compound was known as a psychotomimetic, which is to say, a mind drug
that mimicked psychoses. This was the most obvious and parsimonious interpretation of a psychedelic's effects. Viewed from the outside, people given doses of LSD and, later, psilocybin exhibited many
of the signs of a temporary psychosis. Early researchers reported a
range of disturbing symptoms in their LSD volunteers, including
depersonalization, loss of ego boundaries, distorted body image,
synesthesia (seeing sounds or hearing sights), emotional lability, gig-
gling and weeping, distortion of the sense of time, delirium, hallucinations, paranoid delusions, and, in the words of one writer, "a
tantalizing sense of portentousness." When researchers administered
standardized psychiatric tests to volunteers on LSD—such as the
Rorschach ink blots or the Minnesota Multiphasic Personality In-
ventory test—the results mirrored those of psychotics and, specifically, schizophrenics. Volunteers on LSD appeared to be losing their
minds.

This suggested to some researchers that LSD held promise as a
tool for understanding psychosis, which is precisely how Sandoz initially marketed Delysid. Although the drug might not cure anything,
the resemblance of its effects to the symptoms of schizophrenia
suggested that the mental disorder might have a chemical basis that
LSD could somehow illuminate. For clinicians, the drug promised to

help them better understand and empathize with their schizophrenic patients. That of course meant taking the drug themselves, which seems odd, even scandalous, to us today. But in the years before 1962, when Congress passed a law giving the FDA authority to regulate new "investigational" drugs, this was in fact common practice. Indeed, it was considered the ethical thing to do, for to *not* take the drug yourself was tantamount to treating your patients as guinea pigs. Humphry Osmond wrote that the extraordinary promise of LSD was to allow the therapist who took it to "enter the illness and see with a madman's eyes, hear with his ears, and feel with his skin."

Born in Surrey, England, in 1917, Osmond is a little-known but pivotal figure in the history of psychedelic research,* probably contributing more to our understanding of these compounds and their therapeutic potential than any other single researcher. In the years following World War II, Osmond, a tall reed of a man with raucous teeth, was practicing psychiatry at St. George's Hospital in London when a colleague named John Smythies introduced him to an obscure body of medical literature about mescaline. After learning that mescaline induced hallucinations much like those reported by schizophrenics, the two researchers began to explore the idea that the disease was caused by a chemical imbalance in the brain. At a time when the role of brain chemistry in mental illness had not yet been established, this was a radical hypothesis. The two psychiatrists had observed that the molecular structure of mescaline closely resembled that of adrenaline. Could schizophrenia result from some kind of dysfunction in the metabolism of adrenaline, transforming it into a compound that produced the schizophrenic rupture with reality?

---

* Osmond's story, and the rich Canadian history of psychedelic research, is well told in Erika Dyck, *Psychedelic Psychiatry: LSD from Clinic to Campus* (Baltimore: Johns Hopkins University Press, 2008).

No, as it would turn out. But it was a productive hypothesis even so, and Osmond's research into the biochemical basis of mental illness contributed to the rise of neurochemistry in the 1950s. LSD research would eventually give an important boost to the nascent field. The fact that such a vanishingly small number of LSD molecules could exert such a profound effect on the mind was an important clue that a system of neurotransmitters with dedicated receptors might play a role in organizing our mental experience. This insight eventually led to the discovery of serotonin and the class of antidepressants known as SSRIs.

But the powers that be at St. George's Hospital were unsupportive of Osmond's research on mescaline. In frustration, the young doctor went looking for a more hospitable institution in which to conduct it. This he found in the western Canadian province of Saskatchewan, of all places. Beginning in the mid-1940s, the province's leftist government had instituted several radical reforms in public policy, including the nation's first system of publicly funded health care. (It became the model for the system Canada would adopt in 1966.) Hoping to make the province a center of cutting-edge medical research, the government offered generous funding and a rare degree of freedom to lure researchers to the frozen wastes of the Canadian prairies. After replying to an ad in the *Lancet*, Osmond received an invitation from the provincial government to move his family and his novel research project to the remote agrarian community of Weyburn, Saskatchewan, forty-five miles north of the North Dakota border. The Saskatchewan Mental Hospital in Weyburn would soon become the world's most important hub of research into psychedelics—or rather, into the class of compounds still known as psychotomimetics.

That paradigm still ruled the thinking of Osmond and his new, like-minded colleague and research director, a Canadian psychiatrist named Abram Hoffer, as they began conducting experiments using a supply of LSD-25 obtained from Sandoz. The psychotomimetic model was introduced to the general public in 1953, when *Maclean's*, the popular Canadian magazine, published a harrowing account of a journalist's experience on LSD titled "My 12 Hours as a Madman."

Sidney Katz had become the first "civilian" to participate in one of Osmond and Hoffer's LSD experiments at Weyburn hospital. Katz had been led to expect madness, and madness he duly experienced: "I saw faces of familiar friends turn into fleshless skulls and the heads of menacing witches, pigs and weasels. The gaily patterned carpet at my feet was transformed into a fabulous heaving mass of living matter, part vegetable, part animal." Katz's article, which was illustrated with an artist's rendering of chairs flying through a collapsing room, reads like the work of a fervent anti-LSD propagandist circa 1965: "I was repeatedly held in the grip of a terrifying hallucination in which I could feel and see my body convulse and shrink until all that remained was a hard sickly stone." Yet, curiously, his twelve hours of insanity "were not all filled with horror," he reported. "At times I beheld visions of dazzling beauty—visions so rapturous, so unearthly, that no artist will ever paint them."

During this period, Osmond and Hoffer administered Sandoz LSD to dozens of people, including colleagues, friends, family members, volunteers, and, of course, themselves. Their focus on LSD as a window into the biochemistry of mental illness gradually gave way to a deepening curiosity about the power of the experience itself and whether the perceptual disturbances produced by the drug might themselves confer some therapeutic benefit. During a late night brainstorming session in an Ottawa hotel room in 1953, Osmond and Hoffer noted that the LSD experience appeared to share many features with the descriptions of delirium tremens reported by alcoholics—the hellish, days-long bout of madness alcoholics often suffer while in the throes of withdrawal. Many recovering alcoholics look back on the hallucinatory horrors of the DTs as a conversion experience and the basis of the spiritual awakening that allows them to remain sober.

The idea that an LSD experience could mimic the DTs "seemed so bizarre that we laughed uproariously," Hoffer recalled years later. "But when our laughter subsided, the question seemed less comical and we formed our hypothesis . . . : would a controlled LSD-produced delirium help alcoholics stay sober?"

Here was an arresting application of the psychotomimetic paradigm: use a single high-dose LSD session to induce an episode of madness in an alcoholic that would simulate delirium tremens, shocking the patient into sobriety. Over the next decade, Osmond and Hoffer tested this hypothesis on more than seven hundred alcoholics, and in roughly half the cases, they reported, the treatment worked: the volunteers got sober and remained so for at least several months. Not only was the new approach more effective than other therapies, but it suggested a whole new way to think about psychopharmacology. "From the first," Hoffer wrote, "we considered not the chemical, but the experience as a key factor in therapy." This novel idea would become a central tenet of psychedelic therapy.

The emphasis on what subjects *felt* represented a major break with the prevailing ideas of behaviorism in psychology, in which only observable and measurable outcomes counted and subjective experience was deemed irrelevant. The analysis of these subjective experiences, sometimes called phenomenology, had of course been the basis of Freudian psychoanalysis, which behaviorism had rejected as insufficiently rigorous or scientific. There was no point in trying to get inside the mind; it was, in B. F. Skinner's famous phrase, "a black box." Instead, you measured what you could measure, which was outward behavior. The work with psychedelics would eventually spark a revival of interest in the subjective dimensions of the mind—in consciousness. How ironic that it took, of all things, a chemical—LSD-25—to bring interiority back into psychology.

And yet, successful as the new therapy seemed to be, there was a nagging little problem with the theoretical model on which it was based. When the therapists began to analyze the reports of volunteers, their subjective experiences while on LSD bore little if any resemblance to the horrors of the DTs, or to madness of any kind. To the contrary, their experiences were, for the most part, incredibly—and bafflingly—positive. When Osmond and Hoffer began to catalog their volunteers' session reports, "psychotic changes"—hallucinations, paranoia, anxiety—sometimes occurred, but there were also descriptions of, say, "a transcendental feeling of being united with the world," one of the most common feelings reported. Rather than madness, most volunteers described sensations such as a new ability "to see oneself objectively"; "enhancement in the sensory fields"; profound new understandings "in the field of philosophy or religion"; and "increased sensitivity to the feelings of others."* In spite of the powerful expectancy effect, symptoms that looked nothing like those of insanity were busting through the researchers' preconceptions.

For many of the alcoholics treated at Weyburn hospital, the core of the LSD experience seemed to involve something closer to transcendence, or spiritual epiphany, than temporary psychosis. Osmond and Hoffer began to entertain doubts about their delirium tremens model and, eventually, to wonder if perhaps the whole psychotomimetic paradigm—and name for these drugs—might need retooling.

---

* Duncan C. Blewett and Nick Chwelos, *Handbook for the Therapeutic Use of Lysergic Acid Diethlylamide-25: Individual and Group Procedures* (1959), http://www.maps.org/research-archive/ritesofpassage/lsdhandbook.pdf. Blewett and Chwelos drew heavily on Osmond and Hoffer's case reports for their manual.

They received a strong push in that direction from Aldous Huxley after his mescaline experience, which he declared bore scant resemblance to psychosis. What a psychiatrist might diagnose as depersonalization, hallucinations, or mania might better be thought of as instances of mystical union, visionary experience, or ecstasy. Could it be that the doctors were mistaking transcendence for insanity?

At the same time, Osmond and Hoffer were learning from their volunteers that the environment in which the LSD session took place exerted a powerful effect on the kinds of experiences people had and that one of the best ways to avoid a bad session was the presence of an engaged and empathetic therapist, ideally someone who had had his or her own LSD experience. They came to suspect that the few psychotic reactions they did observe might actually be an artifact of the metaphorical white room and white-coated clinician. Though the terms "set" and "setting" would not be used in this context for several more years (and became closely identified with Timothy Leary's work at Harvard a decade later), Osmond and Hoffer were already coming to appreciate the supreme importance of those factors in the success of their treatment.

But however it worked, it worked, or certainly seemed to: by the end of the decade, LSD was widely regarded in North America as a miracle cure for alcohol addiction. Based on this success, the Saskatchewan provincial government helped develop policies making LSD therapy a standard treatment option for alcoholics in the province. Yet not everyone in the Canadian medical establishment found the Saskatchewan results credible: they seemed too good to be true. In the early 1960s, the Addiction Research Foundation in Toronto, the leading institute of its kind in Canada, set out to replicate the Saskatchewan trials using better controls. Hoping to isolate the effects of the drug from all other variables, clinicians administered LSD to alcoholics in neutral rooms and under instructions not to engage with them during their trips, except to administer an extensive questionnaire. The volunteers were then put in constraints or blindfolded, or both. Not surprisingly, the results failed to match those obtained by Osmond and Hoffer. Worse still, more than a few of the volunteers endured terrifying experiences—bad trips, as they would come to be called. Critics of treating alcoholics with LSD concluded that the treatment didn't work as well under rigorously controlled conditions, which was true enough, while supporters of the practice concluded that attention to set and setting was essential to the success of LSD therapy, which was also true.

o   o   o

IN THE MID-1950S, Bill Wilson, the cofounder of Alcoholics Anonymous, learned about Osmond and Hoffer's work with alcoholics. The idea that a drug could occasion a life-changing spiritual experience was not exactly news to Bill W., as he was known in the fellowship. He credited his own sobriety to a mystical experience he had on belladonna, a plant-derived alkaloid with hallucinogenic properties that was administered to him at Towns Hospital in Manhattan in 1934. Few members of AA realize that the whole idea of a spiritual awakening leading one to surrender to a "higher power"—a cornerstone of Alcoholics Anonymous—can be traced to a psychedelic drug trip.

Twenty years later, Bill W. became curious to see if LSD, this new wonder drug, might prove useful in helping recovering alcoholics have such an awakening. Through Humphry Osmond he got in touch with Sidney Cohen, an internist at the Brentwood VA hospital (and, later, UCLA) who had been experimenting with Sandoz LSD since 1955. Beginning in 1956, Bill W. had several LSD sessions in Los Angeles with Sidney Cohen and Betty Eisner, a young psychologist who had recently completed her doctorate at UCLA. Along with the psychiatrist Oscar Janiger, Cohen and Eisner were by then leading figures in a new hub of LSD research loosely centered on UCLA. By the mid-1950s, there were perhaps a dozen such hubs in North America and Europe; most of them kept in close contact with one another, sharing techniques, discoveries, and, sometimes, drugs, in a spirit that was generally more cooperative than competitive.

Bill W.'s sessions with Cohen and Eisner convinced him that LSD could reliably occasion the kind of spiritual awakening he believed one needed in order to get sober; however, he did not believe the LSD experience was anything like the DTs, thus driving another nail in the coffin of *that* idea. Bill W. thought there might be a place for LSD therapy in AA, but his colleagues on the board of the fellowship strongly disagreed, believing that to condone the use of any mind-altering substance risked muddying the organization's brand and message.

o   o   o

SIDNEY COHEN AND HIS COLLEAGUES in Los Angeles had, like the Canadian group, started out thinking that LSD was a psychotomimetic, but by the mid-1950s Cohen, too, had come to question that model. Born in 1910 in New York City to Lithuanian Jewish immigrants, Cohen, who in photographs looks very distinguished, with thick white hair slicked back, trained in pharmacology at Columbia University and served in the U.S. Army Medical Corps in the South Pacific during World War II. It was in 1953, while working on a review article about chemically induced psychoses—a long-standing research interest—that Cohen first read about a new drug called LSD.

Yet when Cohen finally tried LSD himself in October 1955, he "was taken by surprise." Expecting to find himself trapped inside the mind of a madman, Cohen instead experienced a profound, even transcendent sense of tranquillity, as if "the problems and strivings, the worries and frustrations of everyday life [had] vanished; in their place was a majestic, sunlit, heavenly inner quietude . . . I seemed to have finally arrived at the contemplation of eternal truth." Whatever this was, he felt certain it wasn't a temporary psychosis. Betty Eisner wrote that Cohen came to think of it instead as something he called "unsanity": "a state beyond the control of the ego."

As often happens in science when a theoretical paradigm comes under the pressure of contrary evidence, the paradigm totters for a period of time as researchers attempt to prop it up with various amendments and adjustments, and then, often quite suddenly and swiftly, it collapses as a new paradigm rises to take its place. Such was the fate of the psychotomimetic paradigm in the mid-1950s. Certainly, a number of volunteers were reporting challenging and sometimes even harrowing trips, but remarkably few were having the full-on psychosis the paradigm promised. Even poor Mr. Katz's twelve hours as a madman included passages of indescribable pleasure and insight that could not be overlooked.

As it happened, the psychotomimetic paradigm was replaced not by one but by two distinct new theoretical models: the psycholytic and, later, the psychedelic model. Each was based on a different conception of how the compounds worked on the mind and therefore how they might best be deployed in the treatment of mental illness. The two models weren't at odds with each other, exactly, and some researchers explored both at various times, but they did represent profoundly different approaches to understanding the psyche, as well as to psychotherapy and, ultimately, science itself.

The so-called psycholytic paradigm was developed first and proved especially popular in Europe and with the Los Angeles group identified with Sidney Cohen, Betty Eisner, and Oscar Janiger. Coined by an English psychiatrist named Ronald Sandison, "psycholytic" means "mind loosening," which is what LSD and psilocybin seem to do—at least at low doses. Therapists who administered doses of LSD as low as 25 micrograms (and seldom higher than 150 micrograms) reported that their patients' ego defenses relaxed, allowing them to bring up and discuss difficult or repressed material with relative ease. This suggested that the drugs could be used as an aid to talking therapy, because at these doses the patients' egos remained sufficiently intact to allow them to converse with a therapist and later recall what was discussed.

The supreme virtue of the psycholytic approach was that it meshed so neatly with the prevailing modes of psychoanalysis, a practice that the drugs promised to speed up and streamline, rather than revolutionize or render obsolete. The big problem with psychoanalysis is that the access to the unconscious mind on which the whole approach depends is difficult and limited to two less-than-optimal routes: the patient's free associations and dreams. Freud called dreams "the royal road" to the subconscious, bypassing the gates of both the ego and the superego, yet the road has plenty of ruts and potholes: patients don't always remember their dreams, and when they do recall them, it is often imperfectly. Drugs like LSD and psilocybin promised a better route into the subconscious.

Stanislav Grof, who trained as a psychoanalyst, found that under moderate doses of LSD his patients would quickly establish a strong transference with the therapist, recover childhood traumas, give voice to buried emotions, and, in some cases, actually relive the experience of their birth—our first trauma and, Grof believed (following Otto Rank), a key determinant of personality. (Grof did extensive research trying to correlate his patients' recollections of their birth experience on LSD with contemporaneous reports from medical personnel and parents. He concluded that with the help of LSD many people can indeed recall the circumstances of their birth, especially when it was a difficult one.)

In Los Angeles, Cohen, Eisner, and Janiger began incorporating LSD in their weekly therapeutic sessions, gradually stepping up the dose each week until their patients gained access to subconscious material such as repressed emotions and buried memories of childhood trauma. They mainly treated neurotics and alcoholics and people with minor personality disorders—the usual sorts of patients seen by psychotherapists, functional and articulate people with intact egos and the will to get better. The Los Angeles group also treated hundreds of painters, composers, and writers, on the theory that if the wellspring of creativity was the subconscious, LSD would expand one's access to it.

These therapists and their patients expected the drug to be therapeutic, and, lo and behold, it frequently was: Cohen and Eisner reported that sixteen of their first twenty-two patients showed marked improvement. A 1967 review article summarizing papers about psycholytic therapy published between 1953 and 1965 estimated that the technique's rate of success ranged from 70 percent in cases of anxiety neurosis, 62 percent for depression, and 42 percent for obsessive-compulsive disorder. These results were impressive, yet there were few if any attempts to replicate them in controlled trials.

By the end of the decade, psycholytic LSD therapy was routine practice in the tonier precincts of Los Angeles, such as Beverly Hills. Certainly the business model was hard to beat: some therapists were charging upwards of five hundred dollars a session to administer a drug they were often getting from Sandoz for free. LSD therapy also became the subject of remarkably positive press attention. Articles like "My 12 Hours as a Madman" gave way to the enthusiastic testimonials of the numerous Hollywood celebrities who had had transformative experiences in the offices of Oscar Janiger, Betty Eisner, and Sidney Cohen and a growing number of other therapists. Anaïs Nin, Jack Nicholson, André Previn, James Coburn, and the beat comedian Lord Buckley all underwent LSD therapy, many of them on the couch of Oscar Janiger. But the most famous of these patients was Cary Grant, who gave an interview in 1959 to the syndicated gossip columnist Joe Hyams extolling the benefits of LSD therapy. Grant had more than sixty sessions and by the end declared himself "born again."

"All the sadness and vanities were torn away," the fifty-five-year-old actor told Hyams, in an interview all the more surprising in the light of Cary Grant's image as a reserved and proper Englishman. "I've had my ego stripped away. A man is a better actor without ego, because he has truth in him. Now I cannot behave untruthfully toward anyone, and certainly not to myself." From the sound of it, LSD had turned Cary Grant into an American.

"I'm no longer lonely and I am a happy man," Grant declared. He said the experience had allowed him to overcome his narcissism, greatly improving not only his acting but his relationships with women: "Young women have never before been so attracted to me."

Not surprisingly, Grant's interview, which received boatloads of national publicity, created a surge in demand for LSD therapy, and for just plain LSD. Hyams received more than eight hundred letters from readers eager to know how they might obtain it: "Psychiatrists called, complaining that their patients were now begging them for LSD."

If the period we call "the 1960s" actually began sometime in the 1950s, the fad for LSD therapy that Cary Grant unleashed in 1959 is one good place to mark a shift in the cultural breeze. Years before Timothy Leary became notorious for promoting LSD outside a therapeutic or research context, the drug had already begun "escaping from the lab" in Los Angeles and receiving fervent national press attention. By 1959, LSD was showing up on the street in some places. Several therapists and researchers in Los Angeles and New York began holding LSD "sessions" in their homes for friends and colleagues, though exactly how these sessions could be distinguished from parties is difficult to say. At least in Los Angeles, the premise of "doing research" had become tenuous at best. As one of these putative researchers would later write, "LSD became for us an intellectual fun drug."

Sidney Cohen, who by now was the dean of LSD researchers in Los Angeles, scrupulously avoided this scene and began to have second thoughts about the drug, or at least about the way it was now being used and discussed. According to his biographer, the historian Steven Novak, Cohen was made uncomfortable by the cultishness and aura of religiosity and magic that now wreathed LSD. Sounding a theme that would crop up repeatedly in the history of psychedelic research, Cohen struggled with the tension between the spiritual import of the LSD experience (and the mystical inclinations it brought out in its clinical practitioners) and the ethos of science to which he was devoted. He remained deeply ambivalent: LSD, he wrote in a 1959 letter to a colleague, had "opened a door from which we must not retreat merely because we feel uncomfortably unscientific at the threshold." And yet that is precisely how the LSD work often made him feel: uncomfortably unscientific.

Cohen also began to wonder about the status of the insights that patients brought back from their journeys. He came to believe that "under LSD the fondest theories of the therapist are confirmed by his patient." The expectancy effect was such that patients working with Freudian therapists returned with Freudian insights (framed in terms of childhood trauma, sexual drives, and oedipal emotions), while patients working with Jungian therapists returned with vivid archetypes from the attic of the collective unconscious, and Rankians with recovered memories of their birth traumas.

This radical suggestibility posed a scientific dilemma, surely, but was it necessarily a therapeutic dilemma as well? Perhaps not: Cohen wrote that "any explanation of the patient's problems, if firmly believed by both the therapist and the patient, constitutes insight or is useful as insight." Yet he qualified this perspective by acknowledging it was "nihilistic," which, scientifically speaking, it surely was. For it takes psychotherapy perilously close to the world of shamanism and faith healing, a distinctly uncomfortable place for a scientist to be. And yet as long as it works, as long as it heals people, why should anyone care? (This is the same discomfort scientists feel about using placebos. It suggests an interesting way to think about psychedelics: as a kind of "active placebo," to borrow a term proposed by Andrew Weil in his 1972 book, *The Natural Mind*. They do *something*, surely, but most of what that is may be self-generated. Or as Stanislav Grof put it, psychedelics are "nonspecific amplifiers" of mental processes.)

Cohen's thoughtful ambivalence about LSD, which he would continue to feel until the end of his career, marks him as that rare figure in a world densely populated by psychedelic evangelists: the open-minded skeptic, a man capable of holding contrary ideas in his head. Cohen continued to believe in the therapeutic power of LSD, especially in the treatment of anxiety in cancer patients, which he wrote about, enthusiastically, for *Harper's* in 1965. There, he called it "therapy by self-transcendence," suggesting he saw a role in Western medicine for what would come to be called applied mysticism. Yet Cohen never hesitated to call attention to the abuses and dangers of LSD, or to call out his more fervent colleagues when they strayed too far off the path of science—the path from which the siren song of psychedelics would lure so many.

o   o   o

BACK IN SASKATCHEWAN, Humphry Osmond and Abram Hoffer had taken a very different path after the collapse of the psychotomimetic paradigm, though this path, too, ended up complicating their own relationship to science. Struggling to formulate a new therapeutic model for LSD, they turned to a pair of brilliant amateurs—one a famous author, Aldous Huxley, and the other an obscure former bootlegger and gunrunner, spy, inventor, boat captain, ex-con, and Catholic mystic named Al Hubbard. These two most unlikely nonscientists would help the Canadian psychiatrists reconceptualize the LSD experience and develop the therapeutic protocol that is still in use today.

The name for this new approach, and the name for this class of drugs that would finally stick—psychedelics—emerged from a 1956 exchange of letters between Humphry Osmond and Aldous Huxley. The two had first met in 1953, after Huxley wrote to Osmond expressing interest in trying mescaline; he had read a journal article by Osmond describing the drug's effects on the mind. Huxley had long harbored a lively interest in drugs and consciousness—the plot of his most famous novel, *Brave New World* (1932), turns on a mind-control drug he called soma—as well as mysticism, paranormal perception, reincarnation, UFOs, and so on.

So in the spring of 1953, Humphry Osmond traveled to Los Angeles to administer mescaline to Aldous Huxley, though not without some trepidation. In advance of the session, he confided to a colleague that he did not "relish the possibility, however remote, of finding a small but discreditable niche in literary history as the man who drove Aldous Huxley mad."

He need not have worried. Huxley had a splendid trip, one that would change forever the culture's understanding of these drugs when, the following year, he published his account of his experience in *The Doors of Perception*.

"It was without question the most extraordinary and significant experience this side of the Beatific Vision," Huxley wrote in a letter to his editor shortly after it happened. For Huxley, there was no question but that the drugs gave him access not to the mind of the madman but to a spiritual realm of ineffable beauty. The most mundane objects glowed with the light of a divinity he called "the Mind at Large." Even "the folds of my gray flannel trousers were charged with 'is-ness,'" he tells us, before dilating on the beauty of the draperies in Botticelli's paintings and the "Allness and Infinity of folded cloth." When he gazed upon a small vase of flowers, he saw "what Adam had seen on the morning of his creation—the miracle, moment by moment, of naked existence . . . flowers shining with their own inner light and all but quivering under the pressure of the significance with which they were charged."

"Words like 'grace' and 'transfiguration' came to my mind." For Huxley, the drug gave him unmediated access to realms of existence usually known only to mystics and a handful of history's great visionary artists. This other world is always present but in ordinary moments is kept from our awareness by the "reducing valve" of everyday waking consciousness, a kind of mental filter that admits only "a measly trickle of the kind of consciousness" we need in order to survive. The rest was a gorgeous superfluity, which, like poetry, men die every day for the lack thereof. Mescaline flung open what William Blake had called "the doors of perception," admitting to our conscious awareness a glimpse of the infinite, which is always present all around us—even in the creases in our trousers!—if only we could just *see*.

Like every psychedelic experience before or since, Huxley's did not unfold on a blank slate, de novo, the pure product of the chemical, but rather was shaped in important ways by his reading and the philosophical and spiritual inclinations he brought to the experience. (It was only when I typed his line about flowers "shining with their own inner light" and "all but quivering under the pressure" of their significance that I realized just how strongly Huxley had inflected my own perception of plants under the influence of psilocybin.) The idea of a mental reducing valve that constrains our perceptions, for instance, comes from the French philosopher Henri Bergson. Bergson believed that consciousness was not generated by human brains but rather exists in a field outside us, something like electromagnetic waves; our brains, which he likened to radio receivers, can tune in to different frequencies of consciousness. Huxley also believed that at the base of all the world's religions there lies a common core of mystical experience he called "the Perennial Philosophy." Naturally, Huxley's morning on mescaline confirmed him in all these ideas; as one reviewer of *The Doors of Perception* put it, rather snidely, the book contained "99 percent Aldous Huxley and only one half gram mescaline." But it didn't matter: great writers stamp the world with their minds, and the psychedelic experience will forevermore bear Huxley's indelible imprint.

Whatever else it impressed on the culture, Huxley's experience left no doubt in his mind or Osmond's that the "model psychosis" didn't begin to describe the mind on mescaline or LSD, which Huxley would try for the first time two years later. One person's "depersonalization" could be another's "sense of oneness"; it was all a matter of perspective and vocabulary.

"It will give that elixir a bad name if it continues to be associated, in the public mind, with schizophrenia symptoms," Huxley wrote to Osmond in 1955. "People will think they are going mad, when in fact they are beginning, when they take it, to go sane."

Clearly a new name for this class of drugs was called for, and in a 1956 exchange of letters the psychiatrist and the writer came up with a couple of candidates. Surprisingly, however, it was the psychiatrist, not the writer, who had the winning idea. Huxley's proposal came in a couplet:

> *To make this trivial world sublime,*
> *Take half a gramme of phanerothyme*

His coinage combined the Greek words for "spirit" and "manifesting."

Perhaps wary of adopting such an overtly spiritual term, the scientist replied with his own rhyme:

> *To fathom Hell or go angelic*
> *Just take a pinch of psychedelic*

Osmond's neologism married two Greek words that together mean "mind manifesting." Though by now the word has taken on the Day-Glo coloring of the 1960s, at the time it was the very neutrality of "psychedelic" that commended it to him: the word "had no particular connotation of madness, craziness or ecstasy, but suggested an enlargement and expansion of mind." It also had the virtue of being "uncontaminated by other associations," though that would not remain the case for long.

"Psychedelic therapy," as Osmond and his colleagues practiced it beginning in the mid-1950s, typically involved a single, high-dose session, usually of LSD, that took place in comfortable surroundings, the subject stretched out on a couch, with a therapist (or two) in attendance who says very little, allowing the journey to unfold according to its own logic. To eliminate distractions and encourage an inward journey, music is played and the subject usually wears eyeshades. The goal was to create the conditions for a spiritual epiphany—what amounted to a conversion experience.

But though this mode of therapy would become closely identified with Osmond and Hoffer, they themselves credited someone else for critical elements of its design, a man of considerable mystery with no formal training as a scientist or therapist: Al Hubbard. A treatment space decorated to feel more like a home than a hospital came to be known as a Hubbard Room, and at least one early psychedelic researcher told me that this whole therapeutic regime, which is now the norm, should by all rights be known as "the Hubbard method." Yet Al Hubbard, a.k.a. "Captain Trips" and "the Johnny Appleseed of LSD," is not the kind of intellectual forebear anyone doing serious psychedelic science today is eager to acknowledge, much less celebrate.

o   o   o

AL HUBBARD IS SURELY the most improbable, intriguing, and elusive figure to grace the history of psychedelics, and that's saying a lot. There is much we don't know about him, and many key facts about his life are impossible to confirm, contradictory, or just plain fishy. To cite one small example, his FBI file puts his height at five feet eleven, but in photographs and videos Hubbard appears short and stocky, with a big round head topped with a crew cut; for reasons known only to himself, he often wore a paramilitary uniform and carried a Colt .45 revolver, giving the impression of a small-town sheriff. But based on his extensive correspondence with colleagues and a handful of accounts in the Canadian press and books about the period,* as well as interviews with a handful of people who knew him well, it's possible to assemble a rough portrait of the man, even if it does leave some important areas blurry or blank.

Hubbard was born poor in the hills of Kentucky in either 1901 or

---

* See especially Martin A. Lee and Bruce Shlain, *Acid Dreams: The Complete Social History of LSD* (New York: Grove Press, 1992), and Jay Stevens, *Storming Heaven: LSD and the American Dream* (New York: Grove Press, 1987).

1902 (his FBI file gives both dates); he liked to tell people he was twelve before he owned a pair of shoes. He never got past the third grade, but the boy evidently had a flair for electronics. As a teenager, he invented something called the Hubbard Energy Transformer, a new type of battery powered by radioactivity that "could not be explained by the technology of the day"—this according to the best account we have of his life, a well-researched 1991 *High Times* article by Todd Brendan Fahey. Hubbard sold a half interest in the patent for seventy-five thousand dollars, though nothing ever came of the invention and *Popular Science* magazine once included it in a survey of technological hoaxes. During Prohibition, Hubbard drove a taxi in Seattle, but that appears to have been a cover: in the trunk of his cab he kept a sophisticated ship-to-shore communications system he used to guide bootleggers seeking to evade the Coast Guard. Hubbard was eventually busted by the FBI and spent eighteen months in prison on a smuggling charge.

After his release from prison the trail of Hubbard's life becomes even more difficult to follow, muddied by vague and contradictory accounts. In one of them, Hubbard became involved in an undercover operation to ship heavy armaments from San Diego to Canada and from there on to Britain, in the years before the U.S. entered World War II, when the nation was still officially neutral. (Scouts for the future OSS officer Allen Dulles, impressed by Hubbard's expertise in electronics, may or may not have recruited him for the mission.) But when Congress began investigating the operation, Hubbard fled to Vancouver to avoid prosecution. There he became a Canadian citizen, founded a charter boat business (earning him the title of Captain) and became the science director of a uranium mining company. (According to one account, Hubbard had something to do with supplying uranium to the Manhattan Project.) By the age of fifty, the "barefoot boy from Kentucky" had become a millionaire, owner of a fleet of aircraft, a one-hundred-foot yacht, a Rolls-Royce, and a private island off Vancouver. At some point during the war Hubbard apparently returned to the United States, and he joined the OSS shortly before the wartime intelligence agency became the CIA.

A few other curious facts about the prepsychedelic Al Hubbard: He was an ardent Catholic, with a pronounced mystical bent. And he was unusually flexible in his professional loyalties, working at various times as a rum- and gunrunner as well as an agent for the Bureau of Alcohol, Tobacco, and Firearms. Was he a double agent of some kind? Possibly. At one time or another, he also worked for the Canadian Special Services, the U.S. Department of Justice, and the Food and Drug Administration. His FBI file suggests he had links to the CIA during the 1950s, but the redactions are too heavy for it to reveal much about his role, if any. We know the government kept close tabs on the psychedelic research community all through the 1950s, 1960s, and 1970s (funding university research on LSD and scientific conferences in some cases), and it wouldn't be surprising if, in exchange for information, the government would allow Hubbard to operate with as much freedom as he did. But this remains speculation.

Al Hubbard's life made a right-angled change of course in 1951. At the time, he was hugely successful but unhappy, "desperately searching for meaning in his life"—this according to Willis Harman, one of a group of Silicon Valley engineers to whom Hubbard would introduce LSD later in the decade. As Hubbard told the story to Harman (and Harman told it to Todd Brendan Fahey), he was hiking in Washington State when an angel appeared to him in a clearing. "She told Al that something tremendously important to the future of mankind would be coming soon, and that he could play a role in it if he wanted to. But he hadn't the faintest clue what he was supposed to be looking for."

The clue arrived a year later, in the form of an article in a scientific journal describing the behavior of rats given a newly discovered compound called LSD. Hubbard tracked down the researcher, obtained some LSD, and had a literally life-changing experience. He witnessed the beginning of life on earth as well as his own conception. "It was the deepest mystical thing I've ever seen," he told friends later. "I saw myself as a tiny mite in a big swamp with a spark of intelligence. I saw my mother and father having intercourse." Clearly this was what the angel had foretold—"something tremendously important to the future of mankind." Hubbard realized it was up to him to bring the new gospel of LSD, and the chemical itself, to as many people as he possibly could. He had been given what he called a "special chosen role."

Thus began Al Hubbard's career as the Johnny Appleseed of LSD. Through his extensive connections in both government and business, he persuaded Sandoz Laboratories to give him a mind-boggling quantity of LSD—a liter bottle of it, in one account, forty-three cases in another, six thousand vials in a third. (He reportedly told Albert Hofmann he planned to use it "to liberate human consciousness.") Depending on whom you believe, he kept his supply hidden in a safe-deposit box in Zurich or buried somewhere in Death Valley, but a substantial part of it he carried with him in a leather satchel. Eventually, Hubbard became the exclusive distributor of Sandoz LSD in Canada and, later, somehow secured an Investigational New Drug permit from the FDA allowing him to conduct clinical research on LSD in the United States—this even though he had a third-grade education, a criminal record, and a single, arguably fraudulent scientific credential. (His PhD had been purchased from a diploma mill.) Seeing himself as "a catalytic agent," Hubbard would introduce an estimated six thousand people to LSD between 1951 and 1966, in an avowed effort to shift the course of human history.

Curiously, the barefoot boy from Kentucky was something of a mandarin, choosing as his subjects leading figures in business, government, the arts, religion, and technology. He believed in working from the top down and disdained other psychedelic evangelists, like Timothy Leary, who took a more democratic approach. Members of Parliament, officials of the Roman Catholic Church,* Hollywood actors, government officials, prominent writers and philosophers, university officials, computer engineers, and prominent businessmen were all introduced to LSD as part of Hubbard's mission to shift the course of history from above. (Not everyone Hubbard approached would play: J. Edgar Hoover, whom Hubbard claimed as a close friend, declined.) Hubbard believed that "if he could give the psychedelic experience to the major executives of the Fortune 500 companies," Abram Hoffer recalled, "he would change the whole of society." One of the executives Hubbard turned on in the late 1950s—Myron Stolaroff, assistant to the president for long-term planning at Ampex, at the time a leading electronics firm in Silicon Valley—became "convinced that [Al Hubbard] was the man to bring LSD to planet Earth."

o o o

IN 1953, not long after his psychedelic epiphany, Hubbard invited Humphry Osmond to lunch at the Vancouver Yacht Club. Like so many others, Osmond was deeply impressed by Hubbard's worldliness, wealth, connections, and access to seemingly endless supplies of LSD. The lunch led to a collaboration that changed the course of

---

* Hubbard treasured a 1957 letter he received from a Monsignor Brownmajor in Vancouver endorsing his work: "We therefore approach the study of these psychedelics and their influence on the mind of man anxious to discover whatever attributes they possess, respectfully evaluating their proper place in the Divine Economy."

psychedelic research and, in important ways, laid the groundwork for the research taking place today.

Under the influence of both Hubbard and Huxley, whose primary interest was in the revelatory import of psychedelics, Osmond abandoned the psychotomimetic model. It was Hubbard who first proposed to him that the mystical experience many subjects had on a single high dose of mescaline or LSD might itself be harnessed as a mode of therapy—and that the experience was more important than the chemical. The psychedelic journey could, like the conversion experience, forcibly show people a new, more encompassing perspective on their lives that would help them to change. But perhaps Hubbard's most enduring contribution to psychedelic therapy emerged in, of all places, the treatment room.

It is easier to accumulate facts about Al Hubbard's life than it is to get a steady sense of the character of the man, it was so rife with contradiction. The pistol-packing tough guy was also an ardent mystic who talked about love and the heavenly beatitudes. And the well-connected businessman and government agent proved to be a remarkably sensitive and gifted therapist. Though he never used those terms, Hubbard was the first researcher to grasp the critical importance of set and setting in shaping the psychedelic experience. He instinctively understood that the white walls and fluorescent lighting of the sanitized hospital room were all wrong. So he brought pictures and music, flowers and diamonds, into the treatment room, where he would use them to prime patients for a mystical revelation or divert a journey when it took a terrifying turn. He liked to show people paintings by Salvador Dalí and pictures of Jesus or to ask them to study the facets of a diamond he carried. One patient he treated in Vancouver, an alcoholic paralyzed by social anxiety, recalled Hubbard handing him a bouquet of roses during an LSD session: "He said, 'Now hate them.' They withered and the petals fell off, and I started to cry. Then he said, 'Love them,' and they came back brighter and even more spectacular than before. That meant a lot to me. I realized that you can make your relationships anything you want. The trouble I was having with people was coming from me."

What Hubbard was bringing into the treatment room was something well known to any traditional healer. Shamans have understood for millennia that a person in the depths of a trance or under the influence of a powerful plant medicine can be readily manipulated with the help of certain words, special objects, or the right kind of music. Hubbard understood intuitively how the suggestibility of the human mind during an altered state of consciousness could be harnessed as an important resource for healing—for breaking destructive patterns of thought and proposing new perspectives in their place. Researchers might prefer to call this a manipulation of set and setting, which is accurate enough, but Hubbard's greatest contribution to modern psychedelic therapy was to introduce the tried-and-true tools of shamanism, or at least a Westernized version of it.

o   o   o

WITHIN A FEW YEARS, Hubbard had made the acquaintance of just about everybody in the psychedelic research community in North America, leaving an indelible impression on everyone he met, along with a trail of therapeutic tips and ampules of Sandoz LSD. By the late 1950s, he had become a kind of psychedelic circuit rider. One week he might be in Weyburn, assisting Humphry Osmond and Abram Hoffer in their work with alcoholics, which was earning them international attention. From there to Manhattan, to meet with R. Gordon Wasson, and then a stop on his way back west to administer LSD to a VIP or check in on a research group working in Chicago.

The next week might find him in Los Angeles, conducting LSD sessions with Betty Eisner, Sidney Cohen, or Oscar Janiger, freely sharing his treatment techniques and supplies of LSD. ("We waited for him like the little old lady on the prairie waiting for a copy of the Sears Roebuck catalog," Oscar Janiger recalled years later.) And then it was back to Vancouver, where he had persuaded Hollywood Hospital to dedicate an entire wing to treating alcoholics with LSD.* Hubbard would often fly his plane down to Los Angeles to discreetly ferry Hollywood celebrities up to Vancouver for treatment. It was this sideline that earned him the nickname Captain Trips. Hubbard also established two other alcoholism treatment facilities in Canada, where he regularly conducted LSD sessions and reported impressive rates of success. LSD treatment for alcoholism using the Hubbard method became a business in Canada. But Hubbard believed it was unethical to profit from LSD, which led to tensions between him and some of the institutions he worked with, because they were charging patients upwards of five hundred dollars for an LSD session. For Hubbard, psychedelic therapy was a form of philanthropy, and he drained his fortune advancing the cause.

Al Hubbard moved between these far-flung centers of research like a kind of psychedelic honeybee, disseminating information, chemicals, and clinical expertise while building what became an extensive network across North America. In time, he would add Menlo Park and Cambridge to his circuit. But was Hubbard just spreading information, or was he also collecting it and passing it on to the CIA? Was the pollinator also a spy? It's impossible to say for certain;

---

* Hubbard's name appears on a single scientific paper, written with his colleagues at Hollywood Hospital: "The Use of LSD-25 in the Treatment of Alcoholism and Other Psychiatric Problems," *Quarterly Journal of Studies on Alcohol* 22 (March 1961): 34–45.

some people who knew Hubbard (like James Fadiman) think it's entirely plausible, while others aren't so sure, pointing to the fact the Captain often criticized the CIA for using LSD as a weapon. "The CIA work stinks," he told Oscar Janiger in the late 1970s.

Hubbard was referring to the agency's MK-Ultra research program, which since 1953 had been trying to figure out whether LSD could be used as a nonlethal weapon of war (by, say, dumping it in an adversary's water supply), a truth serum in interrogations, a means of mind control,* or a dirty trick to play on unfriendly foreign leaders, causing them to act or speak in embarrassing ways. None of these schemes panned out, at least as far as we know, and all reflected a research agenda that remained stuck on the psychotomimetic model long after other researchers had abandoned it. Along the way, the CIA dosed its own employees and unwitting civilians with LSD; in one notorious case that didn't come to light until the 1970s, the CIA admitted to secretly giving LSD to an army biological weapons specialist named Frank Olson in 1953; a few days later, Olson supposedly jumped to his death from the thirteenth floor of the Statler Hotel in New York. (Others believe Olson was pushed and that the CIA's admission, embarrassing as it was, was actually a cover-up for a crime far more heinous.) It could be Olson whom Al Hubbard was referring to when he said, "I tried to tell them how to use it, but even when they were killing people, you couldn't tell them a goddamned thing."

A regular stop on Hubbard's visits to Los Angeles was the home of Aldous and Laura Huxley. Huxley and Hubbard had formed the most unlikely of friendships after Hubbard introduced the author to

---

* Sidney Gottlieb, the CIA officer in charge of MK-Ultra, would testify to Congress that its goal was "to investigate whether and how it was possible to modify an individual's behavior by covert means." We would know more about MK-Ultra had Gottlieb not destroyed most of the program's records on the orders of the CIA director Richard Helms.

LSD—and the Hubbard method—in 1955. The experience put the author's 1953 mescaline trip in the shade. As Huxley wrote to Osmond in its aftermath, "What came through the closed door was the realization . . . the direct, total awareness, from the inside, so to say, of Love as the primary and fundamental cosmic fact." The force of this insight seemed almost to embarrass the writer in its baldness: "The words, of course, have a kind of indecency and must necessarily ring false, seem like twaddle. But the fact remains."

Huxley immediately recognized the value of an ally as skilled in the ways of the world as the man he liked to call "the good Captain." As so often seems to happen, the Man of Letters became smitten with the Man of Action.

"What Babes in the Woods we literary gents and professional men are!" Huxley wrote to Osmond about Hubbard. "The great World occasionally requires your services, is mildly amused by mine, but its full attention and deference are paid to Uranium and Big Business. So what extraordinary luck that this representative of both these Higher Powers should (a) have become so passionately interested in mescaline and (b) be such a very nice man."

Neither Huxley nor Hubbard was particularly dedicated to medicine or science, so it's not surprising that over time their primary interest would drift from the treatment of individuals with psychological problems to a desire to treat the whole of society. (This aspiration seems eventually to infect everyone who works with psychedelics, touching scientists, too, including ones as different in temperament as Timothy Leary and Roland Griffiths.) But psychological research proceeds person by person and experiment by experiment; there is no real-world model for using a drug to change all of society as Hubbard and Huxley determined to do, with the result that the scientific method began to feel to them, as it later would to Leary, like a straitjacket.

In the wake of his first LSD experience, Huxley wrote to Osmond suggesting that "who, having once come to the realization of the primordial fact of unity in love, would ever want to return to experimentation on the psychic level? . . . My point is that the opening of the door by mescalin[e] or LSD is too precious an opportunity, too high a privilege to be neglected for the sake of experimentation." Or to be limited to sick people. Osmond was actually sympathetic to this viewpoint—after all, he had administered mescaline to Huxley, hardly a controlled experiment—and he participated in many of Hubbard's sessions turning on the Best and Brightest. But Osmond wasn't prepared to abandon science or medicine for whatever Huxley and Hubbard imagined might lay beyond it.

In 1955, Al Hubbard sought to escape the scientific straitjacket and formalize his network of psychedelic researchers by establishing something he called the Commission for the Study of Creative Imagination. The name reflected his own desire to take his work with psychedelics beyond the limits of medicine and its focus on the ill. To serve on the commission's board, Hubbard recruited Osmond, Hoffer, Huxley, and Cohen, as well as half a dozen other psychedelic researchers, a philosopher (Gerald Heard), and a UN official; he named himself "scientific director."

(What did these people think of Hubbard and his grandiose title, not to mention his phony academic credentials? They were at once indulgent and full of admiration. After Betty Eisner wrote a letter to Osmond expressing discomfort with some of Hubbard's representations, he suggested she think of him as a kind of Christopher Columbus: "Explorers have not always been the most scientific, excellent or wholly detached people.")

It isn't clear how much more there was to the Commission for the Study of Creative Imagination than a fancy letterhead, but its very existence signaled a deepening fissure between the medical and the spiritual approach to psychedelics. (Sidney Cohen, ever ambivalent on questions of science versus mysticism, abruptly resigned in 1957, only a year after joining the board.) His title as "scientific director" notwithstanding, Hubbard himself said during this period, "My regard for science, as an end within itself, is diminishing as time goes on . . . when the thing I want with all of my being, is something that lives far outside and out of reach of empirical manipulation." Long before Leary, the shift in the objective of psychedelic research from psychotherapy to cultural revolution was well under way.

o   o   o

ONE LAST NODE worth visiting in Al Hubbard's far-flung psychedelic network is Silicon Valley, where the potential for LSD to foster "creative imagination" and thereby change the culture received its most thorough test to date. Indeed, the seeds that Hubbard planted in Silicon Valley continue to yield interesting fruit, in the form of the valley's ongoing interest in psychedelics as a tool for creativity and innovation. (As I write, the practice of microdosing—taking a tiny, "subperceptual" regular dose of LSD as a kind of mental tonic— is all the rage in the tech community.) Steve Jobs often told people that his experiments with LSD had been one of his two or three most important life experiences. He liked to taunt Bill Gates by suggesting, "He'd be a broader guy if he had dropped acid once or gone off to an ashram when he was younger." (Gates has said he did in fact try LSD.) It might not be a straight one, but it is possible to draw a line connecting Al Hubbard's arrival in Silicon Valley with his satchelful of LSD to the tech boom that Steve Jobs helped set off a quarter century later.

The key figure in the marriage of Al Hubbard and Silicon Valley was Myron Stolaroff. Stolaroff was a gifted electrical engineer who, by the mid-1950s, had become assistant to the president for strategic planning at Ampex, one of the first technology companies to set up shop in what at the time was a sleepy valley of farms and orchards. (It wouldn't be called Silicon Valley until 1971.) Ampex, which at its peak had thirteen thousand employees, was a pioneer in the development of reel-to-reel magnetic tape for audio, video, and data recording. Born in Roswell, New Mexico, in 1920, Stolaroff studied engineering at Stanford and was one of Ampex's very first employees, a fact that would make him a wealthy man. Nominally Jewish, he was by his thirties a spiritual seeker whose path eventually led him to Gerald Heard, the English philosopher and friend of Aldous Huxley's. Stolaroff was so moved by Heard's description of his LSD experience with Al Hubbard that in March 1956 he traveled to Vancouver for a session with the Captain in his apartment.

Sixty-six micrograms of Sandoz LSD launched Stolaroff on a journey by turns terrifying and ecstatic. Over the course of several hours, he witnessed the entire history of the planet from its formation through the development of life on earth and the appearance of humankind, culminating in the trauma of his own birth. (This seems to have been a common trajectory of Hubbard-guided trips.) "That was a remarkable opening for me," he told an interviewer years later, "a tremendous opening. I relived a very painful birth experience that had determined almost all my personality features. But I also experienced the oneness of mankind, and the reality of God. I knew that from then on . . . I would be totally committed to this work.

"After that first LSD experience, I said, 'this is the greatest discovery man has ever made.'"

Stolaroff shared the news with a small number of his friends and colleagues at Ampex. They began meeting every month or so to discuss spiritual questions and the potential of LSD to help individuals—healthy individuals—realize their full potential. Don Allen, a young Ampex engineer, and Willis Harman, a professor of electrical engineering at Stanford, joined the group, and Al Hubbard began coming down to Menlo Park to guide the members on psychedelic journeys and then train them to guide others. "As a therapist," Stolaroff recalled, "he was one of the best."

Convinced of the power of LSD to help people transcend their limitations, Stolaroff tried for a time, with Hubbard's help, to reshape Ampex as the world's first "psychedelic corporation." Hubbard conducted a series of weekly workshops at headquarters and administered LSD to company executives at a site in the Sierra. But the project foundered when the company's general manager, who was Jewish, objected to the images of Christ, the Virgin Mary, and the Last Supper that Hubbard insisted on bringing into his office. Around the same time, Willis Harman shifted the focus of his teaching at Stanford, offering a new class on "the human potential" that ended with a unit on psychedelics. The engineers were getting religion. (And have it still: I know of one Bay Area tech company today that uses psychedelics in its management training. A handful of others have instituted "microdosing Fridays.")

In 1961, Stolaroff left Ampex to dedicate himself full-time to psychedelic research. With Willis Harman, he established the orotundly titled International Foundation for Advanced Study (IFAS) to explore the potential of LSD to enhance human personality and creativity. Stolaroff hired a psychiatrist named Charles Savage as medical director and, as staff psychologist, a first-year graduate student by the name of James Fadiman. (Fadiman, who graduated from Harvard in 1960, was introduced to psilocybin by Richard Alpert, though not until after his graduation. "The greatest thing in the world has happened to me," Alpert told his former student, "and I want to share it with you.") Don Allen also left his engineering post at Ampex to join IFAS as a screener and guide. The foundation secured a drug research permit from the FDA and a supply of LSD and mescaline from Al Hubbard and began—to use an Al Hubbard term— "processing clients." Over the next six years, the foundation would process some 350 people.

As James Fadiman and Don Allen recall those years at the foundation (both sat for extensive interviews), it was a thrilling and heady time to be working on what they were convinced was the frontier of human possibility. For the most part, their experimental subjects were "healthy normals" or what Fadiman described as "a healthy neurotic outpatient population." Each client paid five hundred dollars for a package that included before-and-after personality testing, a guided LSD session, and some follow-up. Al Hubbard "would float in and out," Don Allen recalls. He "was both our inspiration and our resident expert." James Fadiman says, "He was the hidden force behind the Menlo Park research." From time to time, Hubbard would take members of the staff to Death Valley for training sessions, in the belief that the primordial landscape there was particularly conducive to revelatory experience.

In half a dozen or so papers published in the early 1960s, the foundation's researchers reported some provocative "results." Seventy-eight percent of clients said the experience had increased their ability to love, 71 percent registered an increase in self-esteem, and 83 percent said that during their sessions they had glimpsed "a higher power, or ultimate reality." Those who had such an experience were the ones who reported the most lasting benefits from their session. Don Allen told me that most clients emerged with "notable and fairly sustainable changes in beliefs, attitudes, and behavior, way above statistical probability." Specifically, they became "much less judgmental, much less rigid, more open, and less defended." But it wasn't all sweetness and light: several clients abruptly broke off marriages after their sessions, now believing they were mismatched or trapped in destructive patterns of behavior.

The foundation also conducted studies to determine if LSD could in fact enhance creativity and problem solving. "This wasn't at all obvious," James Fadiman points out, "since the experience is so powerful, you might just wander off and lose track of what you were trying to accomplish." So to test their hypothesis, Fadiman and his colleagues started with themselves, seeing if they could design a credible creativity experiment while on a relatively light dose of LSD—a hundred micrograms. Perhaps not surprisingly, they determined that they could.

Working in groups of four, James Fadiman and Willis Harman administered the same dose of LSD to artists, engineers, architects, and scientists, all of whom were somehow "stuck" in their work on a particular project. "We used every manipulation of set and setting in the book," Fadiman recalled, telling subjects "they would be fascinated by their intellectual capacities and would solve problems as never before." Subjects reported much greater fluidity in their thinking, as well as an enhanced ability to both visualize a problem and recontextualize it. "We were amazed, as were our participants, at how many novel and effective solutions came out of our sessions," Fadiman wrote. Among their subjects were some of the visionaries who in the next few years would revolutionize computers, including William English and Doug Engelbart.* There are all sorts of problems with this study—it was not controlled, it relied on the subjects' own assessments of their success, and it was halted before it could be completed— but it does at least point to a promising avenue for research.

---

* During his LSD session, Engelbart invented a "tinkle toy" to toilet train children, or at least boys: a waterwheel floating in a toilet that could be powered by a stream of urine. He went on to considerably more significant accomplishments, including the computer mouse, the graphical computer interface, text editing, hypertext, networked computers, e-mail, and videoconferencing, all of which he demonstrated in a legendary "mother of all demos" in San Francisco in 1968.

The foundation had closed up shop by 1966, but Hubbard's work in Silicon Valley was not quite over. In one of the more mysterious episodes of his career, Hubbard was called out of semiretirement by Willis Harman in 1968. After IFAS disbanded, Harman had gone to work at the Stanford Research Institute (SRI), a prestigious think tank affiliated with Stanford University and a recipient of contracts from several branches of the federal government, including the military. Harman was put in charge of SRI's Educational Policy Research Center, with a mandate to envision education's future. LSD by now was illegal but still very much in use in the community of engineers and academics in and around Stanford.

Hubbard, who by now was broke, was hired as a part-time "special investigative agent," ostensibly to keep tabs on the use of drugs in the student movement. Harman's letter of employment to Hubbard is both obscure and suggestive: "Our investigations of some of the current social movements affecting education indicate that the drug use prevalent among student members of the New Left is not entirely undesigned. Some of it appears to be present as a deliberate weapon aimed at political change. We are concerned with assessing the significance of this as it impacts on matters of long-range educational policy. In this connection it would be advantageous to have you considered in the capacity of a special investigative agent who might have access to relevant data which is not ordinarily available." Though not mentioned in the letter, Hubbard's services to SRI also included using his extensive government contacts to keep contracts flowing. So Al Hubbard once again donned his khaki security-guard uniform, complete with gold badge, sidearm, and a belt studded with bullets, and got back to work.

But the uniform and the "special agent" title were all a cover, and an audacious one at that.

As a vocal enemy of the rising counterculture, it's entirely possible Hubbard did investigate illegal drug use on campus for SRI (or others*), but if he did, he was once again working both sides of the street. For though the legal status of LSD had changed by 1968, Hubbard and Harman's mission—"to provide the [LSD] experience to political and intellectual leaders around the world"—apparently had not. The work might well have continued, just more quietly and beneath a cover story. For as Willis Harman told Todd Brendan Fahey in a 1990 interview and as a former SRI employee confirmed, "Al never did anything resembling security work.

"Al's job was to run the special sessions for us."

That former SRI employee is Peter Schwartz, an engineer who became a leading futurist; he is currently senior vice president for government relations and strategic planning at Salesforce.com. In 1973, Schwartz went to work for Willis Harman at SRI, his first job out of graduate school. By then, Al Hubbard was more or less retired, and Schwartz was given his office. On the wall above the desk hung a large photograph of Richard Nixon, inscribed "to my good friend, Al, for all your years of service, your friend, Dick." A pile of mail accumulated in the in-box, with letters addressed to A. M. Hubbard from all over the world, including, he recalled, one from George Bush, the future CIA director, who at the time was serving as head of the Republican National Committee.

---

* Hubbard hated the idea of street acid and the counterculture's use of it. According to Don Allen, he played a role in at least one bust of an important underground LSD chemist in 1967. Hubbard sent Don Allen to a meeting to pose as a Canadian buyer looking to purchase "pure LSD" from a Bay Area group that included the notorious LSD chemist (and Grateful Dead sound engineer) Owsley Stanley III. Federal agents tailed the people at the meeting back to Stanley and his lab in Orinda, California; during the bust, they reportedly found 350,000 doses of LSD.

"Who was this fellow?" Schwartz wondered. And then one day this round fellow with a gray crew cut, dressed in a security guard's uniform and carrying a .38, showed up to retrieve his mail.

"'I'm a friend of Willis's,'" Hubbard told Schwartz. "And then he began asking me the strangest questions, completely without context. 'Where do you think you *actually* came from? What do you think about the cosmos?' I learned later this was how he checked people out, to decide whether or not you were a worthy candidate."

Intrigued, Schwartz asked Harman about this mystery man and, piece by piece, began to put together much of the tale of Hubbard's life. The young futurist soon realized that "most of the people I was meeting who had interesting ideas had tripped with Hubbard: professors at Stanford, Berkeley, the staff at SRI, computer engineers, scientists, writers. And all of them had been transformed by the experience." Schwartz said that several of the early computer engineers relied on LSD in designing circuit chips, especially in the years before they could be designed on computers. "You had to be able to visualize a staggering complexity in three dimensions, hold it all in your head. They found that LSD could help."

Schwartz eventually realized that "everyone in that community"—referring to the Bay Area tech crowd in the 1960s and early 1970s, as well as the people in and around Stewart Brand's Whole Earth Network—"had taken Hubbard LSD."

Why were engineers in particular so taken with psychedelics? Schwartz, himself trained as an aerospace engineer, thinks it has to do with the fact that unlike the work of scientists, who can simplify the problems they work on, "problem solving in engineering always involves irreducible complexity. You're always balancing complex variables you can never get perfect, so you're desperately searching to find patterns. LSD shows you patterns.

"I have no doubt that all that Hubbard LSD all of us had taken had a big effect on the birth of Silicon Valley."

Stewart Brand received his own baptism in Hubbard LSD at IFAS in 1962, with James Fadiman presiding as his guide. His first experience with LSD "was kind of a bum trip," he recalls, but it led to a series of other journeys that reshaped his worldview and, indirectly, all of ours. The Whole Earth Network Brand would subsequently gather together (which included Peter Schwartz, Esther Dyson, Kevin Kelly, Howard Rheingold, and John Perry Barlow) and play a key role in redefining what computers meant and did, helping to transform them from a top-down tool of the military-industrial complex—with the computer punch card a handy symbol of Organization Man—into a tool of personal liberation and virtual community, with a distinctly countercultural vibe. How much does the idea of cyberspace, an immaterial realm where one can construct a new identity and merge with a community of virtual others, owe to an imagination shaped by the experience of psychedelics? Or for that matter virtual reality?* The whole notion of cybernetics, the idea that material reality can be translated into bits of information, may also owe something to the experience of LSD, with its power to collapse matter into spirit.

Brand thinks LSD's value to his community was as an instigator of creativity, one that first helped bring the power of networked computers to people (via SRI computer visionaries such as Doug En-

---

* The two best accounts of the counterculture's (and its chemicals') influence on the computer revolution are Fred Turner's *From Counterculture to Cyberculture: Stewart Brand, the Whole Earth Network, and the Rise of Digital Utopianism* (Chicago: University of Chicago Press, 2006) and John Markoff's *What the Dormouse Said: How the Sixties Counterculture Shaped the Personal Computer Industry* (New York: Penguin Books, 2005).

gelbart and the early hacker community), but then was superseded by the computers themselves. ("At a certain point, the drugs weren't getting any better," Brand said, "but the computers were.") After his experience at IFAS, Brand got involved with Ken Kesey and his notorious Acid Tests, which he describes as "a participatory art form that led directly to Burning Man," the annual gathering of the arts, technology, and psychedelic communities in the Nevada desert. In his view, LSD was a critical ingredient in nourishing the spirit of collaborative experiment, and tolerance of failure, that distinguish the computer culture of the West Coast. "It gave us permission to try weird shit in cahoots with other people."

On occasion, the LSD produced genuine insight, as it did for Brand himself one chilly afternoon in the spring of 1966. Bored, he went up onto the roof of his building in North Beach and took a hundred micrograms of acid—Fadiman's creativity dose. As he looked toward downtown while wrapped in a blanket, it appeared that the streets lined with buildings were not quite parallel. This must be due to the curvature of Earth, Brand decided. It occurred to him that when we think of Earth as flat, as we usually do, we assume it is infinite, and we treat its resources that way. "The relationship to infinity is to use it up," he thought, "but a round earth was a finite spaceship you had to manage carefully." At least that's how it appeared to him that afternoon, "from three stories and one hundred mikes up."

It would change everything if he could convey this to people! But how? He flashed on the space program and wondered, "Why haven't we seen a picture of the earth from space? I become fixed on this, on how to get this photo that would revolutionize our understanding of our place in the universe. *I know, I'll make a button!* But what should it say? 'Let's have a photo of the earth from space.' No, it needs to be a question, and maybe a little paranoid—draw on *that* American resource. 'Why haven't we seen a photograph of the whole earth yet?'"

Brand came down from his roof and launched a campaign that eventually reached the halls of Congress and NASA. Who knows if it was the direct result of Brand's campaign, but two years later, in 1968, the Apollo astronauts turned their cameras around and gave us the first photograph of Earth from the moon, and Stewart Brand gave us the first edition of the *Whole Earth Catalog*. Did everything change? The case could be made that it had.

Part II: The Crack-Up

Timothy Leary came late to psychedelics. By the time he launched the Harvard Psilocybin Project in 1960, there had already been a full decade of psychedelic research in North America, with hundreds of academic papers and several international conferences to show for it. Leary himself seldom made reference to this body of work, preferring to give the impression that his own psychedelic research represented a radical new chapter in the annals of psychology. In 1960, the future of psychedelic research looked bright. Yet within the brief span of five years, the political and cultural weather completely shifted, a moral panic about LSD engulfed America, and virtually all psychedelic research and therapy were either halted or driven underground. *What happened?*

"Timothy Leary" is the too-obvious answer to that question. Just about everyone I've interviewed on the subject—dozens of people—has prefaced his or her answer by saying, "It's far too easy to blame Leary," before proceeding to do precisely that. It's hard to avoid the conclusion that the flamboyant psychology professor with a tropism bending him toward the sun of publicity, good *or* bad, did grave damage to the cause of psychedelic research. He did. And yet the social forces unleashed by the drugs themselves once they moved from the laboratory out into the culture were bigger and stronger than any individual could withstand—or take credit for. With or without the heedless, joyful, and amply publicized antics of Timothy Leary, the sheer Dionysian power of LSD was itself bound to shake things up and incite a reaction.

By the time Leary was hired by Harvard in 1959, he had a national reputation as a gifted personality researcher, and yet even then—before his first shattering experience with psilocybin in Cuernavaca during the summer of 1960—Leary was feeling somewhat disenchanted with his field. A few years before, while working as director of psychiatric research at Kaiser Hospital in Oakland, Leary and a colleague had conducted a clever experiment to assess the effectiveness of psychotherapy. A group of patients seeking psychiatric care were divided into two groups; one received the standard treatment of the time, the other (consisting of people on a waiting list) no treatment at all. After a year, one-third of all the subjects had improved, one-third had gotten worse, and one-third remained unchanged—regardless of which group they were in. Whether or not a subject received treatment made no difference whatsoever in the outcome. So what good was conventional psychotherapy? Psychology? Leary had begun to wonder.

Leary quickly established himself at Harvard's Department of Social Relations as a dynamic and charismatic, if somewhat cynical, teacher. The handsome professor was a great talker, in the expansive Irish mode, and could charm the pants off anyone, especially women, for whom he was apparently catnip. Leary had always had a roguish, rebellious streak—he was court-martialed during his time at West Point for violating the honor code and expelled from the University of Alabama for spending the night in a women's dorm—and Harvard-the-institution brought out rebellion in him. Leary would speak cynically of psychological research as a "game." Herbert Kelman, a colleague in the department who later became Leary's chief adversary, recalls the new professor as "personable" (Kelman helped him find his first house) but says, "I had misgivings about him from the beginning. He would often talk out of the top of his head about things he knew nothing about, like existentialism, and he was telling our students psychology was all a game. It seemed to me a bit cavalier and irresponsible."

I met Kelman, now in his nineties, in the small, overstuffed apartment where he lives with his wife in an assisted-living facility in West Cambridge. Kelman displayed no rancor toward Leary yet evinced little respect for him either as a teacher or as a scientist; indeed, he believes Leary had become disenchanted with science well before psychedelics came into his life. In Kelman's opinion, even before the psilocybin, "He was already halfway off the deep end."

Leary's introduction to psilocybin, poolside in Mexico during the summer of 1960, came three years after R. Gordon Wasson published his notorious *Life* magazine article about the "mushrooms that cause strange visions." For Leary, the mushrooms were transformative. In an afternoon, his passion to understand the human mind had been reignited—indeed, had exploded.

"In four hours by the swimming pool in Cuernavaca I learned more about the mind, the brain, and its structures than I did in the preceding fifteen as a diligent psychologist," he wrote later in *Flashbacks*, his 1983 memoir. "I learned that the brain is an underutilized biocomputer . . . I learned that normal consciousness is one drop in an ocean of intelligence. That consciousness and intelligence can be systematically expanded. That the brain can be reprogrammed."

Leary returned from his journey with an irresistible urge to "rush back and tell everyone," as he recalled in *High Priest*, his 1968 memoir. And then in a handful of sentences he slid into a prophetic voice, one in which the whole future trajectory of Timothy Leary could be foretold:

> *Listen! Wake up! You are God! You have the Divine plan engraved in cellular script within you. Listen! Take this sacrament! You'll see! You'll get the revelation! It will change your life!*

But at least for the first year or two at Harvard, Leary went through the motions of doing science. Back in Cambridge that fall, he recruited Richard Alpert, a promising assistant professor who was heir to a railroad fortune, and, having secured the tacit approval of their department chair, David McClelland, the two launched the Harvard Psilocybin Project, operating out of a tiny broom closet of an office in the Department of Social Relations in a house at 5 Divinity Avenue. (I went looking for the house, but it has long since been razed and replaced by a sprawling, block-long brick science building.) Leary, ever the salesman, had convinced Harvard that the research he proposed to undertake was squarely in the tradition of William James, who in the early years of the century had also studied altered states of consciousness and mystical experience at Harvard. The university placed one condition on the research: Leary and Alpert could give the new drugs to graduate students, but not to undergraduates. Before long, an intriguingly titled new seminar showed up in the Harvard course listings:

Experimental Expansion of Consciousness
The literature describing internally and externally induced changes in awareness will be reviewed. The basic elements of mystical experiences will be studied cross-culturally. The members of the seminar will participate in experiences

with consciousness expanding methods and a systematic analysis of attention will be paid to the problems of methodology in this area. This seminar will be limited to advanced graduate students. Admission by consent of the instructor.

"Experimental Expansion of Consciousness" proved to be extremely popular.

o   o   o

IN ITS THREE YEARS of existence, the Harvard Psilocybin Project accomplished surprisingly little, at least in terms of science. In their first experiments, Leary and Alpert administered psilocybin to hundreds of people of all sorts, including housewives, musicians, artists, academics, writers, fellow psychologists, and graduate students, who then completed questionnaires about their experiences. According to "Americans and Mushrooms in a Naturalistic Environment: A Preliminary Report," most subjects had generally very positive and occasionally life-changing experiences.

"Naturalistic" was apt: these sessions took place not in university buildings but in comfortable living rooms, accompanied by music and candlelight, and to a casual observer they would have looked more like parties than experiments, especially because the researchers themselves usually joined in. (Leary and Alpert took a heroic amount of psilocybin and, later, LSD.) At least in the beginning, Leary, Alpert, and their graduate students endeavored to write up accounts of their own and their subjects' psilocybin journeys, as if they were pioneers exploring an unmapped frontier of consciousness and the previous decade of work surveying the psychedelic landscape had never happened. "We were on our own," Leary wrote, somewhat disingenuously. "Western literature had almost no guides, no maps, no texts that even recognized the existence of altered states."

Drawing on their extensive fieldwork, however, Leary did do some original work theorizing the idea of "set" and "setting," deploying the words in this context for the first time in the literature. These useful terms, if not the concepts they denote—for which Al Hubbard deserves most of the credit—may well represent Leary's most enduring contribution to psychedelic science. Leary and Alpert published a handful of papers in the early years at Harvard that are still worth reading, both as well-written and closely observed ethnographies of the experience and as texts in which the early stirrings of a new sensibility can be glimpsed.

Building on the idea that the life-changing experiences of volunteers in the Psilocybin Project might have some broader social application, in 1961 Leary and a graduate student, Ralph Metzner, dreamed up a more ambitious research project. The Concord Prison Experiment sought to discover if the potential of psilocybin to change personality could be used to reduce recidivism in a population of hardened criminals. That this audacious experiment ever got off the ground is a testimony to Leary's salesmanship and charm, for not only the prison psychiatrist but the warden had to sign off on it.

The idea was to compare the recidivism rates of two groups of prisoners in a maximum security prison in Concord, Massachusetts. A group of thirty-two inmates received psilocybin in sessions that took place in the prison, with one member of Leary's team taking the drug with them—so as not to condescend to the prisoners, Leary explained, or treat them like guinea pigs.* The other remained

---

* Leary wrote in *Flashbacks* that he was initially frightened to take psilocybin in a prison with violent criminals. When he confessed his fear to one of the prisoners, the inmate admitted he was afraid too. "Why are you afraid of me?" Leary asked, puzzled. "I'm afraid of you 'cause you're a fucking mad scientist."

straight in order to observe and take notes. A second group of inmates received no drugs or special treatment of any kind. The two groups were then followed for a period of months after their release.

Leary reported eye-popping results: ten months after their release, only 25 percent of the psilocybin recipients had ended up back in jail, while the control group returned at a more typical rate of 80 percent. But when Rick Doblin at MAPS meticulously reconstructed the Concord experiment decades later, reviewing the outcomes subject by subject, he concluded that Leary had exaggerated the data; in fact, there was no statistically significant difference in the rates of recidivism between the two groups. (Even at the time, the methodological shortcomings of the study had prompted David McClelland, the department chair, to write a scathing memo to Metzner.) Of Leary's scientific work, Sidney Cohen, himself a psychedelic researcher, concluded that "it was the sort of research that made scientists wince."

Leary played a more tangential role in one other, much more credible study done in the spring of 1962: the Good Friday Experiment, described in chapter one. Unlike the Concord Prison Experiment, the "Miracle at Marsh Chapel," as it became known, made a good faith effort to honor the conventions of the controlled, double-blind psychology experiment. Neither the investigators nor the subjects—twenty divinity students—were told who had gotten the drug and who had gotten the placebo, which was active. The Good Friday study was far from perfect; Pahnke suppressed the fact that one subject freaked out and had to be sedated. Yet Pahnke's main conclusion—that psilocybin can reliably occasion a mystical experience that is "indistinguishable from, if not identical with," the experiences described in the literature—still stands and helped to inspire the current wave of research, particularly at Johns Hopkins, where it was replicated (roughly speaking) in 2006.

But most of the credit for the Good Friday Experiment rightfully belongs to Walter Pahnke, not Timothy Leary, who was critical of its design from the start; he had told Pahnke it was a waste of time to use a control group or a placebo. "If we learned one thing from that experience," Leary later wrote, "it was how foolish it was to use a double-blind experiment with psychedelics. After five minutes, no one's fooling anyone."

o   o   o

BY NOW, Leary had pretty much lost interest in doing science; he was getting ready to trade the "psychology game" for what he would call the "guru game." (Perhaps Leary's most endearing character trait was never to take himself too seriously—even as a guru.) It had become clear to him that the spiritual and cultural import of psilocybin and LSD far outweighed any therapeutic benefit to individuals. As with Hubbard and Huxley and Osmond before him, psychedelics had convinced Leary that they had the power not just to heal people but to change society and save humankind, and it was his mission to serve as their prophet. It was as though the chemicals themselves had hit upon a brilliant scheme for their own proliferation, by colonizing the brains of a certain type of charismatic and messianic human.

"We were thinking far-out history thoughts at Harvard," Leary later wrote about this period, "believing that it was a time (after the shallow, nostalgic fifties) for far-out visions, knowing that America had run out of philosophy, that a new, empirical, tangible meta-physics was desperately needed." The bomb and the cold war formed the crucial background to these ideas, investing the project with urgency.

Leary was also encouraged in his shift from scientist to evangelist by some of the artists he turned on. In one notable session at his Newton home in December 1960, Leary gave psilocybin to the Beat poet Allen Ginsberg, a man who needed no chemical inducement to play the role of visionary prophet. Toward the end of an ecstatic trip, Ginsberg stumbled downstairs, took off all his clothes, and announced his intention to march naked through the streets of Newton preaching the new gospel.

"We're going to teach people to stop hating," Ginsberg said, "start a peace and love movement." You can almost hear in his words the 1960s being born, the still-damp, Day-Glo chick cracking out of its shell. When Leary managed to persuade Ginsberg not to leave the house (among other issues, it was December), the poet got on the phone and started dialing world leaders, trying to get Kennedy, Khrushchev, and Mao Zedong on the line to work out their differences. In the end, Ginsberg was only able to reach his friend Jack Kerouac, identifying himself as God ("that's G-O-D") and telling him he *must* take these magic mushrooms.

Along with everyone else.

Ginsberg was convinced that Leary, the Harvard professor, was the perfect man to lead the new psychedelic crusade. To Ginsberg, the fact that the new prophet "should emerge from Harvard University," the alma mater of the newly elected president, was a case of "historic comedy," for here was "the one and only Dr. Leary, a respectable human being, a worldly man faced with the task of a Messiah." Coming from the great poet, the words landed like seeds on the fertile, well-watered soil of Timothy Leary's ego. (It is one of the many paradoxes of psychedelics that these drugs can sponsor an ego-dissolving experience that in some people quickly leads to massive ego inflation. Having been let in on a great secret of the universe, the recipient of this knowledge is bound to feel special, chosen for great things.)

Huxley and Hubbard and Osmond shared Leary's sense of historical mission, but they had a very different idea of how best to fulfill it. The three were inclined to a more supply-side kind of spiritualism—first you must turn on the elite, and then let the new consciousness filter down to the masses, who might not be ready to absorb such a shattering experience all at once. Their unspoken model was the Eleusinian mysteries, in which the Greek elite gathered in secret to ingest the sacred *kykeon* and share a night of revelation. But Leary and Ginsberg, both firmly in the American grain, were determined to democratize the visionary experience, make transcendence available to everyone *now*. Surely that was the great blessing of psychedelics: for the first time, there was a technology that made this possible. Years later Lester Grinspoon, a Harvard professor of psychiatry, captured the ethos nicely in a book he wrote with James Bakalar, *Psychedelic Drugs Reconsidered*: "Psychedelic drugs opened to mass tourism mental territories previously explored only by small parties of particularly intrepid adventurers, mainly religious mystics." As well as visionary artists like William Blake, Walt Whitman, and Allen Ginsberg. Now, with a pill or square of blotter paper, anyone could experience firsthand exactly what in the world Blake and Whitman were talking about.

But this new form of spiritual mass tourism had not yet received much advertising or promotion before the spring of 1962. That's when news of controversy surrounding the Harvard Psilocybin Project first hit the newspapers, beginning with Harvard's own student paper, the *Crimson*. Harvard being Harvard, and Leary Leary, the story quickly spread to the national press, turning the psychology professor into a celebrity and hastening his, and Alpert's, departure from Harvard, in a scandal that both prefigured and helped fuel the backlash against psychedelics that would soon close down most research.

Leary and Alpert's colleagues had been uncomfortable about the Harvard Psilocybin Project almost from the start. A 1961 memo from David McClelland had raised questions about the absence of controls in Leary and Alpert's "naturalistic" studies as well as the lack of medical supervision and the fact that the investigators insisted on taking the drugs with their subjects, of whom there were hundreds. ("How often should a person take psilocybin?" he asked, referring to Leary and Alpert.) McClelland also called the two researchers out on their "philosophical naivete."

"Many reports are given of deep mystical experiences," he wrote, "but their chief characteristic is the wonder at one's own profundity." The following year, in a detailed critique of Ralph Metzner's Concord Prison Experiment, McClelland accused the graduate student of failing to "analyz[e] your data objectively and carefully. You know what the conclusions are to be . . . and the data are simply used to support what you already know to be true." No doubt the popularity of the Psilocybin Project among the department's students, as well as its cliquishness, rankled the rest of the faculty, who had to compete with Leary and Alpert and their drugs for a precious academic resource: talented graduate students.

But these grievances didn't leave the premises of 5 Divinity Avenue—not until March 1962. That's when McClelland, responding to a request by Herb Kelman, called a meeting of the faculty and students to air concerns about the Psilocybin Project. Kelman asked for the meeting because he had heard from his graduate students that a kind of cult had formed around Alpert and Leary, and some students felt pressure to participate in the drug taking. Early in the meeting Kelman took the floor: "I wish I could treat this as scholarly disagreement, but this work violates the values of the academic community. The whole program has an anti-intellectual atmosphere. Its emphasis is on pure experience, not on verbalizing findings.

"I'm also sorry to say that Dr. Leary and Dr. Alpert have taken a very nonchalant attitude toward these experiments—especially considering the effects these drugs might have on the subjects.

"What most concerns me," Kelman concluded, "and others who have come to me, is how the hallucinogenic and mental effects of these drugs have been used to form a kind of 'insider' sect within the department. Those who choose not to participate are labeled as 'squares.' I just don't think that kind of thing should be encouraged in this department." Psychedelic drugs had divided a Harvard department just as they would soon divide the culture.

Alpert responded forcefully, claiming the work was "right in the tradition of William James," the department's presiding deity, and that Kelman's critique amounted to an attack on academic freedom. But Leary took a more conciliatory approach, consenting to a few reasonable restrictions on the research. Everyone went home thinking the matter had been closed.

Until the following morning.

The room had been so completely jammed with faculty and students that no one noticed the presence of an undergraduate reporter from the *Crimson* named Robert Ellis Smith, furiously taking notes. The next day's *Crimson* put the controversy on page 1: "Psychologists Disagree on Psilocybin Research." The day after that, the story was picked up by the *Boston Herald*, a Hearst paper, and given a much punchier if not quite as accurate headline: "Hallucination Drug Fought at Harvard—350 Students Take Pills." Now the story was out, and very soon Timothy Leary, always happy to supply a reporter with a delectably outrageous quote, was famous. He delivered a particularly choice one after the university forced him to put his supply of Sandoz psilocybin pills under the control of Health Services: "Psychedelic drugs cause panic and temporary insanity in people who have not taken them."

By the end of the year, Leary and Alpert had concluded that "these materials are too powerful and too controversial to be researched in a university setting." They announced in a letter to the *Crimson* they were forming something called the International Federation for Internal Freedom (IFIF) and henceforth would be conducting research under its umbrella rather than Harvard's. They decried the new restrictions placed on psychedelic research, not only at Harvard, but by the federal government: in the wake of the thalidomide tragedy, in which a new sedative given to pregnant women for morning sickness had caused terrible birth defects in their children, Congress had given the FDA authority to regulate experimental drugs. "For the first time in American history," the IFIF announced, "and for the first time in the Western world since the Inquisition there now exists a scientific underground." They predicted that "a major civil liberties issue of the next decade will be the control and expansion of consciousness."

"Who controls your cortex?" they wrote in their letter to the *Crimson*—which is to say, to students. "Who decides on the range and limits of your awareness? If you want to research your own nervous system, expand your consciousness, who is to decide that you can't and why?"

It's often said that in the 1960s psychedelics "escaped from the laboratory," but it would probably be more accurate to say they were thrown over the laboratory wall, and never with as much loft or velocity as by Timothy Leary and Richard Alpert at the end of 1962. "We're through playing the science game," Leary told McClelland when he returned to Cambridge that fall. Now, Leary and Alpert were playing the game of cultural revolution.

o   o   o

THE LARGER COMMUNITY of psychedelic researchers across
North America reacted to Leary's provocations with dismay and
then alarm. Leary had been in regular contact with the West Coast
and Canadian groups, exchanging letters and visits with his far-flung
colleagues on a fairly regular basis. (He and Alpert had paid a visit to
Stolaroff's foundation in 1960 or 1961; "I think they thought we were
too straitlaced," Don Allen told me.) Soon after arriving at Harvard,
Leary had gotten to know Huxley, who was teaching for a semester
at MIT. Huxley had become extremely fond of the roguish professor,
and shared his aspirations for psychedelics as an agent of cultural
transformation, but worried that Leary was moving too fast and too
flagrantly.* During his last visit to Cambridge (Huxley would die in
Los Angeles in November 1963, on the same day as John F. Kennedy), Huxley felt that Leary "had talked such nonsense . . . that I
became quite concerned. Not about his sanity—because he is perfectly sane—but about his prospects in the world."

Soon after Leary announced the formation of the International
Federation for Internal Freedom, Humphry Osmond traveled to
Cambridge to try to talk some sense into him. He and Abram Hoffer
were worried that Leary's promotion of the drugs outside the context of clinical research threatened to provoke the government and
upend their own research. Osmond also faulted Leary for working
without a psychopharmacologist and for treating these "powerful

---

* In a 1992 letter to Betty Eisner, Humphry Osmond wrote, "Where both Al [Hub-
bard] and Aldous [Huxley] disagreed with Timothy Leary was that they believed
that he had got the time scale wrong, and that the US had a much greater inertia
than he supposed. They both believed for quite different reasons that working in-
conspicuously but determinedly within the system could transform it in the long
run. Timothy believed that it could be taken by storm."

chemicals [as] harmless toys." Hoping to distance serious research from irresponsible use, and troubled that the counterculture was contaminating his formerly neutral term "psychedelic," Osmond tried once again to coin a new one: "psychodelytic." I don't need to tell you it failed to catch on.

"You must face these objections rather than dissipate them with a smile, however cosmic," Osmond told him. There it was again: the indestructible Leary smile! But Osmond got nothing more than that for his troubles.

Myron Stolaroff weighed in with a blunt letter to Leary describing the IFIF as "insane" and accurately prophesying the crack-up to come: It will "wreak havoc on all of us doing LSD work all over the nation . . .

"Tim, I am convinced you are heading for very serious trouble if your plan goes ahead as you have described it to me, and it would not only make a great deal of trouble for you, but for all of us, and may do irreparable harm to the psychedelic field in general."

But what exactly *was* the plan of the IFIF? Leary was happy to state it openly: to introduce as many Americans to "the strong psychedelics" as it possibly could in order to change the country one brain at a time. He had done the math and concluded that "the critical figure for blowing the mind of the American society would be four million LSD users and this would happen by 1969."

As it would turn out, Leary's math was not far off. Though closer to two million Americans had tried LSD by 1969, this cadre had indeed blown the mind of America, leaving the country in a substantially different place.

But perhaps the most violent response to Leary's plans for worldwide mental revolution came from Al Hubbard, who had always had an uneasy relationship with the professor. The two had met soon after Leary got to Harvard, when Hubbard made the drive to

Cambridge in his Rolls-Royce, bringing a supply of LSD he hoped to trade for some of Leary's psilocybin.

"He blew in with that uniform," Leary recalled, "laying down the most incredible atmosphere of mystery and flamboyance, and really impressive bullshit!"—a subject on which Leary was certainly qualified to judge. Hubbard "started name-dropping like you wouldn't believe . . . claimed he was friends with the Pope.

"The thing that impressed me is, on one hand he looked like a carpetbagger con man, and on the other he had these most impressive people in the world in his lap, basically backing him."

But Leary's legendary charm never had much traction with Hubbard, a deeply conservative and devout man who disdained both the glare of publicity and the nascent counterculture. "I liked Tim when we first met," he said years later, "but I warned him a dozen times" about staying out of trouble and the press. "He seemed like a well-intentioned person, but then he went overboard . . . he turned out to be completely no good." Like many of his colleagues, Hubbard strongly objected to Leary's do-it-yourself approach to psychedelics, especially his willingness to dispense with the all-important trained guide. His attitude toward Leary might also have been influenced by his extensive contacts in law enforcement and intelligence, which by now had the professor on their radar.

According to Osmond, the Captain's antipathy toward Leary surfaced alarmingly during a psychedelic session the two shared during this period of mounting controversy. "Al got greatly preoccupied with the idea he ought to shoot Timothy, and when I began to reason with him that this would be a very bad idea . . . I became much concerned he might shoot me."

Hubbard was probably right to think that nothing short of a bullet was going to stop Timothy Leary now. As Stolaroff put the matter in closing his letter to Leary, "I suppose there is little hope that with the bit so firmly in your mouth you can be deterred."

o   o   o

BY THE SPRING OF 1963, Leary had one foot out of Harvard, skipping classes and voicing his intention to leave at the end of the school year, when his contract would be up. But Alpert had a new appointment in the School of Education and planned to stay on— until another explosive article in the *Crimson* got them both fired. This one was written by an undergraduate named Andrew Weil.

Weil had arrived at Harvard with a keen interest in psychedelic drugs—he had devoured Huxley's *Doors of Perception* in high school— and when he learned about the Psilocybin Project, he beat a path to Professor Leary's office door to ask if he could participate.

Leary explained the university rule restricting the drugs to graduate students. Yet, trying to be helpful, he told Weil about a company in Texas where he might order some mescaline by mail (it was still legal at the time), which Weil promptly did (using university stationery). Weil became fascinated with the potential of psychedelics and helped form an undergraduate mescaline group. But he wanted badly to be part of Leary and Alpert's more exclusive club, so when in the fall of 1962 Weil began to hear about other undergraduates who had received drugs from Richard Alpert, he was indignant. He went to his editor at the *Crimson* and proposed an investigation.

Weil developed leads on a handful of fellow students whom Alpert had turned on in violation of university rules. (Weil would later write that "students and others were using hallucinogens for seductions both heterosexual and homosexual.") But there were two problems with his scoop: none of the students to whom Alpert supposedly gave drugs were willing to say so on the record, and the *Crimson*'s lawyers were worried about printing defamatory charges against professors. The lawyers advised Weil to turn over his information to the administration. He could then write a story reporting on whatever actions the university took in response to the charges, thereby reducing the newspaper's legal exposure. But Weil still needed a student to come forward.

He traveled to New York City to meet with the prominent father of one of them—Ronnie Winston—and offered him a deal. As Alpert tells the story,* "He went to Harry Winston"—the famous Fifth Avenue jeweler—"and he said, 'Your son is getting drugs from a faculty member. If your son will admit to that charge, we'll cut out your son's name. We won't use it in the article.'" So young Ronnie went to the dean and, when asked if he had taken drugs from Dr. Alpert, confessed, adding an unexpected fillip: "Yes, sir, I did. And it was the most educational experience I've had at Harvard."

Alpert and Leary appear to be the only Harvard professors fired in the twentieth century. (Technically, Leary wasn't fired, but Harvard stopped paying him several months before his contract ended.) The story became national news, introducing millions of Americans to the controversy surrounding these exotic new drugs. It also earned Andrew Weil a plum assignment from *Look* magazine to write about the controversy, which spread the story still further. Describing the psychedelic scene at Harvard in the third person, Weil alluded to "an undergraduate group . . . conducting covert research with mescaline," neglecting to mention he was a founding member of that group.

This was not, suffice it to say, Andrew Weil's proudest moment, and when I spoke to him about it recently, he confessed that he's felt badly about the episode ever since and had sought to make amends to

---

* In Don Lattin, *The Harvard Psychedelic Club* (New York: HarperOne, 2010), 94.

both Leary and Ram Dass. (Two years after his departure from Harvard, Alpert embarked on a spiritual journey to India and returned as Ram Dass.) Leary readily accepted Weil's apology—the man was apparently incapable of holding a grudge—but Ram Dass refused to talk to Weil for years, which pained him. But after Ram Dass suffered a stroke in 1997, Weil traveled to Hawaii to seek his forgiveness. Ram Dass finally relented, telling Weil that he had come to regard being fired from Harvard as a blessing. "If you hadn't done what you did," he told Weil, "I would never have become Ram Dass."

o   o   o

HERE, UPON THEIR EXIT from Harvard, we should probably take our leave of Timothy Leary and Richard Alpert, even though their long, strange trip through American culture still had a long, strange way to go. The two would now take their show (with its numerous ex-students and hangers-on) on the road, moving the International Federation for Internal Freedom (which would later morph into the League for Spiritual Discovery) from Cambridge to Zihuatanejo, until the Mexican government (under pressure from U.S. authorities) kicked them out, then briefly to the Caribbean island of Dominica, until that government kicked them out, before finally settling for several raucous years in a sixty-four-room mansion in Millbrook, New York, owned by a wealthy patron named Billy Hitchcock.

Embraced by the rising counterculture, Leary was invited (along with Allen Ginsberg) to speak at the first Human Be-In in San Francisco, an event that drew some twenty-five thousand young people to Golden Gate Park in January 1967, to trip on freely distributed LSD while listening to speakers proclaim a new age. The ex-professor, who for the occasion had traded in his Brooks Brothers for white robes and love beads (and flowers in his graying hair), implored the throng of tripping "hippies"—the term popularized that year by the local newspaper columnist Herb Caen—to "turn on, tune in, drop out." The slogan—which he at first said he had thought up in the shower but years later claimed was "given to him" by Marshall McLuhan—would cling to Leary for the rest of his life, earning him the contempt of parents and politicians the world over.

But Leary's story only gets weirder, and sadder. Soon after his departure from Cambridge, the government, alarmed at his growing influence on the country's youth, launched a campaign of harassment that culminated in the 1966 bust in Laredo; he was driving his family to Mexico on vacation, when a border search of his car turned up a small quantity of marijuana. Leary would spend years in jail battling federal marijuana charges and then several more years on the lam as an international fugitive from justice. He acquired this status in 1970 after his bold escape from a California prison, with the help of the Weathermen, the revolutionary group. His comrades managed to spirit Leary out of the country to Algeria, into the arms of Eldridge Cleaver, the Black Panther, who had established a base of operations there. But asylum under Cleaver turned out to be no picnic: the Panther confiscated his passport, effectively holding Leary hostage. Leary had to escape yet again, this time making his way to Switzerland (where he found luxurious refuge in the chalet of an arms dealer), then (after the U.S. government persuaded Switzerland to jail him) on to Vienna, Beirut, and Kabul, where he was finally seized by U.S. agents and remanded to an American prison, now maximum security and, for a time, solitary confinement. But the persecution only fed his sense of destiny.

The rest of his life is an improbable 1960s tragicomedy featuring plenty of courtrooms and jails (twenty-nine in all) but also memoirs and speeches and television appearances, a campaign for governor of California (for which John Lennon wrote, and the Beatles recorded, the campaign song, "Come Together"), and a successful if somewhat pathetic run on the college lecture circuit teamed up with G. Gordon Liddy. Yes, the Watergate burglar, who in an earlier incarnation as Dutchess County assistant DA had busted Leary at Millbrook. Through it all, Leary remains improbably upbeat, never displaying anger or, it would seem from the countless photographs and film clips, forgetting Marshall McLuhan's sage advice to smile always, no matter what.

Meanwhile, beginning in 1965, Leary's former partner in psychedelic research, Richard Alpert, was off on a considerably less hectic spiritual odyssey to the East. As Ram Dass, and the author of the 1971 classic *Be Here Now*, he would put his own lasting mark on American culture, having blazed one of the main trails by which Eastern religion found its way into the counterculture and then the so-called New Age. To the extent that the 1960s birthed a form of spiritual revival in America, Ram Dass was one of its fathers.

But Leary's post-Harvard "antics" are relevant to the extent they contributed to the moral panic that now engulfed psychedelics and doomed the research. Leary became a poster boy not just for the drugs but for the idea that a crucial part of the counterculture's DNA could be spelled out in the letters *LSD*. Beginning with Allen Ginsberg's December 1960 psilocybin trip at his house in Newton, Leary forged a link between psychedelics and the counterculture that has never been broken and that is surely one of the reasons they came to be regarded as so threatening to the establishment. (Could it have possibly been otherwise? What if the cultural identity of the drugs had been shaped by, say, a conservative Catholic like Al Hubbard? It's difficult to imagine such a counter history.)

It didn't help that Leary liked to say things like "LSD is more frightening than the bomb" or "The kids who take LSD aren't going to fight your wars. They're not going to join your corporations." These were no empty words: beginning in the mid-1960s, tens of thousands of American children actually *did* drop out, washing up on the streets of Haight-Ashbury and the East Village.* And young men were refusing to go to Vietnam. The will to fight and the authority of Authority had been undermined. These strange new drugs, which seemed to change the people who took them, surely had *some*thing to do with it. Timothy Leary had said so.

But this upheaval would almost certainly have happened without Timothy Leary. He was by no means the only route by which psychedelics were seeping into American culture; he was just the most notorious. In 1960, the same year Leary tried psilocybin and launched his research project, Ken Kesey, the novelist, had his own mind-blowing LSD experience, a trip that would inspire him to spread the psychedelic word, and the drugs themselves, as widely and loudly as he could.

It is one of the richer ironies of psychedelic history that Kesey had his first LSD experience courtesy of a government research program conducted at the Menlo Park Veterans Hospital, which paid him seventy-five dollars to try the experimental drug. Unbeknownst to Kesey, his first LSD trip was bought and paid for by the CIA, which had sponsored the Menlo Park research as part of its MK-Ultra program, the agency's decade-long effort to discover whether LSD could somehow be weaponized.

With Ken Kesey, the CIA had turned on exactly the wrong man. In what he aptly called "the revolt of the guinea pigs," Kesey proceeded to organize with his band of Merry Pranksters a series of "Acid Tests" in which thousands of young people in the Bay Area were given LSD in an effort to change the mind of a generation. To

---

* One could argue that the LSD dropout problem began back in the 1950s, when successful engineers like Myron Stolaroff, Willis Harman, and Don Allen left Ampex and Stanford to tune in to psychedelics.

the extent that Ken Kesey and his Pranksters helped shape the new zeitgeist, a case can be made that the cultural upheaval we call the 1960s began with a CIA mind-control experiment gone awry.

o   o   o

IN RETROSPECT, the psychiatric establishment's reaction was probably unavoidable the moment that Humphry Osmond, Al Hubbard, and Aldous Huxley put forward their new paradigm for psychedelic therapy in 1956–1957. The previous theoretical models used to make sense of these drugs were, by comparison, easy to fold into the field's existing frameworks without greatly disturbing the status quo. "Psychotomimetics" fit nicely into the standard psychiatric understanding of mental illness—the drugs' effects resembled familiar psychoses—and "psycholytics" could be incorporated into both the theory and the practice of psychoanalysis as a useful adjunct to talking therapy. But the whole idea of psychedelic therapy posed a much stiffer challenge to the field and the profession. Instead of interminable weekly sessions, the new mode of therapy called for only a single high-dose session, aimed at achieving a kind of conversion experience in which the customary roles of both patient and therapist had to be reimagined.

Academic psychiatrists were also made uncomfortable by the spiritual trappings of psychedelic therapy. Charles Grob, the UCLA psychiatrist who would play an important role in the revival of research, wrote in a 1998 article on the history of psychedelics that "by blurring the boundaries between religion and science, between sickness and health, and between healer and sufferer, the psychedelic model entered the realm of applied mysticism"—a realm where psychiatry, increasingly committed to a biochemical understanding of the mind, was reluctant to venture. With its emphasis on set and setting—what Grob calls "the critical extra-pharmacological variables"—psychedelic therapy was also a little too close to shaman-ism for comfort. For so-called shrinks not entirely secure in their identity as scientists (the slang is short for "headshrinkers," conjur-ing images of witch doctors in loincloths), this was perhaps too far to go. Another factor was the rise of the placebo-controlled double-blind trial as the "gold standard" for testing drugs in the wake of the thalidomide scandal, a standard difficult for psychedelic research to meet.

By 1963, leaders of the profession had begun editorializing against psychedelic research in their journals. Roy Grinker, the editor of the *Archives of General Psychiatry*, lambasted researchers who were ad-ministering "the drugs to themselves and . . . [had become] enam-ored with the mystical hallucinatory state," thus rendering them "disqualified as competent investigators." Writing the following year in the *Journal of the American Medical Association* (*JAMA*), Grinker deplored the practice of investigators taking the drugs themselves, thereby "rendering their conclusions biased by their own ecstasy." An unscientific "aura of magic" surrounded the new drugs, another critic charged in *JAMA* in 1964. (It didn't help that some psychedelic therapists, like Betty Eisner, celebrated the introduction of "the transcendental into psychiatry" and developed an interest in para-normal phenomenon.)

But although there is surely truth to the charge that researchers were often biased by their own experiences using the drugs, the ob-vious alternative—abstinence—posed its own set of challenges, with the result that the loudest and most authoritative voices in the debate over psychedelics during the 1960s were precisely the people who knew the least about them. To psychiatrists with no personal experi-ence of psychedelics, their effects were bound to look a lot more like psychoses than transcendence. The psychotomimetic paradigm had returned, now with a vengeance.

After quantities of "bootleg LSD" showed up on the street in 1962–1963 and people in the throes of "bad trips" began appearing in emergency rooms and psych wards, mainstream psychiatry felt compelled to abandon psychedelic research. LSD was now regarded as a cause of mental illness rather than a cure. In 1965, Bellevue Hospital in Manhattan admitted sixty-five people for what it called LSD-induced psychoses. With the media now in full panic mode, urban legends about the perils of LSD spread more rapidly than facts.* The same was often true in the case of ostensibly scientific findings. In one widely publicized study, a researcher reported in *Science* that LSD could damage chromosomes, potentially leading to birth defects. But when the study was later discredited (also in *Science*), the refutation received little attention. It didn't fit the new public narrative of LSD as a threat.

Yet it was true that the mid-1960s saw a surge of people on LSD showing up in emergency rooms with acute symptoms of paranoia, mania, catatonia, and anxiety, as well as "acid flashbacks"—a spontaneous recurrence of symptoms days or weeks after ingesting LSD. Some of these patients were having genuine psychotic breaks. Especially in the case of young people at risk for schizophrenia, an LSD trip can trigger their first psychotic episode, and sometimes

---

* Several of these urban legends have been traced to their source and discredited. For example, a 1967 *Newsweek* story about six college students tripping on LSD who went blind after staring into the sun turned out to be a hoax concocted by Pennsylvania's state commissioner for the blind, Dr. Norman Yoder. According to the governor, who disclosed the hoax, Yoder had "attended a lecture on the use of LSD by children and became concerned and emotionally involved." Yet once introduced into the culture, these urban legends survive and, on occasion, go on to become "true" when people tripping on LSD are inspired to imitate them, as has happened in the case of the staring-into-the-sun story. See David Presti and Jerome Beck, "Strychnine and Other Enduring Myths: Expert and User Folklore Surrounding LSD," in *Psychoactive Sacramentals: Essays on Entheogens and Religion*, ed. Thomas B. Roberts (San Francisco: Council on Spiritual Practices, 2001).

did. (It should be noted that any traumatic experience can serve as such a trigger, including the divorce of one's parents or graduate school.) But in many other cases, doctors with little experience of psychedelics mistook a panic reaction for a full-blown psychosis. Which usually made things worse.

Andrew Weil, who as a young doctor volunteered in the Haight-Ashbury Free Clinic in 1968, saw a lot of bad trips and eventually developed an effective way to "treat" them. "I would examine the patient, determine it was a panic reaction, and then tell him or her, 'Will you excuse me for a moment? There's someone in the next room who has a serious problem.' They would immediately begin to feel much better."

The risks of LSD and other psychedelic drugs were fiercely debated during the 1960s, both among scientists and in the press. Voices on both sides of this debate typically cherry-picked evidence and anecdotes to make their case, but Sidney Cohen was an exception, approaching the question with an open mind and actually conducting research to answer it. Beginning in 1960, he published a series of articles that track his growing concerns. For his first study, Cohen surveyed forty-four researchers working with psychedelics, collecting data on some five thousand subjects taking LSD or mescaline on a total of twenty-five thousand occasions. He found only two credible reports of suicide in this population (a low rate for a group of psychiatric patients), several transient panic reactions, but "no evidence of serious prolonged physical side effects." He concluded that when psychedelics are administered by qualified therapists and researchers, complications were "surprisingly infrequent" and that LSD and mescaline were "safe."

Leary and others often cited Cohen's 1960 paper as an exoneration of psychedelics. Yet in a follow-up article published in the *Journal of the American Medical Association* in 1962, Cohen reported new and "alarming" developments. The casual use of LSD outside the clinical setting, and in the hands of irresponsible therapists, was leading to "serious complications" and occasional "catastrophic reactions." Alarmed that physicians were losing control of the drug, Cohen warned that "the dangers of suicide, prolonged psychotic reactions and antisocial acting out behavior exist." In another paper published in the *Archives of General Psychiatry* the following year, he reported several cases of psychotic breaks and an attempted suicide and presented an account of a boy who, after ingesting a sugar cube laced with LSD that his father, a detective, had confiscated from a "pusher," endured more than a month of visual distortions and anxiety before recovering. It was this article that inspired Roy Grinker, the journal's editor, to condemn psychedelic research in an accompanying commentary, even though Cohen himself continued to believe that psychedelics in the hands of responsible therapists had great potential. A fourth article that Cohen published in 1966 reported still more LSD casualties, including two accidental deaths associated with LSD, one from drowning and the other from walking into traffic shouting, "Halt."

But balanced assessments of the risks and benefits of psychedelics were the exception to what by 1966 had become a full-on moral panic about LSD. A handful of headlines from the period suggests the mood: "LSD-Use Charged with Killing Teacher"; "Sampled LSD, Youth Plunges from Viaduct"; "LSD Use Near Epidemic in California"; "Six Students Blinded on LSD Trip in Sun"; "Girl, 5, Eats LSD and Goes Wild"; "Thrill Drug Warps Mind, Kills"; and "A Monster in Our Midst—a Drug Called LSD." Even *Life* magazine, which had helped ignite public interest in psychedelics just nine years before with R. Gordon Wasson's enthusiastic article on psilocybin, joined the chorus of condemnation, publishing a feverish cover story titled "LSD: The Exploding Threat of the Mind Drug That Got out of

Control." Never mind that the magazine's publisher and his wife had recently had several positive LSD experiences themselves (under the guidance of Sidney Cohen); now the kids were doing it, and it had gotten "out of control." With pictures of crazed people cowering in corners, the story warned that "an LSD trip is not always a round trip" but rather could be "a one-way trip to an asylum, a prison or a grave."* As Clare Boothe Luce wrote to Sidney Cohen in 1965, "LSD *has* been your Frankenstein monster."

o   o   o

OTHER POWERFUL DRUGS subject to abuse, such as the opiates, have managed to maintain a separate identity as a legitimate tool of medicine. Why not psychedelics? The story of Timothy Leary, the most famous psychedelic researcher, made it difficult to argue that a bright line between the scientific and the recreational use of psychedelics could be drawn and patrolled. The man had deliberately—indeed gleefully—erased all such lines. But the "personality" of the drug may have as much to do with the collapse of such distinctions as the personalities of people like Timothy Leary or the flaws in their research.

What doomed the first wave of psychedelic research was an irrational exuberance about its potential that was nourished by the drugs themselves—that, and the fact that these chemicals are what today we would call disruptive technologies. For people working with these powerful molecules, it was impossible not to conclude that—like that divinity student running down Commonwealth Avenue—you were

---

* There are quotations in this piece that should have set off any editor's bullshit detector. "When my husband and I want to take a trip together," says the psychedelic mother of four, "I just put a little acid in the kids' orange juice in the morning and let them spend the day freaking out in the woods."

suddenly in possession of news with the power to change not just individuals but the world. To confine these drugs to the laboratory, or to use them only for the benefit of the sick, became hard to justify, when they could do so much for everyone, including the researchers themselves!

Leary might have made his more straitlaced colleagues cringe at his lack of caution, yet most of them shared his exuberance and had come to more or less the same conclusions about the potential of psychedelics; they were just more judicious when speaking about them in public.

Who among the first generation of psychedelic researchers would dispute a word of this classic gust of Leary exuberance, circa 1963: "Make no mistake: the effect of consciousness-expanding drugs will be to transform our concepts of human nature, of human potentialities, of existence. The game is about to be changed, ladies and gentlemen. Man is about to make use of that fabulous electrical network he carries around in his skull. Present social establishments had better be prepared for the change. Our favorite concepts are standing in the way of a floodtide, two billion years building up. The verbal dam is collapsing. Head for the hills, or prepare your intellectual craft to flow with the current."*

So perhaps Leary's real sin was to have the courage of his convictions—his and everyone else's in the psychedelic research community. It's often said that a political scandal is what happens when someone in power inadvertently speaks the truth. Leary was

---

* Originally published in *Harvard Review* (Summer 1963) and reprinted in Timothy Leary and James Penner, *Timothy Leary, The Harvard Years: Early Writings on LSD and Psilocybin with Richard Alpert, Huston Smith, Ralph Metzner, and Others* (Rochester, Vt.: Park Street Press, 2014). The paragraph also appears in the transcript of a 1966 Senate hearing on federal regulation of LSD by the Senate Subcommittee on Executive Reorganization, p. 141.

all too often willing to say out loud to anyone in earshot what everyone else believed but knew better than to speak or write about candidly. It was one thing to use these drugs to treat the ill and maladjusted—society will indulge any effort to help the wayward individual conform to its norms—but it is quite another to use them to treat society itself as if it were sick and to turn the ostensibly healthy into wayward individuals.

The fact is that whether by their very nature or the way that first generation of researchers happened to construct the experience, psychedelics introduced something deeply subversive to the West that the various establishments had little choice but to repulse. LSD truly was an acid, dissolving almost everything with which it came into contact, beginning with the hierarchies of the mind (the superego, ego, and unconscious) and going on from there to society's various structures of authority and then to lines of every imaginable kind: between patient and therapist, research and recreation, sickness and health, self and other, subject and object, the spiritual and the material. If all such lines are manifestations of the Apollonian strain in Western civilization, the impulse that erects distinctions, dualities, and hierarchies and defends them, then psychedelics represented the ungovernable Dionysian force that blithely washes all those lines away.

But it surely is not the case that the forces unleashed by these chemicals are *necessarily* ungovernable. Even the most powerful acids can be carefully handled and put to use as tools for accomplishing important things. What is the story of the first-wave researchers if not a story about searching for an appropriate container for these powerful chemicals? They tested several different possibilities: the psychotomimetic, the psycholytic, the psychedelic, and, still later, the entheogenic. None were perfect, but each represented a different way to regulate the power of these compounds, by proposing a set of protocols for their use as well as a theoretical framework. Where

Leary and the counterculture ultimately parted ways with the first generation of researchers was in deciding that no such container—whether medical, religious, or scientific—was needed and that an unguided, do-it-yourself approach to psychedelics was just fine. This is risky, as it turns out, and probably a mistake. But how would we ever have discovered this, without experimenting? Before 1943, our society had never had such powerful mind-changing drugs available to it.

Other societies have had long and productive experience with psychedelics, and their examples might have saved us a lot of trouble had we only known and paid attention. The fact that we regard many of these societies as "backward" probably kept us from learning from them. But the biggest thing we might have learned is that these powerful medicines can be dangerous—both to the individual and to the society—when they don't have a sturdy social container: a steadying set of rituals and rules—protocols—governing their use, and the crucial involvement of a guide, the figure that is usually called a shaman. Psychedelic therapy—the Hubbard method—was groping toward a Westernized version of this ideal, and it remains the closest thing we have to such a protocol. For young Americans in the 1960s, for whom the psychedelic experience was new in every way, the whole idea of involving elders was probably never going to fly. But this is, I think, the great lesson of the 1960s experiment with psychedelics: the importance of finding the proper context, or container, for these powerful chemicals and experiences.

Speaking of lines, psychedelics in the 1960s did draw at least one of them, and it has probably never before been quite so sharp or bright: the line, I mean, between generations. Saying exactly how or what psychedelics contributed to the counterculture of the 1960s is not an easy task, there were so many other forces at work. With or without psychedelics, there probably would have been a counterculture; the Vietnam War and the draft made it more than likely. But the forms the counterculture took and its distinctive styles—of music, art, writing, design, and social relations—would surely have been completely different were it not for these chemicals. Psychedelics also contributed to what Todd Gitlin has called the "as if" mood of 1960s politics—the sense that everything now was up for grabs, that nothing given was inviolate, and that it might actually be possible to erase history (there was that acid again) and start the world over again from scratch.

But to the extent that the upheaval of the 1960s was the result of an unusually sharp break between generations, psychedelics deserve much of the blame—or credit—for creating this unprecedented "generation gap." For at what other time in history did a society's young undergo a searing rite of passage with which the previous generation was utterly unfamiliar? Normally, rites of passage help knit societies together as the young cross over hurdles and through gates erected and maintained by their elders, coming out on the other side to take their place in the community of adults. Not so with the psychedelic journey in the 1960s, which at its conclusion dropped its young travelers onto a psychic landscape unrecognizable to their parents. That this won't ever happen again is reason to hope that the next chapter in psychedelic history won't be quite so divisive.

So maybe this, then, is the enduring contribution of Leary: by turning on a generation—the generation that, years later, has now taken charge of our institutions—he helped create the conditions in which a revival of psychedelic research is now possible.

o o o

BY THE END OF 1966, the whole project of psychedelic science had collapsed. In April of that year, Sandoz, hoping to distance itself from the controversy engulfing the drug that Albert Hofmann would come to call his "problem child," withdrew LSD-25 from circulation, turning over most of its remaining stocks to the U.S. government and leading many of the seventy research programs then under way to shut down.

In May of that year, the Senate held hearings about the LSD problem. Timothy Leary and Sidney Cohen both testified, attempting valiantly to defend psychedelic research and draw lines between legitimate use and a black market that the government was now determined to crush. They found a surprisingly sympathetic ear in Senator Robert F. Kennedy, whose wife, Ethel, had reportedly been treated with LSD at Hollywood Hospital in Vancouver—one of Al Hubbard's outposts. Grilling the FDA regulators about their plans to cancel many of the remaining research projects, Kennedy demanded to know, "Why if [these projects] were worthwhile six months ago, why aren't they worthwhile now?" Kennedy said it would be a "loss to the nation" if psychedelics were banned from medicine because of illicit use. "Perhaps we have lost sight of the fact that [they] can be very, very helpful in our society if used properly."

But Kennedy got nowhere. Leary, and perhaps the drugs themselves, had made drawing such distinctions impossible. In October, some sixty psychedelic researchers scattered across the United States received a letter from the FDA ordering them to stop their work.

James Fadiman, the psychologist conducting experiments on creativity at the International Foundation for Advanced Study in Menlo Park, remembers the day well. The letter revoking FDA approval of the project arrived at the very moment he had finished dosing four of his problem-solving creatives to begin their session. As he read the letter, sprawled on the floor in the next room, "four men lay, their minds literally expanding." Fadiman said to his colleagues, "I think we need to agree that we got this letter tomorrow." And so it was not until the following day that the research program of the International Foundation for Advanced Study, along with virtually every other research program then under way in the United States, closed down.

One psychedelic research program survived the purge: the Maryland Psychiatric Research Center at Spring Grove. Here, researchers such as Stanislav Grof, Bill Richards, Richard Yensen, and, until his death in 1971, Walter Pahnke (the Good Friday researcher) continued to explore the potential of psilocybin and LSD to treat alcoholism, schizophrenia, and the existential distress of cancer patients, among other indications. It remains something of a mystery why this large psychedelic research program was allowed to continue—as it did until 1976—when dozens of others were being closed down. Some researchers who weren't so fortunate speculate that Spring Grove might have been making psychedelic therapy available to powerful people in Washington who recognized its value or hoped to learn from the research or perhaps wanted to retain their own access to the drugs. But the former staff members at the center I spoke to doubt this was the case. They did confirm, however, that the center's director, Albert Kurland, MD, besides having a sterling reputation among federal officials, was exceptionally well connected in Washington and used his connections to keep the lights on—and obtain LSD, some of it from the government—for a decade after they had been switched off everywhere else.

Yet it turns out that the events of neither 1966 nor 1976 put an end to psychedelic research and therapy in America. Moving now underground, it went on, quietly and in secret.

Coda

In February 1979, virtually all the important figures in the first wave of American psychedelic research gathered for a reunion in Los Angeles at the home of Oscar Janiger. Someone made a videotape of the event, and though the quality is poor, most of the conversation is audible. Here in Janiger's living room we see Humphry Osmond, Sidney Cohen, Myron Stolaroff, Willis Harman, Timothy Leary, and, sitting on the couch next to him, looking distinctly uncomfortable, Captain Al Hubbard. He's seventy-seven (or eight), and he's traveled from Casa Grande, Arizona, where he lives in a trailer park. He's wearing his paramilitary getup, though I can't tell if he's carrying a sidearm.

The old men reminisce, a bit stiffly at first. Some hard feelings hang in the air. But Leary, still charming, is remarkably generous, working to put everyone at ease. Their best days are behind them; the great project to which they devoted their lives lay in ruins. But something important was accomplished, they all believe—else they wouldn't be here at this reunion. Sidney Cohen, dressed in a jacket and tie, asks the question on everyone's mind—"What does it all mean?"—and then ventures an answer: "It stirred people up. It cracked their frame of reference by the thousands—millions perhaps. And anything that does that is pretty good I think."

It's Leary, of all people, who asks the group, "Does anyone here feel that mistakes were made?"

Osmond, the unfailingly polite Englishman, his teeth now in full revolt, declines to use the word "mistake." "What I would say is . . . you could have seen other ways of doing it." Someone I don't recognize cracks, "There *was* a mistake made: nobody gave it to Nixon!"

It's Myron Stolaroff who finally confronts the elephant in the room, turning to Leary to say, "We were a little disturbed at some of the things you were doing that [were] making it more difficult to carry on legitimate research." Leary reminds him that as he told them then, he had a different role to play: "Let us be the far-out explorers. The farther out we go, the more ground it gives the people at Spring Grove to denounce us." And so appear responsible.

"And I just wish, I hope we all understand that we've all been playing parts that have been assigned to us, and there's no good-guy/bad-guy, or credit or blame, whatever . . ."

"Well, I think we need people like Tim and Al," Sidney Cohen offers, genially accepting Leary's framing. "They're absolutely necessary to get out, way out, too far out in fact—in order to move the ship . . . [turn] things around." Then, turning to Osmond: "And we need people like you, to be reflective about it and to study it. And little by little, a slight movement is made in the totality. So, you know, I can't think of how it could have worked out otherwise."

Al Hubbard listens intently to all this but has little to add; he fiddles with a hardback book in his lap. At one point, he pipes up to suggest the work should go on, drug laws be damned: We should "just keep on doing it. Wake people up! Let them see for themselves what they are. I think old Carter could stand a good dose!" Carter's defense secretary, Harold Brown, and CIA director, Stansfield Turner, too. But Hubbard's not at all sure he wants to be on this couch with Timothy Leary and is less willing than the others to let bygones be bygones, or Leary off the hook, no matter how solicitous he is of the Captain.

"Oh, Al! I owe everything to you," Leary offers at one point, beaming his most excellent smile at Hubbard. "The galactic center sent you down just at the right moment."

Hubbard doesn't crack a smile. And then, a few minutes later:

"You sure as heck contributed your part."

CHAPTER FOUR

TRAVELOGUE

Journeying Underground

My plan had been to volunteer for one of the Hopkins or NYU experimental trials. If I was going to have my own guided psychedelic journey, a harrowing prospect under any circumstances, I very much liked the idea of traveling in the company of trained professionals close by a hospital emergency room. But the above-ground researchers were no longer working with "healthy normals." This meant that if I hoped to have the journey I had heard so much about, it would have to take place underground. Could I find a guide willing to work with a writer who planned to publish an account of his journey, and would that person be someone I felt sufficiently comfortable with and confident in to entrust with my mind? The whole endeavor was fraught with uncertainty and entailed risks of several kinds—legal, ethical, psychological, and even literary. For how do you put into words an experience said to be ineffable?

"Curiosity" is an accurate but tepid word for what drove me. By now, I had interviewed at length more than a dozen people who had gone on guided psychedelic journeys, and it was impossible to listen to their stories without wondering what the journey would be like for one's self. For many of them, these were among the two or three most profound experiences of their lives, in several cases changing them in positive and lasting ways. To become more "open"—especially at this age, when the grooves of mental habit have been etched so deep as to seem inescapable—was an appealing prospect. And then there was the possibility, however remote, of having some kind of spiritual epiphany. Many of the people I'd interviewed had started out stone-cold materialists and atheists, no more spiritually developed than I, and yet several had had "mystical experiences" that left them with the unshakable conviction that there was something more to this world than we know—a "beyond" of some kind that transcended the material universe I presume to constitute the whole shebang. I thought often about one of the cancer patients I interviewed, an avowed atheist who had nevertheless found herself "bathed in God's love."

Yet not everything I'd heard from these people made me eager to follow them onto the couch. Many had been borne by psilocybin deep into their pasts, a few of them traveling all the way back to scenes of unremembered childhood trauma. These journeys had been wrenching, shaking the travelers to their core, but they had been cathartic too. Clearly these medicines—as guides both above- and belowground invariably call the drugs they administer—powerfully stir the psychic pot, surfacing all sorts of repressed material, some of it terrifying and ugly. Did I *really* want to go there?

No!—to be perfectly honest. You should know I have never been one for deep or sustained introspection. My usual orientation is more forward than back, or down, and I generally prefer to leave my psychic depths undisturbed, assuming they exist. (There's quite enough to deal with up here on the surface; maybe that's why I became a journalist rather than a novelist or poet.) All that stuff down there in the psychic basement has been stowed there for a reason, and unless you're looking for something specific to help solve a problem, why would anyone willingly go down those steps and switch on that light?

People generally think of me as a fairly even-keeled and psychologically sturdy person, and I've played that role for so long now—in my family as a child, in my family as an adult, with my friends, and with my colleagues—that it's probably an accurate enough characterization. But every so often, perhaps in the wee-hour throes of insomnia or under the influence of cannabis, I have found myself tossed in a psychic storm of existential dread so dark and violent that the keel comes off the boat, capsizing this trusty identity. At such times, I begin seriously to entertain the possibility that somewhere deep beneath the equable presence I present, there exists a shadow me made up of forces roiling, anarchic, and potentially mad. Just how thin is the skin of my sanity? There are times when I wonder. Perhaps we all do. But did I really want to find out? R. D. Laing once said there are three things human beings are afraid of: death, other people, and their own minds. Put me down as two for three. But there are moments when curiosity gets the better of fear. I guess for me such a moment had arrived.

o   o   o

By "psychedelic underground," I don't mean the shadowy world of people making, selling, and using psychedelic drugs illegally. I have in mind a specific subset of that world, populated by perhaps a couple hundred "guides," or therapists, working with a variety of psychedelic substances in a carefully prescribed manner, with the intention of healing the ill or bettering the well by helping them fulfill their spiritual, creative, or emotional potential. Many of these guides are credentialed therapists, so by doing this work they are risking not only their freedom but also their professional licenses. I met one who was a physician and heard about another. Some are religious professionals—rabbis and ministers of various denominations; a few call themselves shamans; one described himself as a druid. The rest are therapists trained in dizzying combinations of alternative schools: I met Jungians and Reichians, Gestalt therapists and "transpersonal" psychologists; energy healers; practitioners of aura work, breathwork, and bodywork; EST, past-life, and family constellation therapists, vision questers, astrologers, and meditation teachers of every stripe—a shaggy reunion of that whole 1970s class of alternative "modalities" that usually get lumped together under the rubric of the "human potential movement" and that has as its world headquarters Esalen. The New Age terminology can be a little off-putting; there were times when I felt I was listening to people whose language and vocabulary had stopped evolving sometime in the early 1970s, at the very moment when psychedelic therapy was forced underground, freezing a subculture in time.

I tracked down several of these people in the Bay Area, which probably has the largest concentration of underground guides in the country, without much difficulty. Asking around, I soon discovered that a friend had a friend who worked with a guide down in Santa Cruz, doing an annual psilocybin journey on the occasion of his birthday. I also soon discovered that the membrane between the aboveground and the belowground psychedelic worlds is permeable in certain places; a couple of the people I befriended while reporting on the university psilocybin trials were willing to introduce me to "colleagues" who worked underground. One introduction led to another as people came to trust my intentions. By now, I've interviewed fifteen underground guides and have worked with five.

Considering the risks involved, I found most of these people unexpectedly open, generous, and trusting. Although the authorities have so far shown no interest in going after people practicing psychedelic-assisted therapy, the work remains illegal and so is dangerous to share with a journalist without taking precautions. All the guides asked me not to disclose their names or locations and to take whatever other measures I could to protect them. With that in mind, I have changed not only their names and locations but also certain other identifying details in each of their stories. But all the people you are about to meet are real individuals, not composites or fictions.

Virtually all of the underground guides I met are descended in one way or another from the generation of psychedelic therapists working on the West Coast and around Cambridge during the 1950s and 1960s when this work was still legal. Indeed, just about everyone I interviewed could trace a professional lineage reaching back to Timothy Leary (often through one of his graduate students), Stanislav Grof, Al Hubbard, or a Bay Area psychologist named Leo Zeff. Zeff, who died in 1988, was one of the earliest underground therapists, and certainly the most well-known; he claims to have "processed" (Al Hubbard's term) three thousand patients and trained 150 guides during his career, including several of the ones I met on the West Coast.

Zeff also left a posthumous (and anonymous) account of his work, in the form of a 1997 book called *The Secret Chief,* a series of interviews with a therapist called Jacob conducted by his close friend Myron Stolaroff. (In 2004, Zeff's family gave Stolaroff permission to disclose his identity and republish the book as *The Secret Chief Revealed.*) On the evidence of his interviews, Zeff is in many ways typical of the underground therapists I met, in both his approach and his manner; he comes across rather as folksy, or *haimish,* to use a Yiddish word Zeff would have appreciated, rather than as a renegade, guru, or hippie. In a photograph included in the 2004 edition, a smiling Zeff, wearing a big pair of aviator glasses and a sweater vest over his shirtsleeves, looks more like a favorite uncle than either an outlaw or mystic. Yet he was both.

Zeff was a forty-nine-year-old Jungian therapist practicing in Oakland in 1961 when he had his first trip, on a hundred micrograms of LSD. (It might have been Stolaroff himself who first "tripped him," to borrow one of Zeff's locutions.) The guide had asked him to bring along an object of personal significance, so Zeff brought his Torah. After the effects of the LSD had come on, his guide "laid the Torah across my chest and I immediately went into the lap of God. He and I were One."

Zeff soon began incorporating a range of different psychedelics in his practice and found that the medicines helped his patients break through their defenses, bringing buried layers of unconscious material to the surface, and achieve spiritual insights, often in a single session. The results were so "fantastic," he told Stolaroff, that when the federal government put psychedelics on schedule I in 1970, prohibiting their use for any purpose, Zeff made the momentous decision to continue his work underground.

This was not easy. "Many times I'd be in much agony falling asleep, and wake up in the morning and have it hit me," he told Stolaroff. "'Jacob [his pseudonym], for Christ's sake what are you exposing yourself to all this shit for? You don't need it.' Then I'd look and I'd say, 'Look at the people. Look what's happening to them.' I'd say, 'Is it worth it?' . . . Inevitably I'd come back with 'Yeah, it's worth it' . . . Whatever you have to go through. It's *worth* it to produce these results!"

During his long career, Zeff helped codify many of the protocols of underground therapy, setting forth the "agreements" guides typically make with their clients—regarding confidentiality (strict), sexual contact (forbidden), obedience to the therapist's instructions during the session (absolute), and so on—and developing many of the ceremonial touches, such as having participants take the medicine from a cup: "a very important symbol of the transformation experience." Zeff also described the departures from conventional therapeutic practice common among psychedelic guides. He believed it was imperative that guides have personal experience of any medicine they administer. (Aboveground guides either don't seek such experience or don't admit to it.) He came to believe that guides should not try to direct or manipulate the psychedelic journey, allowing it instead to find its own course and destination. ("*Just leave 'em alone!*" he tells Stolaroff.) Guides should also be willing to drop the analyst's mask of detachment, offering their personalities and emotions, as well as a comforting touch or hug to the client undergoing a particularly challenging trip.

In his introduction to *The Secret Chief Revealed*, Myron Stolaroff sketched the influence of underground guides like Leo Zeff on the field as a whole, suggesting that the legitimate psychedelic research that resumed in the late 1990s, when he was writing, had "evolved as a result of anecdotal evidence from underground therapists" like Zeff, as well as from the first wave of psychedelic research done in the 1950s and 1960s. Psychedelic researchers working in universities today are understandably reluctant to acknowledge it, but there is a certain amount of traffic between the two worlds, and a small number of figures who move, somewhat gingerly, back and forth between them. For example, some prominent underground therapists have been recruited to help train a new cohort of psychedelic guides to work in university trials of psychedelic drugs. When the Hopkins team wanted to study the role of music in the guided psilocybin session, it reached out to several underground guides, surveying their musical practices.

No one had any idea how many underground guides were working in America, or exactly what that work consisted of, until 2010. That was the year James Fadiman, the Stanford-trained psychologist who took part in psychedelic research at the International Foundation for Advanced Study in Menlo Park in the early 1960s, attended a conference on psychedelic science in the Bay Area. The conference was organized by MAPS, with sponsorship from Heffter, the Beckley Foundation, and Bob Jesse's Council on Spiritual Practices, the three other nonprofits that funded most of the psychedelic research under way at the time. In a Holiday Inn in San Jose, the conference brought together more than a thousand people, including several dozen scientists (who presented their research, complete with Power-Point slides), a number of guides drawn from both the university trials and the underground, and a great many more "psychonauts"— people of all ages who make regular use of psychedelics in their lives, whether for spiritual, therapeutic, or "recreational" purposes. (As Bob Jesse is always quick to remind me whenever I use that word, "recreational" doesn't necessarily mean frivolous, careless, or lacking in intention. Point taken.)

James Fadiman came to the MAPS conference "on the science track," to give a talk about the value of the guided entheogenic journey. He wondered if there were many underground guides in the audience, so at the end of his talk he announced that there would be a meeting of guides at 8:00 the following morning.

"I dragged myself out of bed at 7:30 expecting to see maybe five people, but a hundred showed up! It was staggering."

It would probably be too strong to describe this far-flung and disparate group as a community, much less an organization, yet my interviews with more than a dozen of them suggest they are professionals who share an outlook, a set of practices, and even a code of conduct. Soon after the meeting in San Jose, a "wiki" appeared on the Internet—a collaborative website where individuals can share documents and together create new content. (Fadiman included the URL in his 2011 book, *The Psychedelic Explorer's Guide*.) Here, I found two items of particular interest, as well as several sub-wikis—documents under development—that hadn't had a new entry for several years; it could be that public disclosure of the site in Fadiman's book had led the creators to abandon it or move elsewhere online.

The first item was a draft charter: "to support a category of profound, prized experiences becoming more available to more people." These experiences are described as "unitive consciousness" and "non-dual consciousness," among other terms, and several non-pharmacological modalities for achieving these states are mentioned, including meditation, breathwork, and fasting. "A principal tool of the Guides is the judicious use of a class of psychoactive substances" known to be "potent spiritual catalysts."

The website offers would-be guides links to printable forms for legal releases, ethical agreements, and medical questionnaires. ("We don't have very good insurance," one guide told me, with a sardonic smile. "So we're very careful.") There's also a link to a thoughtful "Code of Ethics for Spiritual Guides," which acknowledges the psychological and physical risks of journeying and emphasizes the guide's ultimate responsibility for the well-being of the client. Recognizing that during "primary religious practices" "participants may be especially open to suggestion, manipulation, and exploitation," the code states that it is incumbent upon the guide to disclose all risks, obtain consent, guarantee confidentiality, protect the safety and health of participants at all times, "safeguard against . . . ambition" and self-promotion, and accommodate clients "without regard to their ability to pay."

Perhaps the most useful document on the website is the "Guidelines for Voyagers and Guides."* The guidelines represent a compendium of half a century's accumulated knowledge and wisdom about how best to approach the psychedelic journey, whether as a participant or as a guide. It covers the basics of set and setting; mental and physical preparation for the session; potential drug interactions; the value of formulating an intention; what to expect during the experience, both good and bad; the stages of the journey; what can go wrong and how to deal with frightening material; the supreme importance of post-session "integration"; and so on.

For me, standing on the threshold of such an experience, it was reassuring to learn that the underground community of psychedelic guides, which I had assumed consisted of a bunch of individuals all doing pretty much their own thing, operated like professionals, working from a body of accumulated knowledge and experience and in a set of traditions that had been handed down from psychedelic pioneers such as Al Hubbard, Timothy Leary, Myron Stolaroff, Stan Grof, and Leo Zeff. They had rules and codes and agreements, and many elements of the work had been more or less institutionalized.

Stumbling upon the website also made me appreciate just how far the culture of psychedelics has evolved since the 1950s and 1960s. Implicit in these documents, it seemed to me, was the recognition that these powerful, anarchic medicines can and have been misused and that if they are to do more good than harm, they require a cultural vessel of some kind: protocols, rules, and rituals that together form a kind of Apollonian counterweight to contain and channel their sheer Dionysian force. Modern medicine, with its controlled

---

* A version of the guidelines can also be found in James Fadiman's book *The Psychedelic Explorer's Guide: Safe, Therapeutic, and Sacred Journeys* (Rochester, Vt.: Park Street Press, 2011).

trials and white-coated clinicians and *DSM* diagnoses, offers one such container; the underground guides offer another.

o   o   o

YET THE FIRST COUPLE of guides I interviewed did not fill me with confidence. Maybe it was because I was so new to the territory, and nervous about the contemplated journey, but I kept hearing things in their spiels that set off alarm bells and made me want to run in the opposite direction.

Andrei, the first guide I interviewed, was a gruff Romanian-born psychologist in his late sixties with decades of experience; he had worked with a friend of a friend of a friend. We met at his office in a modest neighborhood of small bungalows and neat lawns in a city in the Pacific Northwest. A hand-lettered sign on the door instructed visitors to remove their shoes and come upstairs to the dimly lit waiting room. A kilim rug had been pinned to the wall.

Instead of a table piled with old copies of *People* or *Consumer Reports*, I found a small shrine populated with spiritual artifacts from a bewildering variety of traditions: a Buddha, a crystal, a crow's wing, a brass bowl for burning incense, a branch of sage. At the back of the shrine stood two framed photographs, one of a Hindu guru I didn't recognize and the other of a Mexican *curandera* I did: María Sabina.

This was not the last time I would encounter such a confusing tableau. In fact every guide I met maintained some such shrine in the room where he or she worked, and clients were often asked to contribute an item of personal significance before embarking on their journeys. What I was tempted to dismiss as a smorgasbord of equal-opportunity New Age tchotchkes, I would eventually come to regard more sympathetically, as the material expression of the syncretism prevalent in the psychedelic community. Members of this community tend to be more spiritual than religious in any formal sense, focused on the common core of mysticism or "cosmic consciousness" that they believe lies behind all the different religious traditions. So what appeared to me as a bunch of conflicting symbols of divinity are in fact different means of expressing or interpreting the same underlying spiritual reality, "the perennial philosophy" that Aldous Huxley held to undergird all religions and to which psychedelics supposedly can offer direct access.

After a few minutes, Andrei bounded into the room, and when I stood to offer my hand, he surprised me with a bear hug. A big man with a full head of hastily combed gray hair, Andrei was wearing a blue-checked button-down over a yellow T-shirt that struggled to encompass the globe of his belly. Speaking with a thick accent, he managed to seem both amiable and disconcertingly blunt.

Andrei had his first experience with LSD at twenty-one, soon after he came out of the army; a friend had sent it from America, and the experience transformed him. "It made me realize we live a very limited version of what life is." That realization propelled him on a journey through Eastern religion and Western psychology that eventually culminated in a doctorate in psychology. When military service threatened to interrupt his psycho-spiritual journey, he "decided I have to make my own choices" and deserted.

Andrei eventually left Bucharest for San Francisco, bound for what he had heard was "the first New Age graduate school"—the California Institute of Integral Studies. Founded in 1968, the institute specializes in "transpersonal psychology," a school of therapy with a strong spiritual orientation rooted in the work of Carl Jung and Abraham Maslow as well as the "wisdom traditions" of the East and the West, including Native American healing and South American shamanism. Stanislav Grof, a pioneer of both transpersonal and psychedelic therapy, has been on the faculty for many years. In 2016, the institute began offering the nation's first certificate program in psychedelic therapy.

As part of his degree program, Andrei had to undergo psychotherapy and found his way to a Native American "doing medicine work" in the Four Corners as well as the Bay Area. "Whoopee!" he recalled thinking. "Because of my LSD experience, I knew it was viable." Medicine work became his vocation.

"I help people find out who they are so they can live their lives fully. I used to work with whoever came to me, but some were too fucked up. If you're on the edge of psychosis, this work can push you over. You need a strong ego in order to let go of it and then be able to spring back to your boundaries." He mentioned he'd once been sued by a troubled client who blamed him for a subsequent breakdown. "So I decided, I don't work with crazies anymore. And as soon as I made this statement to the universe, they stopped coming." These days he works with a lot of young people in the tech world. "I'm the dangerous virus of Silicon Valley. They come to me wondering, 'What am I doing here, chasing the golden carrot in the golden cage?' Many of them go on to do something more meaningful with their lives. [The experience] opens them up to the spiritual reality."

It's hard to say exactly what put me off working with Andrei, but oddly enough it was less the New Agey spiritualism than his nonchalance about a process I still found exotic and scary. "I don't play the psychotherapy game," he told me, as blasé as a guy behind a deli counter wrapping and slicing a sandwich. "None of that blank screen. In mainstream psychology, you don't hug. I hug. I touch them. I give advice. I have people come stay with us in the forest." He works with clients not here in the office but in a rural location deep in the woods of the Olympic Peninsula. "Those are all big no-no's." He shrugged as if to say, so what?

I shared some of my fears. He'd heard it all before. "You may not get what you want," he told me, "but you'll get what you need." I gulped mentally. "The main thing is to surrender to the experience, even when it gets difficult. Surrender to your fear. The biggest fears that come up are the fear of death and the fear of madness. But the only thing to do is surrender. So surrender!" Andrei had named my two biggest fears, but his prescription seemed easier said than done.

I was hoping for a guide who exuded perhaps a little more tenderness and patience, I realized, yet I wasn't sure I should let Andrei's gruff manner put me off. He was smart, he had loads of experience, and he was willing to work with me. Then he told a story that decided the matter.

It was about working with a man my age who became convinced during his psilocybin journey he was having a heart attack. "'I'm dying,' he said, 'call 911! I feel it, my heart!' I told him to surrender to the dying. That Saint Francis said that in dying you gain eternal life. When you realize death is just another experience, there's nothing more to worry about."

Okay, but what if it had been a real heart attack? Out there in the woods in the middle of the Olympic Peninsula? Andrei mentioned that an aspiring guide he was training had "once asked me, 'What do you do if someone dies?'" I don't know what I expected him to say, but Andrei's reply, delivered with one of his most matter-of-fact shrugs, was not it.

"You bury him with all the other dead people."

I told Andrei I would be in touch.

The psychedelic underground was populated with a great many such vivid characters, I soon discovered, but not necessarily the kinds of characters to whom I felt I could entrust my mind—or for that matter any part of me. Immediately after my session with Andrei, I had a meeting with a second prospective guide, a brilliant psychologist in his eighties who had been a student of Timothy Leary's at Harvard. His knowledge of psychedelics was deep; his credentials impressive; he had been highly recommended by people I respected. Yet when over lunch at a Tibetan restaurant near his office he removed his bolo tie and suspended it over the menu, I began to lose confidence that this was my man. He explained that he relied on the energies released by the pendulum swing of the silver clasp to choose the entrée most likely to agree with his temperamental digestion. I forget what his tie ordered for lunch, but even before he began dilating on the evidence that 9/11 was an inside job, I knew my search for a guide was not over quite yet.

o   o   o

ONE NOTABLE DIFFERENCE about doing psychedelics at sixty, as opposed to when you're eighteen or twenty, is that at sixty you're more likely to have a cardiologist you might want to consult in advance of your trip. That was me. A year before I had decided to embark on this adventure, my heart, the reliable operations of which I had taken completely for granted to that point, had suddenly made its presence felt and, for the first time in my life, demanded my attention. While sitting at my computer one afternoon, I was suddenly made aware of a pronounced and crazily syncopated new rhythm in my chest.

"Atrial fibrillation" was the name the doctor gave the abnormal squiggles that appeared on my EKG. The danger of AFib is not a heart attack, he said to my (short-lived) relief, but a heightened risk of stroke. "My cardiologist"—the unfortunate phrase had suddenly joined my vocabulary, probably for the duration—put me on a couple of meds to calm the heart rhythms and lower the blood pressure, plus a daily baby aspirin to thin my blood. And then he told me not to worry about it.

I followed all of his advice except the last bit. Now I couldn't help but think about my heart constantly. All of its operations that had previously taken place completely outside my conscious awareness suddenly became salient: something I could hear and feel whenever I thought to check in, which now was incessantly. Months later, the AFib had not recurred, but my surveillance of my poor heart had gotten out of control. I checked my blood pressure daily and listened for signs of ventricular eccentricity every time I got into bed. It took months of not having a stroke before I could once again trust my heart to go about its business without my supervision. Gradually, thankfully, it retreated once again to the background of my attention.

I tell you all this by way of explaining why I felt I should talk to my cardiologist before embarking on a psychedelic journey. My cardiologist was my age, so not likely to be shocked by the word "psilocybin" or "LSD" or "MDMA." I told him what I had in mind and asked if any of the drugs in question were contraindicated, given my coronary issues, or if there was any risk of an interaction with the meds he had prescribed. He was not overly concerned about the psychedelics—most of them concentrate their effects in the mind with remarkably little impact on the cardiovascular system—but one of the drugs I mentioned he advised I avoid. This was MDMA, also known as Ecstasy or Molly, which has been on schedule I since the mid-1980s, when it emerged as a popular rave drug.

The drug 3,4-methylenedioxymethamphetamine is not a classical psychedelic (it works on different brain receptors and doesn't have strong visual effects), yet several of the guides I was interviewing had told me it was part of their regimen. Sometimes called an empathogen, MDMA lowers psychological defenses and helps to swiftly build a bond between patient and therapist. (Leo Zeff was one of the first therapists to use MDMA in the 1970s, after the compound was popularized by his friend the legendary Bay Area chemist Sasha Shulgin and his wife, the therapist Ann Shulgin.) Guides told me MDMA was a good way to "break the ice" and establish trust before the psychedelic journey. (One said, "It condenses years of psychotherapy into an afternoon.") But as its scientific name indicates, MDMA is an amphetamine, and so, chemically, it implicates the heart in a way psychedelics don't. I was disappointed my cardiologist had taken MDMA off the table but pleased that he had more or less given me a green light on the rest of my travel plans.

Trip One: LSD

At least on paper, nothing about the first guide I chose to work with sounds auspicious. The man lived and worked so far off the grid, in the mountains of the American West, that he had no phone service, generated his own electricity, pumped his own water, grew his own food, and had only the spottiest satellite Internet. I could just forget about the whole idea of being anywhere in range of a hospital emergency room. Then there was the fact that while I was a Jew from a family that had once been reluctant to buy a German car, this fellow was the son of a Nazi—a German in his midsixties whose father had served in the SS during World War II. After I had heard so much about the importance of both set and setting, none of these details augured especially well.

Yet I liked Fritz from the moment he came out to greet me, offering a broad grin and a warm hug (I was getting used to these) when I pulled my rental into his remote camp. This consisted of a tidy village of structures—a handmade house and a couple of smaller cabins, an octagonal yurt, and two gaily painted outhouses set out in a clearing on the crest of a heavily wooded mountain. Following the hand-drawn map Fritz had sent me (the area was terra incognita for

GPS), I drove for miles on a dusty dirt road that passed through the blasted landscape of an abandoned mine before rising into a dark forest of cypress and ponderosa pine, with a dense understory of manzanitas, their smooth bark the color of fresh blood. I had come to the middle of nowhere.

Fritz was a tangle of contradiction and yet manifestly a warm and seemingly happy man. At sixty-five, he resembled a European movie actor gone slightly to seed, with thick gray hair parted in the middle and a blocky, muscular frame just beginning to yield. Fritz grew up in Bavaria, the son of a raging alcoholic who had served in the SS as a bodyguard for the cultural attaché responsible for producing operas and other entertainments for the troops—the Nazis' USO. Later, his father fought on the Russian front and survived Stalingrad but came home from the war shell-shocked. Fritz grew up in the dense shade of his misery, sharing the shame and anger of so many in his postwar generation.

"When the military came for me [to serve his period of conscription]," he said, as we sat at his kitchen table sipping tea on a sunny spring afternoon, "I told them to fuck themselves and they threw me into prison." Forced eventually to serve in the army, Fritz was court-martialed twice—once for setting his uniform on fire. He spent time in solitary confinement reading Tolstoy and Dostoyevsky and plotting revolution with the Maoist in the next cell, with whom he communicated through the prison plumbing. "My proudest moment was the time I gave all the guards Orange Sunshine that I had gotten from a friend in California."

At university, he studied psychology and took a lot of LSD, which he obtained from the American troops stationed in Germany. "Compared to LSD, Freud was a joke. For him biography was everything. He had no use for mystical experience." Fritz moved on to Jung and Wilhelm Reich, "my hero." Along the way, he discovered that LSD was a powerful tool for exploring the depths of his own psyche, allowing him to reexperience and then let go of the anger and depression that hobbled him as a young man. "There was more light in my life after that. Something shifted."

As it had for many of the guides I had met, the mystical experience Fritz had on psychedelics launched him on a decades-long spiritual quest that eventually "blew my linear, empirical mind," opening him up to the possibility of past lives, telepathy, precognition, and "synchronicities" that defy our conceptions of space and time. He spent time on an ashram in India, where he witnessed specific scenes that had been prefigured in his psychedelic journeys. Once, making love to a woman in Germany (the two were practicing Tantrism), he and she shared an out-of-body experience that allowed them to observe themselves from the ceiling. "These medicines have shown me that something quote-unquote impossible exists. But I don't think it's magic or supernatural. It's a technology of consciousness we don't understand yet."

Normally when people start talking about transpersonal dimensions of consciousness and "morphogenetic fields," I have little (if any) patience, but there was something about Fritz that made such talk, if not persuasive, then at least . . . provocative. He managed to express the most far-fetched ideas in a disarmingly modest, even down-to-earth way. I had the impression he had no agenda beyond feeding his own curiosity, whether with psychedelics or books on paranormal phenomena. For some people, the privilege of having had a mystical experience tends to massively inflate the ego, convincing them they've been granted sole possession of a key to the universe. This is an excellent recipe for creating a guru. The certitude and condescension for mere mortals that usually come with that key can render these people insufferable. But that wasn't Fritz. To the contrary. His otherworldly experiences had humbled him, opening him up to possibilities and mysteries without closing him to skepticism—or to the pleasures of everyday life on *this* earth. There was nothing ethereal about him. I surprised myself by liking Fritz as much as I did.

After five years spent living on a commune in Bavaria ("we were all trying to undo some of the damage done to the postwar generation"), in 1976 he met a woman from California while hiking in the Himalayas and followed her back to Santa Cruz. There he fell into the whole Northern California human potential scene, at various times running a meditation center for an Indian guru named Rajneesh and doing bodywork (including deep-tissue massage and Rolfing), Gestalt and Reichian therapy, and some landscaping to pay the bills. When in 1982, soon after his father's death, he met Stan Grof at a breathwork course at Esalen, he felt he had at last found his rightful father. During the workshop, Fritz "had an experience as powerful as any psychedelic. Out of the blue, I experienced myself being born—my mother giving birth to me. While this was happening, I watched the goddess Shiva on a gigantic IMAX screen, creating worlds and destroying worlds. Everyone in the group wanted what I had!" He now added Holotropic Breathwork to his bodywork practice.

Eventually, Fritz did an intensive series of multiyear trainings with Grof in Northern California and British Columbia. At one of them, he met his future wife, a clinical psychologist. Grof was ostensibly teaching Holotropic Breathwork, the non-pharmacological modality he had developed after psychedelics were made illegal. But Fritz said that Grof also shared with this select group his deep knowledge about the practice of psychedelic therapy, discreetly passing on his methods to a new generation. Several people in the workshop, Fritz and his future wife among them, went on to become underground guides. She works with the women who find their way up the mountain, he with the men.

"You don't make a lot of money," Fritz told me. Indeed, he charged only nine hundred dollars for a three-day session, which included room and board. "It's illegal and dangerous. You can have a person go psychotic. And you *really* don't make a lot of money. But I'm a healer and these medicines work." It was abundantly clear he had a calling and loved what he did—loved witnessing people undergo profound transformations before his eyes.

o   o   o

FRITZ TOLD ME what to expect if I were to work with him. It would mean returning here for three days, sleeping in the eight-sided yurt, where we would also do "the work." The first afternoon would be a warm-up or get-acquainted session, using either MDMA or breathwork. (I explained why in my case it would have to be breathwork.) This would give him a chance to observe how I handled an altered state of consciousness before sending me on an LSD journey the morning of the second day; it would also help him determine a suitable dose.

I asked him how he could be sure of the purity and quality of the medicines he uses, since they come from chemists working illicitly. Whenever he receives a new shipment, he explained, "I first test it for purity, and then I take a heroic dose to see how it feels before I give it to anyone." Not exactly FDA approval, I thought to myself, but better than nothing.

Fritz doesn't take any medicine himself while he's working but often gets "a contact high" from his clients. During the session he takes notes, selects the music, and checks in every twenty minutes or so. "I'll ask you not how you are but where you are.

"I'm here just for you, to hold the space, so you don't have to worry about anything or anyone else. Not the wife, not the child. So you can really let go—and *go*." This, I realized, was another reason I was eager to work with a guide. When Judith and I had our magic mushroom day the previous summer, the simmer of worry about her welfare kept intruding on my journey, forcing me to stay close to the surface. Much as I hated the psychobabble-y locution, I loved the idea of someone "holding space" for me.

"That night I'll ask you to make some notes before you go to sleep. On your last morning, we'll compare notes and try to integrate and make sense of your experience. Then I'll cook you a big breakfast to get you ready to face the interstate!"

We scheduled a time for me to come back.

o   o   o

THE FIRST THING I learned about myself that first afternoon, working with Fritz in the yurt, is that I am "easy to put under"— susceptible to trance, a mental space completely new to me and accessible by nothing more than a shift in the pattern of one's breathing. It was the damnedest thing.

Fritz's instructions were straightforward: *Breathe deeply and rapidly while exhaling as strongly as you can.* "At first it will feel unnatural and you'll have to concentrate to maintain the rhythm, but after a few minutes your body will take over and do it automatically." I stretched out on the mattress and donned a pair of eyeshades while he put on some music, something generically tribal and rhythmic, dominated by the pounding of a drum. He placed a plastic bucket at my side, explaining that occasionally people throw up.

It was hard work at first, to breathe in such an exaggerated and unnatural way, even with Fritz's enthusiastic coaching, but then all at once my body took over, and I found that no thought was required to maintain the driving pace and rhythm. It was as if I had broken free from gravity and settled into an orbit: the big deep breaths just came, automatically. Now I felt an uncontrollable urge to move my legs and arms in sync with the pounding of the drums, which resonated in my rib cage like a powerful new heartbeat. I felt possessed, both my body and my mind. I can't remember many thoughts except "Hey, this is working, whatever it is!"

I was flat on my back yet dancing wildly, my arms and legs moving with a will of their own. All control of my body I had surrendered to the music. It felt a little like speaking in tongues, or what I imagine that to be, with some external force taking over the mind and body for its own obscure purpose.

There wasn't much visual imagery, just the naked sensation of exhilaration, until I began to picture myself on the back of a big black horse, galloping headlong down a path through a forest. I was perched up high on its shoulders, like a jockey, holding on tight as the beast scissored its great muscles forward and back with each long stride. As my rhythm synced with that of the horse, I could feel myself absorbing the animal's power. It felt fantastic to so fully inhabit my body, as if for the first time. And yet because I am not a very confident rider (or dancer!), it also felt precarious, as if were I to miss a breath or beat I might tumble off.

I had no idea how long the trance lasted, time was utterly lost on me, but when Fritz gently brought me back to the present moment and the reality of the room, simply by encouraging me to slow and relax my breathing, he reported I had been "in it" for an hour and fifteen minutes. I felt flushed and sweaty and triumphant, as if I had run a marathon; Fritz said I looked "radiant"—"young like a baby."

"You had no resistance," he said approvingly; "that's a good sign for tomorrow." I had no idea what had just happened, could recall little more of the hour than riding the horse, but the episode seemed to have involved a terrific physical release of some kind. *Some*thing had let go of me or been expunged, and I felt buoyant. And humbled by the mystery of it. For here was (to quote William James) one of the "forms of consciousness entirely different" from the ordinary and yet so close by—separated from normal waking consciousness by . . . what? A handful of exhalations!

Then something frightening happened. Fritz had gone up to the house to prepare our dinner, leaving me to make some notes about the experience on my laptop, when all at once I felt my heart surge and then begin to dance madly in my chest. I immediately recognized the sensation of turbulence as AFib, and when I took my pulse, it was chaotic. A panicky bird was trapped in my rib cage, throwing itself against the bars in an attempt to get out. And here I was, a dozen miles off the grid smack in the middle of nowhere.

It went on like that for two hours, straight through a subdued and anxious dinner. Fritz seemed concerned; in all the hundreds of breathwork sessions he had led or witnessed, he had never seen such a reaction. (He had mentioned earlier a single fatality attributed to Holotropic Breathwork: a man who had had an aneurism.) Now I was worried about tomorrow, and I think he was too. Though he also wondered if perhaps what I was feeling in my heart might reflect some psychic shift or "heart opening." I resisted the implied metaphor, holding firm to the plane of physiology: *the heart is a pump, and this one is malfunctioning.* We discussed tomorrow's plan. Maybe we want to go with a lower dose, Fritz suggested; "you're so susceptible you might not need very much to journey." I told him I might bail out altogether. And then, as suddenly as it had come on, I felt my heart slip back into the sweet groove of its accustomed rhythm.

I got little sleep that night as a debate raged in my head about whether or not I was crazy to proceed in the morning with LSD at any dose. *I could die up here and wouldn't that be stupid?* But was I really in any danger? Now my heart felt fine, and from everything I read, the effects of LSD were confined to the brain, more or less, leaving the cardiovascular system unaffected. In retrospect, it made perfect sense that a process as physically arduous as Holotropic Breathwork would discombobulate the heart.* Yes, I could take a rain check on my LSD journey, but even the thought of that option landed like a crushing disappointment. I had come this far, and I had had this intriguing glimpse into a state of consciousness that for all my trepidations I was eager to explore more deeply.

This went on all night, back and forth, pro and con, but by the time the sun came up, the earliest rays threading the needles of the eastern pines, I was resolved. At breakfast, I told Fritz I felt good and wanted to proceed. We agreed, however, to go with a modest dose— a hundred micrograms, with "a booster" after an hour or two if I wanted one.

Fritz sent me out on a walk to clear my head and think about my intention while he did the dishes and readied the yurt for my journey. I hiked for an hour on a trail through the forest, which had been refreshed overnight by a rain shower; the cleansed air held the scent of cedar, and the barkless red limbs of the manzanita were glowing. Fritz had told me to look for an object to put on the altar. While I was looking and walking, I decided I would ask Fritz to give me his pledge that if anything whatsoever went wrong, he would call 911 for help regardless of the personal risk.

I returned to the yurt around ten with a manzanita leaf and a smooth black stone in my pocket and a straightforward intention: to

---

* I subsequently learned that hyperventilation, which plays a role in breathwork, changes the CO_2 levels of the blood, which in turn can alter the rhythms of the heart in some people. What I assumed was a physiologically benign alternative to MDMA turns out to be nothing of the kind; even without a drug, it is possible to change one's blood chemistry in ways that can affect heart rhythms.

learn whatever the journey had to teach me about myself. Fritz had lit a fire in the woodstove, and the room was beginning to give up its chill. He had moved the mattress across the room so my head would be close to the speakers. In somber tones, he talked about what to expect and how to handle various difficulties that might arise: "paranoia, spooky places, the feeling you're losing your mind or that you are dying.

"It's like when you see a mountain lion," he suggested. "If you run, it will chase you. So you must stand your ground." I was reminded of the "flight instructions" that the guides employed at Johns Hopkins: instead of turning away from any monster that appears, move toward it, stand your ground, and demand to know, "What are you doing in my mind? What do you have to teach me?"

I added my stone and leaf to the altar, which held a bronze Buddha surrounded by the items of many previous travelers. "Something hard and something soft," Fritz observed. I asked for the assurances I needed to proceed and received them. Now he handed me a Japanese teacup at the bottom of which lay a tiny square of blotter paper and the torn scraps of a second square—the booster. One side of the blotter paper had a Buddha printed on it, the other a cartoon character I didn't recognize. I put the square on my tongue and, taking a sip of water, swallowed. Fritz didn't perform much of a ceremony, but he did talk about the "sacred tradition" I was now joining, the lineage of all the tribes and peoples down through time and around the world who used such medicines in their rites of initiation. Here I was, in range of my sixtieth birthday, taking LSD for the first time. It did feel something like a rite of passage, but a passage to where, exactly?

While waiting for the LSD to come on, we sat on the wooden skirt of decking that circled the yurt, chatting quietly about this and that. Life up here on the mountain; the wildlife that shared the property with him because he didn't keep a dog: there were mountain lions, bears, coyotes, foxes, and rattlesnakes. Jittery, I tried to change the subject; as it was, I'd been afraid during the night to visit the outhouse, choosing instead to pee off the porch. Lions and bears and snakes were the last thing I wanted to think about just now.

Around eleven, I told Fritz I was starting to feel wobbly. He suggested I lie down on the mattress and put on my eyeshades. As soon as he started the music—something Amazonian in flavor, gently rhythmic with traditional instruments but also nature sounds (rain showers and crickets) that created a vivid dimensional sense of outdoor space—I was off, traveling somewhere in my mind, in a fully realized forest landscape that the music had somehow summoned into being. It made me realize what a powerful little technology a pair of eyeshades could be, at least in this context: it was like donning a pair of virtual reality goggles, allowing me immediately to take leave of this place and time.

I guessed I was hallucinating, yet this was not at all what I expected an LSD hallucination to be, which was overpowering. But Fritz had told me that the literal meaning of the word is to wander in one's mind, and that was exactly what I was doing, with the same desultory indifference to agency the wanderer feels. Yet I still *had* agency: I could change at will the contents of my thoughts, but in this dreamy state, so wide open to suggestion, I was happy to let the terrain, and the music, dictate my path.

And for the next several hours the music did just that, summoning into existence a sequence of psychic landscapes, some of them populated by the people closest to me, others explored on my own. A lot of the music was New Age drivel—the sort of stuff you might hear while getting a massage in a high-end spa—yet never had it sounded so evocative, so beautiful! Music had become something much greater and more profound than mere sound. Freely trespassing the borders of the other senses, it was palpable enough to touch, forming three-dimensional spaces I could move through.

The Amazonian-tribal song put me on a trail that ascended steeply through redwoods, following a ravine notched into a hillside by the silvery blade of a powerful stream. *I know this place*: it was the trail that rises from Stinson Beach to Mount Tamalpais. But as soon as I secured that recognition, it morphed into something else entirely. Now the music formed a vertical architecture of wooden timbers, horizontals and verticals and diagonals that were being magically craned into place, forming levels that rose one on top of the other, ever higher into the sky like a multistoried tree house under construction, yet a structure as open to the air and its influences as a wind chime.

I saw that each level represented another phase in my life with Judith. There we were, ascending stage by stage through our many years together, beginning as kids who met in college, falling in love, living together in the city, getting married, having our son, Isaac, becoming a family, moving to the country. Now, here at the top, I watched a new, as yet inchoate stage being constructed as indeed one now is: whatever this life together is going to be now that Isaac has grown up and left home. I looked hard, hoping for some clue about what to expect, but the only thing I could see clearly was that this new stage was being built on the wooden scaffolding of earlier ones and therefore promised to be sturdy.

So it went, song by song, for hours. Something aboriginal, with the deep spooky tones of a didgeridoo, put me underground, moving somehow through the brownish-black rootscape of a forest. I tensed momentarily: Was this about to get terrifying? Have I died and been interred? If so, I was fine with it. I got absorbed watching a white tracery of mycelium threading among the roots and linking the trees in a network intricate beyond comprehension. I knew all about this mycelial network, how it forms a kind of arboreal Internet allowing the trees in a forest to exchange information, but now what had been merely an intellectual conceit was a vivid, felt reality of which I had become a part.

When the music turned more masculine or martial, as it now did, sons and then fathers filled my mental field. I watched a swiftly unfolding biopic of Isaac's life to this point—his struggles as an exquisitely sensitive boy, and how those sensitivities had turned into strengths, making him who he is. I thought about things I needed to tell him—about the surging pride I felt as he embarked on his adult life and made his way in a new city and career, but also my fervent hope that he not harden himself in success or disown his vulnerabilities and his sweetness.

I felt something on my eyeshades and realized I had wet them with my tears.

I was already feeling wide open and undefended when it dawned on me that I wasn't talking to Isaac, or not only to him, but to myself as well. *Something hard and something soft:* the paired terms kept turning over like a coin. The night before coming to Fritz's place, I had spoken to two thousand people in a concert hall, tracked across a stage by a spotlight as I played the role of the man with the answers, the one people could depend on to explain things. This was much the same role I played in my family growing up, not only for my younger sisters, but, in times of crisis, for my parents too. (Even now, my sisters stubbornly refuse ever to accept from me the words "I don't know.") "So now look at me!" I thought, a smile blooming on my face: this grown man blindfolded and laid out on the floor of a psychedelic therapist's yurt, chasing after my mind as it wandered heedlessly through the woods of my life, warm tears—of *what?* I didn't know!—sliding down my cheeks.

This was unfamiliar territory for me and not at all where I

expected to find myself on LSD. I hadn't traveled very far from home. Instead of the demons and angels and various other entities I was expecting to meet, I was having a series of encounters with the people in my family. I visited each of them in turn, the music setting the tone, and the emotions came over me in great waves, whether of admiration (for my sisters and mother, whom I pictured seated around a horseshoe-shaped table—like the UN!—each of them representing a different ideal of feminine strength); gratitude; or compassion, especially for my father, a man both driven and pursued for much of his life, and someone whom before this moment I'd never before fully imagined as a son, and a son of ferociously demanding parents.

The flood tide of compassion overflowed its banks and leaked into some unexpected places, like my fourth-grade music class. Here I inexplicably encountered poor Mr. Roper, this earnest young man in a cheap suit who in spite of heroic efforts could not get us to give a shit about the sections of an orchestra he mapped on the board or the characters of the various instruments, no matter how many times he played *Peter and the Wolf* for us. As he paced the classroom in his excitement, we would wait in breathless suspense for him to step on one of the upturned thumbtacks we placed in his path, a thrill for which we were willing to risk staying after school in detention. But who was this Mr. Roper, really? Why couldn't we see that behind the cartoon figure we tortured so mercilessly was, no doubt, a decent guy who wanted nothing more than to ignite in us his passion for music? *The unthinking cruelty of children* sent a quick shiver of shame through me. But then: What a surfeit of compassion I must be feeling, to spare that much for Mr. Roper!

And cresting over all these encounters came a cascading dam break of love, love for Judith and Isaac and everyone in my family, love even for my impossible grandmother and her long-suffering husband. The next day, during our integration session, Fritz read from his notes two things I apparently said aloud during this part of the journey: "I don't want to be so stingy with my feelings." And, "All this time spent worrying about my heart. What about all the other hearts in my life?"

It embarrasses me to write these words; they sound so thin, so banal. This is a failure of my language, no doubt, but perhaps it is not only that. Psychedelic experiences are notoriously hard to render in words; to try is necessarily to do violence to what has been seen and felt, which is in some fundamental way pre- or post-linguistic or, as students of mysticism say, ineffable. Emotions arrive in all their new-born nakedness, unprotected from the harsh light of scrutiny and, especially, the pitiless glare of irony. Platitudes that wouldn't seem out of place on a Hallmark card glow with the force of revealed truth.

Love is everything.

Okay, but what else did you learn?

No—you must not have heard me: it's *every*thing!

Is a platitude so deeply felt still just a platitude? No, I decided. A platitude is precisely what is left of a truth after it has been drained of all emotion. To resaturate that dried husk with feeling is to see it again for what it is: the loveliest and most deeply rooted of truths, hidden in plain sight. A spiritual insight? Maybe so. Or at least that's how it appeared in the middle of my journey. Psychedelics can make even the most cynical of us into fervent evangelists of the obvious.

You could say the medicine makes you stupid, but after my journey through what must sound like a banal and sentimental landscape, I don't think that's it. For what after all is the sense of banality, or the ironic perspective, if not two of the sturdier defenses the adult ego deploys to keep from being overwhelmed—by our emotions, certainly, but perhaps also by our senses, which are liable at any time to astonish us with news of the sheer wonder of the world. If we are ever to get through the day, we need to put most of what we perceive into boxes neatly labeled "Known," to be quickly shelved with little thought to the marvels therein, and "Novel," to which, understandably, we pay more attention, at least until it isn't that anymore. A psychedelic is liable to take *all* the boxes off the shelf, open and remove even the most familiar items, turning them over and imaginatively scrubbing them until they shine once again with the light of first sight. Is this reclassification of the familiar a waste of time? If it is, then so is a lot of art. It seems to me there is great value in such renovation, the more so as we grow older and come to think we've seen and felt it all before.

Yet one hundred micrograms of LSD had surely not propelled me into the lap of God, as it had Leo Zeff; even after the booster (another fifty micrograms, which I was eager to take, in hopes of going deeper and longer). I never achieved a transcendent, "non-dual" or "mystical-like" experience, and as I recapped the journey with Fritz the following morning, I registered a certain disappointment. But the novel plane of consciousness I'd spent a few hours wandering on had been interesting and pleasurable and, I think, useful to me. I would have to see if its effects endured, but it felt as though the experience had opened me up in unexpected ways.

Because the acid had not completely dissolved my ego, I never completely lost the ability to redirect the stream of my consciousness or the awareness it was in fact mine. But the stream itself felt distinctly different, less subject to will or outside interference. It reminded me of the pleasantly bizarre mental space that sometimes opens up at night in bed when we're poised between the states of being awake and falling asleep—so-called hypnagogic consciousness. The ego seems to sign off a few moments before the rest of the mind does, leaving the field of consciousness unsupervised and vulnerable to gentle eruptions of imagery and hallucinatory snatches of narrative. Imagine that state extended indefinitely, yet with some ability to direct your attention to this or that, as if in an especially vivid and absorbing daydream. Unlike a daydream, however, you are fully present to the contents of whatever narrative is unfolding, completely inside it and beyond the reach of distraction. I had little choice but to obey the daydream's logic, its ontological and epistemological rules, until, either by force of will or by the fresh notes of a new song, the mental channel would change and I would find myself somewhere else entirely.

This, I guess, is what happens when the ego's grip on the mind is relaxed but not eliminated, as a larger dose would probably have done. "For the moment that interfering neurotic who, in waking hours, tries to run the show, was blessedly out of the way," as Aldous Huxley put it in *The Doors of Perception*. Not entirely out of the way in my case, but the LSD had definitely muffled that controlling voice, and in that lightly regulated space all sorts of interesting things could bubble up, things that any self-respecting ego would probably have kept submerged.

I had had a psycholytic dose of LSD, one that allowed the patient to explore his psyche in an unconstrained but still deliberate manner while remaining sufficiently combobulated to talk about it. For me it felt less like a drug experience—the LSD feels completely transparent, with none of the physiological noise I associate with other psychoactive drugs—than a novel mode of cognition, falling somewhere between intellection and feeling. I had conjured several of the people closest to me, and in the presence of each of them had come stronger emotions than I had felt in some time. A dam had been breached, and the sensation of release felt wonderful. Too, a few genuine insights had emerged from these encounters, like the one about my father as a son, which turned on an act of imagination (of empathy) that even grown children seldom have sufficient distance to perform. During our integration session, Fritz mentioned that some people on LSD have an experience that in content and character is more like MDMA than a classic psychedelic trip; maybe what I had had was the MDMA session I'd had to pass up. The notion of a few years of psychotherapy condensed into several hours seemed about right, especially after Fritz and I spent that morning unpacking the scenes from my journey.

As I steered my rental car down the mountain and toward the airport for the flight home, I was relieved that the experience had been so benign (I had survived! Had roused no sleeping monsters in my unconscious!) and grateful it had been productive. All that day and well into the next, a high-pressure system of well-being dominated my psychological weather. Judith found me unusually chatty and available; my usual impatience was in abeyance, and I could outlast her at the table after dinner, being in no hurry to get up and do the dishes so I could move on to the next thing and then the thing after that. I guessed this was the afterglow I'd read about, and for a few days it cast a pleasantly theatrical light over everything, italicizing the ordinary in such a way as to make me feel uncommonly . . . appreciative.

It didn't last, however, and in time I grew disappointed that the experience hadn't been more transformative. I had been granted a taste of a slightly other way to be—less defended, I would say, and so more present. And now that I had acquainted myself with the territory and returned from this first foray more or less intact, I decided it was time to venture farther out.

Trip Two: Psilocybin

My second journey began around an altar, in the middle of a second-story loft in a suburb of a small city on the Eastern Seaboard. The altar was being prayed over by an attractive woman with long blond hair parted in the middle and high cheekbones that I mention only because they would later figure in her transformation into a Mexican Indian. Seated across the altar from me, Mary's eyes were closed as she recited a long and elaborate Native American prayer. She invoked in turn the power of each of the cardinal directions, the four elements, and the animal, plant, and mineral realms, the spirits of which she implored to help guide me on my journey.

My eyes were closed too, but now and again I couldn't resist peeking out to take in the scene: the squash-colored loft with its potted plants and symbols of fertility and female power; the embroidered purple fabric from Peru that covered the altar; and the collection of items arrayed across it, including an amethyst in the shape of a heart, a purple crystal holding a candle, little cups filled with water, a bowl holding a few rectangles of dark chocolate, the two "sacred items" she had asked me to bring (a bronze Buddha a close friend had brought back from a trip to the East; the psilocybin coin Roland Griffiths had given me at our first meeting), and, squarely before me, an antique plate decorated in a grandmotherly floral pattern that held the biggest psilocybin mushroom I had ever seen. It was hard to believe I was about to eat the whole thing.

The crowded altar also held a branch of sage and a stub of Palo Santo, a fragrant South American wood that Indians burn ceremonially, and the jet-black wing of a crow. At various points in the ceremony, Mary lit the sage and the Palo Santo, using the wing to "smudge" me with the smoke—guide the spirits through the space around my head. The wing made an otherworldly whoosh as she flicked it by my ear, the spooky sound of a large bird coming too close for comfort, or a dark spirit being shooed away from a body.

The whole thing must sound ridiculously hokey, I know, but the conviction Mary brought to the ceremony, together with the aromas of the burning plants and the sounds of the wing pulsing the air—

plus my own nervousness about the journey in store—cast a spell that allowed me to suspend my disbelief. I had decided to give myself up to this big mushroom, and for Mary, the guide to whom I had entrusted my psyche for this journey, ceremony counted for as much as chemistry. In this she was acting more like a shaman than a psychologist.

Mary had been recommended by a guide I'd interviewed on the West Coast, a rabbi who had taken an interest in my psychedelic education. Mary, who was my age, had trained with the eighty-something student of Timothy Leary whom I had interviewed and decided was a little too far out there for me. One might think the same of Mary, on paper, but something about her manner, her sobriety, and her evident compassion made me more comfortable in her presence.

Mary had practiced the whole grab bag of New Age therapies, from energy healing to spiritual psychology to family constellation therapy,* before being introduced to medicine work when she was fifty. ("It created the glue that brought together all this other work I'd been doing.") At the time, Mary had used a psychedelic only once and long ago: at her twenty-first birthday party while in college. A friend had given her a jar of honey laced with psilocybin mushrooms. Mary immediately went up to her room, ate two or three spoonfuls, "and had the most profound experience of being with God. I was God and God was me." Friends who had been partying downstairs came up to knock at her door, but Mary was gone.

As a child growing up outside Providence, Mary had been an enthusiastic Catholic, until "I realized I was a girl"—a fact that would

---

* Family constellation therapy, which was founded by a German therapist named Bert Hellinger, focuses on the hidden role of ancestors in shaping our lives and works to help us make peace with these ghostlike presences.

disqualify her from ever performing the ceremonies she cherished. Mary's religiosity lay dormant until that taste of honey, which "catapulted me into a huge change," she told me the first time we met. "I dropped into something I hadn't felt connected to since I was a little girl." The reawakening of her spiritual life led her onto the path of Tibetan Buddhism and eventually to take the vow of an initiate: "'To assist all sentient beings in their awakening and their enlightenment.' Which is still my vocation."

And now sitting before her in her treatment room was me, the next sentient being on deck, hoping to be wakened. I shared my intention: to learn what I could about myself and also about the nature of consciousness—my own but also its "transpersonal" dimension, if such a dimension exists.

"The mushroom teacher helps us to see who we really are," Mary said, "brings us back to our soul's purpose for being here in this lifetime." I can imagine how these words might sound to an outsider. But by now I was inured to the New Age lingo, perhaps because I had glimpsed the potential for something meaningful behind the well-worn words. I'd also been impressed by Mary's intelligence and her professionalism. In addition to having me consent to the standard "agreements" (bowing to her authority for the duration; remaining in the room until she gave me permission to leave; no sexual contact; and so on), she had me fill out a detailed medical form, a legal release, and a fifteen-page autobiographical questionnaire that took me the better part of a day to complete. All of which made me feel I was in good hands—even when those hands were flapping a crow's wing around my head.

Yet, as I sat there before the altar, it seemed doubtful I could choke down that whole mushroom. It had to be five or six inches long, with a cap the size of a golf ball. I asked her if I could crumble it into a glass of hot water, make a tea, and drink it.

"Better to be fully conscious of what you're doing," she said, "which is eating a mushroom that came from the earth, one bite at a time. Examine it first, closely, then start at the cap." She offered me a choice of honey or chocolate to help get it down; I went with the chocolate. Mary had told me that a friend of hers grows the psilocybin and had learned the craft years ago in a mushroom cultivation workshop taught by Paul Stamets. It seems there is only one or two degrees of separation between any two people in this world.

On the tongue, the mushroom was dry as the desert and tasted like earth-flavored cardboard, but alternating each bite with a nibble of the chocolate helped. Except for the gnarly bit at the very base of the stipe, I ate all of it, which amounted to two grams. Mary planned to offer me another two grams along the way, for a total of four. This would roughly approximate the dose being given to volunteers in the NYU and Hopkins trials and was equivalent to roughly three hundred micrograms of LSD—twice as much as I had taken with Fritz.

We chatted quietly for twenty minutes or so before Mary noticed my face was flushed and suggested I lie down and put on eyeshades. I chose a pair of high-tech black plastic ones, which in retrospect might have been a mistake. The perimeters were lined with soft black foam rubber, allowing the wearer to open his eyes to pitch darkness. Called the Mindfold Relaxation Mask, Mary told me, it had been expressly designed for this purpose by Alex Grey, the psychedelic artist.

As soon as Mary put on the first song—a truly insipid New Age composition by someone named Thierry David (an artist thrice nominated, I would later learn, in the category of Best Chill/Groove Album)—I was immediately propelled into a nighttime urban landscape that appeared to have been generated by a computer. Once again, sound begat space ("in the beginning was the note," I remember thinking, with a sense of profundity), and what I took to be Thierry's electronica conjured a depopulated futuristic city, with each note forming another soft black stalagmite or stalactite that together resembled the high-relief soundproofing material used to line recording studios. (The black foam forming this high-relief landscape, I realized later, was the same material lining my eyeshades.) I moved effortlessly through this digital nightscape as if within the confines of a video-game dystopia. Though the place wasn't particularly frightening, and it had a certain sleek beauty, I hated being in it and wished to be somewhere else, but it went on seemingly forever and for hours, with no way out. I told Mary I didn't like the electronic music and asked her to put on something else, but though the feeling tone changed with the new music, I was still stuck in this sunless computer world. Why, oh, why couldn't I be outside! In nature? Because I had never much enjoyed video games, this seemed cruel, an expulsion from the garden: no plants, no people, no sunlight.

Not that the computer world wasn't an interesting place to explore. I watched in awe as, one by one, musical notes turned into palpable forms before my eyes. Annoying music was the presiding deity of the place, the generative force. Even the most spa-appropriate New Age composition had the power to spawn fractal patterns in space that grew and branched and multiplied to infinity. Weirdly, everything in my visual field was black, but in so many different shades that it was easy to see. I was traversing a world generated by mathematical algorithms, and this gave it a certain alienated, lifeless beauty. But whose world was it? Not mine, and I began to wonder, whose brain am I in? (Please, not Thierry David's!)

"This could easily take a terrifying turn," it occurred to me, and with that a dim tide of anxiety began to build. Recalling the flight instructions, I told myself there was nothing to do but let go and surrender to the experience. *Relax and float downstream.* This was not at all like previous trips, which had left me more or less the captain of my attention, able to direct it this way or that and change the mental channel at will. No, this was more like being strapped into the front car of a cosmic roller coaster, its heedless headlong trajectory determining moment by moment what would appear in my field of consciousness.

Actually, this is not completely accurate: all I had to do was to remove my eyeshades and reality, or at least something loosely based on it, would reconstitute itself. This is what I now did, partly to satisfy myself that the world was still existing but mostly because I badly had to pee.

Sunlight and color flooded my eyes, and I drank it in greedily, surveying the room for the welcome signifiers of non-digital reality: walls, windows, plants. But all of it appeared in a new aspect: jeweled with light. I realized I should probably put on my glasses, which partly domesticated the scene, but only partly: objects continued to send their sparkles of light my way. I got up carefully from the mattress, first onto one knee, then, unsteadily, onto my feet. Mary took me by the elbow, geriatrically, and together we made the journey across the room. I avoided looking at her, uncertain what I might see in her face or betray in mine. At the bathroom door she let go of my elbow.

Inside, the bathroom was a riot of sparkling light. The arc of water I sent forth was truly the most beautiful thing I had ever seen, a waterfall of diamonds cascading into a pool, breaking its surface into a billion clattering fractals of light. This went on for a pleasant eternity. When I was out of diamonds, I went to the sink and splashed my face with water, making sure not to catch sight of myself in the mirror, which seemed like a psychologically risky thing to do. I made my unsteady way back to the mattress and lay down.

Speaking softly, Mary asked if I wanted a booster. I did and sat up to receive it. Mary was squatting next to me, and when I finally looked up into her face, I saw she had turned into María Sabina, the Mexican *curandera* who had given psilocybin to R. Gordon Wasson in that dirt basement in Huautla de Jiménez sixty years ago. Her hair was black, her face, stretched taut over its high cheekbones, was anciently weathered, and she was wearing a simple white peasant dress. I took the dried mushroom from the woman's wrinkled brown hand and looked away as I chewed. I didn't think I should tell Mary what had happened to her. (Later, when I did, she was flattered: María Sabina was her hero.)

○   ○   ○

BUT THERE WAS SOMETHING I needed to do before putting my eyeshades back on and going back under, a little experiment I had told Mary I wanted to perform on myself during my trip. I wasn't sure if in my condition I could pull it off, but I'd found that even in the middle of the journey it was possible to summon oneself to a semblance of normality for a few moments at a time.

Loaded on my laptop was a brief video of a rotating face mask, used in a psychological test called the binocular depth inversion illusion. As the mask rotates in space, its convex side turning to reveal its concave back, something remarkable happens: the hollow mask appears to pop out to become convex again. This is a trick performed by the mind, which assumes all faces to be convex, and so automatically corrects for the seeming error—unless, as a neuroscientist had told me, one was under the influence of a psychedelic.

This auto-correct feature is a hallmark of our perception, which in the sane, adult mind is based as much on educated guesswork as the raw data of the senses. By adulthood, the mind has gotten very good at observing and testing reality and developing confident predictions about it that optimize our investments of energy (mental and otherwise) and therefore our survival. So rather than starting from scratch to build a new perception from every batch of raw data delivered by the senses, the mind jumps to the most sensible conclusion based on past experience combined with a tiny sample of that data. Our brains are prediction machines optimized by experience, and when it comes to faces, they have boatloads of experience: faces are always convex, so this hollow mask must be a prediction error to be corrected.

These so-called Bayesian inferences (named for Thomas Bayes, the eighteenth-century English philosopher who developed the mathematics of probability, on which these mental predictions are based) serve us well most of the time, speeding perception while saving effort and energy, but they can also trap us in literally preconceived images of reality that are simply false, as in the case of the rotating mask.

Yet it turns out that Bayesian inference breaks down in some people: schizophrenics and, according to some neuroscientists, people on high doses of psychedelics drugs, neither of whom "see" in this predictive or conventionalized manner. (Nor do young children, who have yet to build the sort of database necessary for confident predictions.) This raises an interesting question: Is it possible that the perceptions of schizophrenics, people tripping on psychedelics, and young children are, at least in certain instances, more accurate— less influenced by expectation and therefore more faithful to reality—than those of sane and sober adults?

Before we started, I had cued up the video on my laptop, and now I clicked to run it. The mask on the screen, gray against a black ground, was clearly the product of computer animation and was uncannily consistent with the visual style of the world I'd been in. (During my integration session with Mary the next day, she suggested that it might have been this image on my laptop that had conjured the computer world and trapped me in it. Could there be a better demonstration of the power of set and setting?) As the convex face rotated to reveal its concave back, the mask popped back out, only a bit more slowly than it did before I ate the mushroom. Evidently, Bayesian inference was still operational in my brain. I'd try again later.

○   ○   ○

WHEN I PUT MY EYESHADES back on and lay down, I was disappointed to find myself back in computer world, but something had changed, no doubt the result of the stepped-up dose. Whereas before I navigated this landscape as myself, taking in the scene from a perspective recognizable as my own, with my attitudes intact (highly critical of the music, for instance, and anxious about what demons might appear), now I watched as that familiar self began to fall apart before my eyes, gradually at first and then all at once.

"I" now turned into a sheaf of little papers, no bigger than Post-its, and they were being scattered to the wind. But the "I" taking in this seeming catastrophe had no desire to chase after the slips and pile my old self back together. No desires of any kind, in fact. Whoever I now was was fine with whatever happened. *No more ego?* That was okay, in fact the most natural thing in the world. And then I looked and saw myself out there again, but this time spread over the landscape like paint, or butter, thinly coating a wide expanse of the world with a substance I recognized as me.

But who was this "I" that was able to take in the scene of its own dissolution? Good question. It wasn't *me*, exactly. Here, the limits of our language become a problem: in order to completely make sense of the divide that had opened up in my perspective, I would need a whole new first-person pronoun. For what was observing the scene was a vantage and mode of awareness entirely distinct from my accustomed self; in fact I hesitate to use the "I" to denote the presiding awareness, it was so different from my usual first person. Where that self had always been a subject encapsulated in this body, this one seemed unbounded by *any* body, even though I now had access to its perspective. That perspective was supremely indifferent, neutral on all questions of interpretation, and unperturbed even in the face of what should by all rights have been an unmitigated personal disaster. Yet the "personal" had been obliterated. Everything I once was and called me, this self six decades in the making, had been liquefied and dispersed over the scene. What had always been a thinking, feeling, perceiving subject based in here was now an object out there. I was paint!

The sovereign ego, with all its armaments and fears, its backward-looking resentments and forward-looking worries, was simply no more, and there was no one left to mourn its passing. Yet something had succeeded it: this bare disembodied awareness, which gazed upon the scene of the self's dissolution with benign indifference. I was present to reality but as something other than my self. And although there was no self left to feel, exactly, there was a feeling tone, which was calm, unburdened, content. There was life after the death of the ego. This was big news.

When I think back on this part of the experience, I've occasionally wondered if this enduring awareness might have been the "Mind at Large" that Aldous Huxley described during his mescaline trip in 1953. Huxley never quite defined what he meant by the term—except to speak of "the totality of the awareness belonging to Mind at Large"—but he seems to be describing a universal, shareable form of consciousness unbounded by any single brain. Others have called it cosmic consciousness, the Oversoul, or Universal Mind. This is supposed to exist outside our brains—as a property of the universe, like light or gravity, and just as pervasive. Constitutive too. Certain individuals at certain times gain access to this awareness, allowing them to perceive reality in its perfected light, at least for a time.

Nothing in my experience led me to believe this novel form of consciousness originated outside me; it seems just as plausible, and surely more parsimonious, to assume it was a product of my brain, just like the ego it supplanted. Yet this by itself strikes me as a remarkable gift: that we can let go of so much—the desires, fears, and defenses of a lifetime!—without suffering complete annihilation. This might not come as a surprise to Buddhists, transcendentalists, or experienced meditators, but it was sure news to me, who has never felt anything but identical to my ego. Could it be there is another ground on which to plant our feet? For the first time since embarking on this project, I began to understand what the volunteers in the cancer-anxiety trials had been trying to tell me: how it was that a psychedelic journey had granted them a perspective from which the very worst life can throw at us, up to and including death, could be regarded objectively and accepted with equanimity.

○ ○ ○

ACTUALLY, this understanding arrived a little later, during the last part of my psilocybin trip, when the journey took a darker turn. After spending an unknown number of hours in computer world—for time was completely lost on me—I registered the desire to check back in on reality, and to pee again. Same deal: Mary guided me to the bathroom by the elbow, geriatrically, and left me there to produce another spectacular crop of diamonds. But this time I dared to look in the mirror. What looked back at me was a human skull, but for the thinnest, palest layer of skin stretched over it, tight as a drum. The bathroom was decorated in a Mexican folk art theme, and the head/skull immediately put me in mind of the Day of the Dead. With its deep sockets and lightning bolt of vein zigzagging down its temple on one side, I recognized this ashen head/skull as my own but at the same time as my dead grandfather's.

This was surprising, if only because Bob, my father's father, is not someone with whom I ever felt much in common. In fact I loved him for all the ways he seemed *unlike* me—or anyone else I knew. Bob was a preternaturally sunny and seemingly uncomplicated man incapable of thinking ill of anyone or seeing evil in the world. (His wife, Harriet, amply compensated for his generosity of spirit.) Bob had a long career as a liquor salesman, making the weekly rounds of the nightclubs in Times Square for a company that everyone but he knew was owned by the mob. Upon reaching the age I am now, he retired to become a painter of lovely naive landscapes and abstractions in spectacular colors; I'd brought one of them with me to Mary's room, along with a watercolor of Judith's. Bob was a genuinely happy, angst-free man who lived to be ninety-six, his paintings becoming ever more colorful, abstract, and free toward the end.

To see him so vividly in my reflection was chilling. A few years before, visiting Bob in the nursing home in the Colorado desert where he would soon die, I'd watched what had been a fit and vigorous man (it had been his habit to stand on his head every day well into his eighties) contract into a parenthesis of skin and bones marooned in a tiny bed. The esophageal muscles required to swallow had given out, and he was tethered to a feeding tube. By then, his situation was pitiful in so many respects, but for some reason I fixed on the fact that never again would a taste of food ever cross his lips.

I splashed cold water on our joint face and made my unsteady way back to Mary.

Risking another glance at her, this time I was rewarded by the sight of a ravishing young woman, blond once again but now in the full radiance of youth. Mary was so beautiful I had to look away.

She gave me another small mushroom—gram number four—and

a piece of chocolate. Before I put on my eyeshade, I attempted to conduct the rotating mask test a second time . . . and it was a complete bust, neither confirming nor disproving the hypothesis. As the mask began to rotate, gradually bringing its back side into view, the whole thing dissolved into a gray jelly that slid down the screen of my laptop before I could determine whether the melting mask I was watching was convex or concave. So much for conducting psychological experiments while tripping.

I put on my eyeshades and sank back down into what now became a cracked and parched desert landscape dense with artifacts and images of death. Bleached skulls and bones and the faces of the familiar dead passed before me, aunts and uncles and grandparents, friends and teachers and my father-in-law—with a voice telling me I had failed to properly mourn all of them. It was true. I had never really reckoned the death of *any*one in my life; something had always gotten in the way. I could do it here and now and did.

I looked hard at each of their faces, one after another, with a pity that seemed bottomless but with no fear whatsoever. Except once, when I came to my aunt Ruthellen and watched, horrified, as her face slowly transformed into Judith's. Ruthellen and Judith were both artists, and both had been diagnosed with breast cancer around the same time. The cancer had killed Ruthellen and spared Judith. So what was Judith doing down here among the unmourned dead? Had I been defending myself against that possibility all this time? Heart wide open, defenses melting, the tears began to flow.

o   o   o

I'VE LEFT OUT one important part of my journey to the underworld: the soundtrack. Before going back under for this last passage, I had asked Mary to please stop playing spa music and put on something classical. We settled on the second of Bach's unaccompanied cello suites, performed by Yo-Yo Ma. The suite in D minor is a spare and mournful piece that I'd heard many times before, often at funerals, but until this moment I had never truly listened to it.

Though "listen" doesn't begin to describe what transpired between me and the vibrations of air set in motion by the four strings of that cello. Never before has a piece of music pierced me as deeply as this one did now. Though even to call it "music" is to diminish what now began to flow, which was nothing less than the stream of human consciousness, something in which one might glean the very meaning of life and, if you could bear it, read life's last chapter. (A question formed: Why don't we play music like this at births as well as funerals? And the answer came immediately: there is too much life-already-lived in this piece, and poignancy for the passing of time that no birth, no beginning, could possibly withstand it.)

Four hours and four grams of magic mushroom into the journey, this is where I lost whatever ability I still had to distinguish subject from object, tell apart what remained of me and what was Bach's music. Instead of Emerson's transparent eyeball, egoless and one with all it beheld, I became a transparent ear, indistinguishable from the stream of sound that flooded my consciousness until there was nothing else in it, not even a dry tiny corner in which to plant an I and observe. Opened to the music, I became first the strings, could feel on my skin the exquisite friction of the horsehair rubbing over me, and then the breeze of sound flowing past as it crossed the lips of the instrument and went out to meet the world, beginning its lonely transit of the universe. Then I passed down into the resonant black well of space inside the cello, the vibrating envelope of air formed by the curves of its spruce roof and maple walls. The instrument's wooden interior formed a mouth capable of unparalleled eloquence—indeed, of articulating everything a human could conceive. But the cello's interior also formed a room to write in and a skull in which to think and I was now it, with no remainder.

So I became the cello and mourned with it for the twenty or so minutes it took for that piece to, well, change everything. Or so it seemed; now, its vibrations subsiding, I'm less certain. But for the duration of those exquisite moments, Bach's cello suite had had the unmistakable effect of reconciling me to death—to the deaths of the people now present to me, Bob's and Ruthellen's and Roy's, Judith's father's, and so many others, but also to the deaths to come and to my own, no longer so far off. Losing myself in this music was a kind of practice for that—for losing myself, period. Having let go of the rope of self and slipped into the warm waters of this worldly beauty—Bach's sublime music, I mean, and Yo-Yo Ma's bow caressing those four strings suspended over that envelope of air—I felt as though I'd passed beyond the reach of suffering and regret.

o   o   o

THAT WAS MY PSILOCYBIN JOURNEY, as faithfully as I can recount it. As I read those words now, doubt returns in full force: "Fool, you were on drugs!" And it's true: you *can* put the experience in that handy box and throw it away, never to dwell on it again. No doubt this has been the fate of countless psychedelic journeys that their travelers didn't quite know what to do with, or failed to make sense of. Yet though it is true that a chemical launched me on this journey, it is also true that everything I experienced I experienced: these are events that took place in my mind, psychological facts that were neither weightless nor evanescent. Unlike most dreams, the traces these experiences inscribed remain indelible and accessible.

The day after my journey I was glad for the opportunity to return to Mary's room for a couple of hours of "integration." I hoped to make sense of what happened by telling the story of my trip and hearing her thoughts about it. What you've just read is the result, and the beneficiary, of that work, for immediately after the journey I was much more confused by it than I am now. What now reads like a reasonably coherent narrative highlighting certain themes began as a jumble of disjointed images and shards of sense. To put words to an experience that was in fact ineffable at the time, and then to shape them into sentences and then a story, is inevitably to do it a kind of violence. But the alternative is, literally, unthinkable.

Mary had taken apart the altar, but we sat in the same chairs, facing each other across a small table. Twenty-four hours later, what had I learned? That I had had no reason to be afraid: no sleeping monsters had awakened in my unconscious and turned on me. This was a deep fear that went back several decades, to a terrifying moment in a hotel room in Seattle when, alone and having smoked too much cannabis, I had had to marshal every last ounce of will to keep myself from doing something deeply crazy and irrevocable. But here in this room I had let down my guard completely, and nothing terrible had happened. The serpent of madness that I worried might be waiting had not surfaced or pulled me under. Did this mean it didn't exist, that I was psychologically sturdier than I believed? Maybe *that's* what the episode with Bob was all about: maybe I was more like him than I knew, and not nearly as deep or complicated as I liked to think. (Can a recognition of one's shallowness qualify as a profound insight?) Mary wasn't so sure: "You bring a different self to the journey every time." The demons might rouse themselves the next time.

That I could survive the dissolution of my ego without struggle or turning into a puddle was something to be grateful for, but even better was the discovery that there might be another vantage—one less neurotic and more generous—from which to take in reality.

"That alone seems worth the price of admission," Mary offered, and I had to agree. Yet, twenty-four hours later, my old ego was back in uniform and on patrol, so what long-term good was that beguiling glimpse of a loftier perspective? Mary suggested that having had a taste of a different, less defended way to be, I might learn, through practice, to relax the ego's trigger-happy command of my reactions to people and events. "Now you have had an experience of another way to react—or not react. That can be cultivated." Meditation, she suggested, was one way to do that.

It is, I think, precisely this perspective that had allowed so many of the volunteers I interviewed to overcome their fears and anxieties, and in the case of the smokers, their addictions. Temporarily freed from the tyranny of the ego, with its maddeningly reflexive reactions and its pinched conception of one's self-interest, we get to experience an extreme version of Keats's "negative capability"—the ability to exist amid doubts and mysteries without reflexively reaching for certainty. To cultivate this mode of consciousness, with its exceptional degree of selflessness (literally!), requires us to transcend our subjectivity or—it comes to the same thing—widen its circle so far that it takes in, besides ourselves, other people and, beyond that, all of nature. Now I understood how a psychedelic could help us to make precisely that move, from the first-person singular to the plural and beyond. Under its influence, a sense of our interconnectedness—*that* platitude—is felt, becomes flesh. Though this perspective is not something a chemical can sustain for more than a few hours, those hours *can* give us an opportunity to see how it might go. And perhaps to practice being there.

I left Mary's loft in high spirits, but also with the feeling I was holding on to something precious by the thinnest, most tenuous of threads. It seemed doubtful I could maintain my grip on this outlook for the rest of the day, much less the rest of my life, but it also seemed worth trying.

Trip Three: 5-MeO-DMT (or, The Toad)

Yes, "the toad," or to be more precise, the smoked venom of the Sonoran Desert toad (*Incilius alvarius*), also called the Colorado River toad, which contains a molecule called 5-MeO-DMT that is one of the most potent and fast-acting psychotropic drugs there is. No, I had never heard of it either. It is so obscure, in fact, that the federal government did not list 5-MeO-DMT as a controlled substance until 2011.

The opportunity to smoke the toad popped up suddenly, giving me very little time to decide if doing so was crazy or not. I got a call from one of my sources, a woman who was training to become a certified psychedelic guide, inviting me to meet her friend Rocío, a thirty-five-year-old Mexican therapist whom she described as "probably the world's leading expert on the toad." (Though how intense, really, could the competition for that title be?) Rocío is from the state of Sonora, in northern Mexico, where she collects the toads and milks their venom; she administers the medicine to people both in Mexico, where its legal status is gray, and in the United States, where it isn't. (It doesn't appear to be on the official radar, however.)

Rocío worked in a clinic in Mexico that treated drug addicts with a combination of *iboga*, a psychedelic plant from Africa, and 5-MeO-DMT—apparently with striking rates of success. In recent years, she's become the Johnny Appleseed of toad, traveling all over North America with her capsules of crystallized venom and her vaporizer. As my circle of psychonauts expanded, most anyone I met who'd had an encounter with the toad had been introduced to it by Rocío.

The first time I met Rocío, at a small dinner organized by our mutual friend, she told me about the toad and what I might expect from it. Rocío was petite, pretty, and fashionably dressed, her shoulder-length black hair cut to frame her face with bangs. She has an easy smile that brings out a dimple on one cheek. Not at all what I expected, Rocío looked less the part of a shaman or *curandera* than that of an urban professional.

After going to college and working for a few years in the United States, five years ago Rocío found herself back at home in Mexico living with her parents and without direction. Online, she found a manual about the toad, which she learned was native to the local desert. (Its habitat extends the length of the Sonoran Desert north into Arizona.) Nine months of the year, the toad lives underground, protected from the desert sun and heat, but when the winter rains come, it emerges at night from its burrow for a brief orgy of eating and copulation. Following the instructions spelled out in the manual, Rocío strapped on a headlamp and went hunting for toads.

"They're not very hard to catch," she told me. "They freeze in the beam of light so you can just grab them." The toads, which are warty, sand colored, and roughly the size of a man's hand, have a large gland on each side of their necks, and smaller ones on their legs. "You gently squeeze the gland while holding a mirror in front of it to catch the spray." The toad is apparently none the worse for being milked. Overnight, the venom dries on the glass, turning into flaky crystals the color of brown sugar.

In its natural state, the venom is toxic—a defense chemical sprayed by the toad when it feels threatened. But when the crystals are volatilized, the toxins are destroyed, leaving behind the 5-MeO-DMT. Rocío vaporizes the crystals in a glass pipe while the recipient inhales; before you've had a chance to exhale, you are gone. "The toad comes on quickly, and at first it can be unbelievably intense." I noticed that

Rocío personified the toad and seldom called the medicine by its molecular name. "Some people remain perfectly still. Other people scream and flail, especially when the toad brings out traumas, which it can do. A few people will vomit. And then after twenty or thirty minutes, the toad is all done and it leaves."

My first instinct when facing such a decision is to read as much about it as I can, and later that night Rocío e-mailed me a few articles. But the pickings were slim. Unlike most other psychedelics, which by now have been extensively studied by scientists and, in many cases, in use for hundreds if not thousands of years, the toad has been known to Western science only since 1992. That's when Andrew Weil and Wade Davis published a paper called "Identity of a New World Psychoactive Toad." They had been inspired to look for such a fantastical creature by the images of frogs in Mayan art. But the only psychoactive toad they could find lives far to the north of Mayan civilization. It's possible that these toads became an item of trade, but as yet there is no proof that the practice of smoking toad venom has any antiquity whatsoever. However, 5-MeO-DMT also occurs in a handful of South American plants, and there are several Amazonian tribes who pound these plants into a snuff for use in shamanic rituals. Among some of these tribes, these snuffs are known as the "semen of the sun."

I couldn't find much in the way of solid medical information about potential side effects or dangerous drug interactions; little research has been done. What I did find were plenty of trip reports online, and many of these were terrifying. I also learned there was someone in town, a friend of a friend I had met a few times at dinner parties, who had tried 5-MeO-DMT—not the toad but a synthetic version of the active ingredient. I took her out to lunch to see what I could learn.

"This is the Everest of psychedelics," she began, portentously, putting a steadying hand on my forearm. Olivia is in her early fifties, a management consultant with a couple of kids; I had vaguely known she was into Eastern religion but had no idea she was a psychonaut, too.

"You need to be prepared." Over grilled cheeses, she described a harrowing onset. "I was shot out into an infinite realm of pure being. There were no figures in this world, no entities of any kind, just pure being. And it was huge; I didn't know what infinity was before this. But it was a two-dimensional realm, not three, and after the rush of liftoff, I found myself installed in this infinite space as a star. I remember thinking, if this is death, I'm fine with it. It was . . . bliss. I had the feeling—no, the *knowledge*—that every single thing there is is made of love.

"After what seemed like an eternity but was probably only minutes, you start to reassemble and come back into your body. I had the thought, 'There are children to raise. And there is an infinite amount of time to be dead.'"

I asked her the question that gnawed at me whenever someone recounted such a mystical experience: "How can you be sure this was a genuine spiritual event and not just a drug experience?"

"It's an irrelevant question," she replied coolly. "This was something being revealed to me."

There it was: the noetic sense William James had described as a mark of the mystical experience. I envied Olivia's certainty. Which I suppose is the reason I decided I would smoke the toad.

o o o

THE NIGHT BEFORE my date with Rocío was, predictably, sleepless. Yes, I'd come through these first two trips intact, grateful, even, for having gone on them, and had come away with the idea I was stronger, physically and mentally, than I had previously thought. But now all the old fears rushed back, assailing me through the long fit-ful night. *Everest!* Could my heart take the intensity of those first harrowing moments of ascent? What were the chances I'd go mad? Slim, perhaps, but surely not zero. So was this an absolutely insane thing to do? On the plus side, I figured, whatever happened, it would all be over in half an hour. On the negative side, *every*thing might be over in half an hour.

As the sun came up, I decided I would decide when I got there. Rocío, whom I'd made aware of my trepidations, had offered to let me watch her work with someone else before it was my turn. This proved reassuring, as she knew it would. The guy before me, a su-premely low-affect college student who had done the toad once be-fore, took a puff from Rocío's pipe, lay back on a mattress, and embarked on what appeared to be a placid thirty-minute nap, during which he exhibited no signs of distress, let alone existential terror. After it was over, he seemed perfectly fine. A great deal had gone on in his mind, he indicated, but from the looks of it, his body had scarcely been perturbed. Okay then. Death or madness seemed much less likely. I could do this.

After positioning me on the mattress just so, Rocío had me sit up while she loaded a premeasured capsule of the crystals into a glass vial that she then screwed onto the barrel of the pipe. She asked me to give thanks to the toad and think about my intention. (Something fairly generic about learning whatever the toad had to teach me.) Rocío lit a butane flame underneath the vial and instructed me to draw on the pipe in short sips of air as the white smoke swirled and then filled the glass. "Then one big final draw that I want you to hold as long as you can."

I have no memory of ever having exhaled, or of being lowered onto the mattress and covered with a blanket. All at once I felt a tremendous rush of energy fill my head accompanied by a punishing roar. I managed, barely, to squeeze out the words I had prepared, "trust" and "surrender." These words became my mantra, but they seemed utterly pathetic, wishful scraps of paper in the face of this category 5 mental storm. Terror seized me—and then, like one of those flimsy wooden houses erected on Bikini Atoll to be blown up in the nuclear tests, "I" was no more, blasted to a confetti cloud by an explosive force I could no longer locate in my head, because it had exploded that too, expanding to become all that there was. Whatever this was, it was not a hallucination. A hallucination implies a reality and a point of reference and an entity to have it. None of those things remained.

Unfortunately, the terror didn't disappear with the extinction of my "I." Whatever allowed me to register this experience, the postegoic awareness I'd first experienced on mushrooms, was now consumed in the flames of terror too. In fact every touchstone that tells us "I exist" was annihilated, and yet I remained conscious. "Is this what death feels like? Could this be it?" That was the thought, though there was no longer a thinker to have it.

Here words fail. In truth, there were no flames, no blast, no thermonuclear storm; I'm grasping at metaphor in the hope of forming some stable and shareable concept of what was unfolding in my mind. In the event, there was no coherent thought, just pure and terrible sensation. Only afterward did I wonder if this was what the mystics call the *mysterium tremendum*—the blinding unendurable mystery (whether of God or some other Ultimate or Absolute) before which humans tremble in awe. Huxley described it as the fear "of being overwhelmed, of disintegrating under a pressure of reality greater than a mind, accustomed to living most of the time in a cosy world of symbols, could possibly bear."

Oh, to be back in the cozy world of symbols!

After the fact I kept returning to one of two metaphors, and while they inevitably deform the experience,* as any words or metaphors or symbols must, they at least allow me to grasp hold of a shadow of it and, perhaps, share it. The first is the image of being on the outside of a rocket after launch. I'm holding on with both hands, legs clenched around it, while the rapidly mounting g-forces clutch at my flesh, pulling my face down into a taut grimace, as the great cylinder rises through successive layers of clouds, exponentially gaining speed and altitude, the fuselage shuddering on the brink of self-destruction as it strains to break free from Earth's grip, while the friction it generates as it crashes through the thinning air issues in a deafening roar.

It was a little like *that*.

The other metaphor was the big bang, but the big bang run in reverse, from our familiar world all the way back to a point before there was anything, no time or space or matter, only the pure unbounded energy that was all there was then, before an imperfection, a ripple in its waveform, caused the universe of energy to fall into time, space, and matter. Rushing backward through fourteen billion years, I watched the dimensions of reality collapse one by one until there was nothing left, not even being. Only the all-consuming roar.

It was just horrible.

And then suddenly the devolution of everything into the noth-

---

* Henri Michaux, a contemporary of Huxley's who also wrote about his psychedelic experiences, took a very different tack, refusing the offer of metaphor to make sense of something he believed was beyond comprehension. In his book *Miserable Miracle*, he aimed to be "attentive to what's going on—as it *is*—without trying to deform it and imagine it otherwise in order to make it more interesting to me." Or sensible to his readers: the book is intermittently brilliant but for long stretches unreadable. "I had no longer any authority over words. I no longer knew how to manage them. Farewell to writing!" I know what he means, but I've elected to resist, even if that means tolerating some measure of deformation in my account.

ingness of pure force reverses course. One by one, the elements of our universe begin to reconstitute themselves: the dimensions of time and space returned first, blessing my still-scattered confetti brain with the cozy coordinates of place; *this is somewhere!* And then I slipped back into my familiar "I" like an old pair of slippers and soon after felt something I recognized as my body begin to reassemble. The film of reality was now running in reverse, as if all the leaves that the thermonuclear blast had blown off the great tree of being and scattered to the four winds were suddenly to find their way back, fly up into the welcoming limbs of reality, and reattach. The order of things was being restored, me notably included. *I was alive!*

The descent and reentry into familiar reality was swifter than I expected. Having undergone the shuddering agony of launch, I had expected to be deposited, weightless, into orbit—my installation in the firmament as a blissed-out star! Alas. Like those first Mercury astronauts, my flight remained suborbital, describing an arc that only kissed the serenity of infinite space before falling back down to Earth.

And yet as I felt myself reconstitute as a self and then a body, something for which I now sought confirmation by running my hands along my legs and squirming beneath the blanket, I felt ecstatic—as happy as I can remember ever feeling. But this ecstasy was not sui generis, not exactly. It was more like the equal and opposite reaction to the terror I had just endured, less of a divine gift than the surge of pleasure that comes from the cessation of unendurable pain. But a sense of relief so vast and deep as to be cosmic.

With the rediscovery of my body, I felt an inexplicable urge to lift my knees, and as soon as I raised them, I felt something squeeze out from between my legs, but easily and without struggle or pain. It was a boy: the infant me. That seemed exactly right: having died, I was now being reborn. Yet as soon as I looked closely at this new being, it morphed smoothly into Isaac, my son. And I thought, how fortunate—how astounding!—for a father to experience the perfect physical intimacy that heretofore only mothers have ever had with their babies. Whatever space had ever intervened between my son and me now closed, and I could feel the warm tears sliding down my cheeks.

Next came an overwhelming wave of gratitude. For what? For once again existing, yes, for the existence of Isaac and Judith too, but also for something even more fundamental: I felt for the first time gratitude for the very fact of being, that there is anything whatsoever. Rather than being necessarily the case, this now seemed quite the miracle, and something I resolved never again to take for granted. Everybody gives thanks for "being alive," but who stops to offer thanks for the bare-bones gerund that comes before "alive"? I had just come from a place where being was no more and now vowed never to forget what a gift (and mystery) it is, that there is something rather than nothing.

I had entered a familiar and more congenial mental space, one in which I was still tripping but could put together thoughts and direct them here or there. (I make no claims as to their quality.) Before I drew the smoke into my lungs, Rocío had asked me, as she asks everyone who meets the toad, to search the experience for a "peace offering"—some idea or resolution I could bring back and put to good use in my life. Mine, I decided, had to do with this question of being and what I took to be its opposite term, "doing." I meditated on this duality, which came to seem momentous, and concluded that I was too much occupied with the latter term in my life and not enough with the former.

True, one had to favor doing in order to get anything done, but wasn't there also a great virtue and psychic benefit in simply being? In contemplation rather than action? I decided I needed to practice being with stillness, being with other people as I find them (imperfect), and being with my own unimproved self. To savor whatever is at this very moment, without trying to change it or even describe it. (Huxley struggled with the same aspiration during his mescaline journey: "If one always saw like this, one would never want to do anything else.") Even now, borne along on this pleasant contemplative stream, I had to resist the urge to drag myself onto shore and tell Rocío about my big breakthrough. No! I had to remind myself: just be with it.

Judith and I had had a fight the previous night that, I realized, turned on this distinction, and on my impatience with being. She was complaining about something she doesn't like about her life, and rather than simply commiserate, being with her and her dilemma, I immediately went to the checklist of practical things she might do to fix it. But this was not at all what she wanted or needed, and she got angry. Now I could see with perfect clarity why my attempt to be helpful had been so hurtful.

So that was my peace offering: to be more and do less. But as soon as I put it that way, I realized there was a problem—a big problem, in fact. For wasn't the very act of resolving to favor being a form of doing? A betrayal of the whole idea? A true connoisseur of being would never dream of making resolutions! I had tied myself up in a philosophical knot, constructed a paradox or koan I was clearly not smart enough or sufficiently enlightened to untangle. And so what had begun as one of the most shattering experiences of my life ended half an hour later with a wan smile.

o   o   o

EVEN NOW, many months later, I still don't know exactly what to make of this last trip. Its violent narrative arc—that awful climax followed so swiftly by such a sweet denouement—upended the form of a story or journey. It lacked the beginning, middle, and end that all my previous trips had had and that we rely on to make sense of experience. That and its mind-bending velocity made it difficult to extract much information or knowledge from the journey, except for the (classic) psychedelic platitude about the importance of being. (A few days after my encounter with the toad, I happened on an old e-mail from James Fadiman that ended, uncannily, with these words, which you should picture arranged on the screen like a poem: "I hope whatever you're doing, / you're stopping now and then / and / not doing it at all.")

The integration had been cursory, leaving me to puzzle out the toad's teachings, such as they were, on my own. Had I had any sort of a spiritual or mystical experience? Or was what took place in my mind merely the epiphenomenon of these strange molecules? (Or was it both?) Olivia's words echoed: "It's an irrelevant question. This was something being revealed to me." What, if anything, had been revealed to me?

Not sure exactly where to begin, I realized it might be useful to measure my experiences against those of the volunteers in the Hopkins and NYU studies. I decided to fill out one of the Mystical Experience Questionnaires (MEQs)* that the scientists had their subjects complete, hoping to learn if mine qualified.

The MEQ asked me to rank a list of thirty mental phenomena—thoughts, images, and sensations that psychologists and philosophers regard as typical of a mystical experience. (The questionnaire draws on the work of William James, W. T. Stace, and Walter Pahnke.) "Looking back on the entirety of your session, please rate the degree to which at any time . . . you experienced the following phenomena"

---

* Specifically, I took the Revised Mystical Experience Questionnaire, or MEQ30.

using a six-point scale. (From zero, for "none at all," to five, for extreme: "more than any other time in my life.")

Some items were easy to rate: "Loss of your usual sense of time." Check; five. "Experience of amazement." Uh-huh. Another five. "Sense that the experience cannot be described adequately in words." Yup. Five again. "Gain of insightful knowledge experienced at an intuitive level." Hmmm. I guess the platitude about being would qualify. Maybe a three? But I was unsure what to do with this one: "Feeling that you experienced eternity or infinity." The language implies something more positive than what I felt when time vanished and terror took hold; NA, I decided. The "experience of the fusion of your personal self into a larger whole" also seemed like an overly nice way to put the sensation of becoming one with a nuclear blast. It seemed less fusion than fission, but okay. I gave it a four.

And what to do with this one? "Certainty of encounter with ultimate reality (in the sense of being able to 'know' and 'see' what is really real at some point in your experience)." I might have emerged from the experience with certain convictions (the one about being and doing, say), but these hardly seemed like encounters with "ultimate reality," whatever that is. Similarly, a few other items made me want to throw up my hands: "Feeling that you experienced something profoundly sacred and holy" (No) or "Experience of the insight 'all is One'" (Yes, but not in a good way; in the midst of that all-consuming mind storm, there was nothing I missed *more* than differentiation and multiplicity). Struggling to assign ratings to a handful of such items, I felt the survey pulling me in the direction of a conclusion that was not at all consistent with what I felt.

But when I tallied my score, I was surprised: I had scored a sixty-one, one point over the threshold for a "complete" mystical experience. I had squeaked through. So *that* was a mystical experience? It didn't feel at all like what I expected a mystical experience to be. I concluded that the MEQ was a poor net for capturing my encounter with the toad. The result was psychological bycatch, I decided, and should probably be tossed out.

Yet I wonder if my dissatisfaction with the survey had something to do with the intrinsic nature—the sheer intensity and bizarre shape—of the toad experience, for which it wasn't designed, after all. Because when I used the same survey to evaluate my psilocybin journey, the fit seemed much better and rating the phenomena much easier. Reflecting just on the cello interlude, for example, I could easily confirm the "fusion of [my] personal self into a larger whole," as well as the "feeling that [I] experienced something profoundly sacred and holy" and "of being at a spiritual height" and even the "experience of unity with ultimate reality." Yes, yes, yes, and yes—provided, that is, my endorsement of those loaded adjectives doesn't imply any belief in a supernatural reality.

My psilocybin journey with Mary yielded a sixty-six on the Mystical Experience Questionnaire. For some reason, I felt stupidly proud of my score. (There I was again, doing being.) It had been my objective to have such an experience, and at least according to the scientists a mystical experience I had had. Yet it had brought me no closer to a belief in God or in a cosmic form of consciousness or in anything magical at all—all of which I might have been, unreasonably, expecting (hoping?) it might do.

Still, there was no question that something novel and profound had happened to me—something I am prepared to call spiritual, though only with an asterisk. I guess I've always assumed that spirituality implied a belief or faith I've never shared and from which it supposedly flows. But now I wondered, is this always or necessarily the case?

Only in the wake of my journeys have I been able to unravel the paradox that had so perplexed me when I interviewed Dinah Bazer, a NYU cancer patient who began and ended her psilocybin experience an avowed atheist. During the climax of a journey that extinguished her fear of death, Bazer described "being bathed in God's love," and yet she emerged with her atheism intact. How could someone hold those two warring ideas in the same brain? I think I get it now. Not only was the flood of love she experienced ineffably powerful, but it was unattributable to any individual or worldly cause, and so was purely gratuitous—a form of grace. So how to convey the magnitude of such a gift? "God" might be the only word in the language big enough.

Part of the problem I was having evaluating my own experience had to do with another big and loaded word—"mystical"—implying as it does an experience beyond the reach of ordinary comprehension or science. It reeks of the supernatural. Yet I think it would be wrong to discard the mystical, if only because so much work has been done by so many great minds—over literally thousands of years—to find the words for this extraordinary human experience and make sense of it. When we read the testimony of these minds, we find a striking commonality in their descriptions, even if we civilians can't *quite* understand what in the world (or out of it) they're talking about.

According to scholars of mysticism, these shared traits generally include a vision of unity in which all things, including the self, are subsumed (expressed in the phrase "All is one"); a sense of certainty about what one has perceived ("Knowledge has been revealed to me"); feelings of joy, blessedness, and satisfaction; a transcendence of the categories we rely on to organize the world, such as time and space or self and other; a sense that whatever has been apprehended is somehow sacred (Wordsworth: "Something far more deeply interfused" with meaning) and often paradoxical (so while the self may vanish, awareness abides). Last is the conviction that the experience is ineffable, even as thousands of words are expended in the attempt to communicate its power. (Guilty.)

Before my journeys, words and phrases such as these left me cold; they seemed utterly opaque, so much quasi-religious mumbo jumbo. Now they paint a recognizable reality. Likewise, certain mystical passages from literature that once seemed so overstated and abstract that I read them indulgently (if at all), now I can read as a subspecies of journalism. Here are three nineteenth-century examples, but you can find them in any century.

Ralph Waldo Emerson crossing a wintry New England commons in "Nature":

> Standing on the bare ground,—my head bathed by the blithe air, and uplifted into infinite space,—all mean egotism vanishes. I become a transparent eye-ball. I am nothing. I see all. The currents of the Universal Being circulate through me; I am part or particle of God.

Or Walt Whitman, in the early lines of the first (much briefer and more mystical) edition of *Leaves of Grass*:

> Swiftly arose and spread around me the peace and joy and knowledge that pass all the art and argument of the earth;
>   And I know that the hand of God is the elderhand of my own,
>   And I know that the spirit of God is the eldest brother of my own,
>   And that all the men ever born are also my brothers
>   . . . and the women my sisters and lovers,
>   And that a kelson* of the creation is love.

---

* "Kelson" is a nautical term for a structural member in the hull of a boat.

And here is Alfred, Lord Tennyson, describing in a letter the "waking trance" that descended upon him from time to time since his boyhood:

> All at once, as it were out of the intensity of the consciousness of individuality, the individuality itself seemed to dissolve and fade into boundless being; and this was not a confused state, but the clearest of the clearest, the surest of the surest; utterly beyond words, where death was an almost laughable impossibility; the loss of personality (if so it were) seeming no extinction, but the only true life.

What had changed for me was that now I understood exactly what these writers were talking about: their own mystical experiences, however achieved, however interpreted. Formerly inert, their words now emitted a new ray of relation, or at least I was now in a position to receive it. Such emissions had always been present in our world, flowing through literature and religion, but like electromagnetic waves they couldn't be understood without some kind of receiver. I had become such a one. A phrase like "boundless being," which once I might have skated past as overly abstract and hyperbolic, now communicated something specific and even familiar. A door had opened for me onto a realm of human experience that for sixty years had been closed.*

But had I earned the right to go through that door, enter into that conversation? I don't know about Emerson's mystical experience (or Whitman's or Tennyson's), but mine owed to a chemical. Wasn't that cheating? Perhaps not: it seems likely that all mental experiences are

---

* Or at least fifty-five years, because I think young children have ready access to these kinds of experiences, as we will see in the next chapter.

mediated by chemicals in the brain, even the most seemingly "transcendent." How much should the genealogy of these chemicals matter? It turns out the very same molecules flow through the natural world and the human brain, linking us all together in a vast watershed of tryptamines. Are these exogenous molecules any less miraculous? (When they come from a mushroom or a plant or *a toad*!) It's worth remembering that there are many cultures where the fact that the inspiration for visionary experiences comes from nature, is the gift of other creatures, renders them *more* meaningful, not less.

My own interpretation of what I experienced—my now officially verified mystical experience—remains a work in progress, still in search of the right words. But I have no problem using the word "spiritual" to describe elements of what I saw and felt, as long as it is not taken in a supernatural sense. For me, "spiritual" is a good name for some of the powerful mental phenomena that arise when the voice of the ego is muted or silenced. If nothing else, these journeys have shown me how that psychic construct—at once so familiar and on reflection so strange—stands between us and some striking new dimensions of experience, whether of the world outside us or of the mind within. The journeys have shown me what the Buddhists try to tell us but I have never really understood: that there is much more to consciousness than the ego, as we would see if it would just shut up. And that its dissolution (or transcendence) is nothing to fear; in fact, it is a prerequisite for making any spiritual progress.

But the ego, that inner neurotic who insists on running the mental show, is wily and doesn't relinquish its power without a struggle. Deeming itself indispensable, it will battle against its diminishment, whether in advance or in the middle of the journey. I suspect that's exactly what mine was up to all through the sleepless nights that preceded each of my trips, striving to convince me that I was risking everything, when really all I was putting at risk was its sovereignty.

When Huxley speaks of the mind's "reducing valve"—the faculty that eliminates as much of the world from our conscious awareness as it lets in—he is talking about the ego. That stingy, vigilant security guard admits only the narrowest bandwidth of reality, "a measly trickle of the kind of consciousness which will help us to stay alive." It's really good at performing all those activities that natural selection values: getting ahead, getting liked and loved, getting fed, getting laid. Keeping us on task, it is a ferocious editor of anything that might distract us from the work at hand, whether that means regulating our access to memories and strong emotions from within or news of the world without.

What of the world it does admit it tends to objectify, for the ego wants to reserve the gifts of subjectivity to itself. That's why it fails to see that there is a whole world of souls and spirits out there, by which I simply mean subjectivities other than our own. It was only when the voice of my ego was quieted by psilocybin that I was able to sense that the plants in my garden had a spirit too. (In the words of R. M. Bucke, a nineteenth-century Canadian psychiatrist and mystic, "I saw that the universe is not composed of dead matter, but is, on the contrary, a living Presence.") "Ecology" and "coevolution" are scientific names for the same phenomena: every species a subject acting on other subjects. But when this concept acquires the flesh of feeling, becomes "more deeply interfused," as it did during my first psilocybin journey, I'm happy to call it a spiritual experience. So too my various psychedelic mergings: with Bach's cello suite, with my son, Isaac, with my grandfather Bob, all spirits directly apprehended and embraced, each time with a flood of feeling.

So perhaps spiritual experience is simply what happens in the space that opens up in the mind when "all mean egotism vanishes." Wonders (and terrors) we're ordinarily defended against flow into our awareness; the far ends of the sensory spectrum, which are normally invisible to us, our senses can suddenly admit. While the ego sleeps, the mind plays, proposing unexpected patterns of thought and new rays of relation. The gulf between self and world, that no-man's-land which in ordinary hours the ego so vigilantly patrols, closes down, allowing us to feel less separate and more connected, "part and particle" of some larger entity. Whether we call that entity Nature, the Mind at Large, or God hardly matters. But it seems to be in the crucible of that merging that death loses some of its sting.

CHAPTER FIVE

THE NEUROSCIENCE
Your Brain on Psychedelics

WHAT JUST HAPPENED in my brain?

A molecule had launched me on each of these trips, and I returned from my travels intensely curious to learn what the chemistry could tell me about consciousness and what that might reveal about the brain's relationship to the mind. How do you get from the ingestion of a compound created by a fungus or a toad (or a human chemist) to a novel state of consciousness with the power to change one's perspective on things, not just during the journey, but long after the molecule has left the body?

Actually, there were three different molecules in question—psilocin, LSD, and 5-MeO-DMT—but even a casual glance at their structures (and I say this as someone who earned a D in high school chemistry) indicates a resemblance. All three molecules are tryptamines. A tryptamine is a type of organic compound (an indole, to be exact) distinguished by the presence of two linked rings, one of them with six atoms and the other with five. Living nature is awash in tryptamines, which show up in plants, fungi, and animals, where they typically act as signaling molecules between cells. The most famous tryptamine in the human body is the neurotransmitter serotonin, the chemical name of which is 5-hydroxytryptamine. It is no coincidence that this molecule has a strong family resemblance with the psychedelic molecules.

Serotonin might be famous, as neurotransmitters go, yet much about it remains a mystery. For example, it binds with a dozen or so different receptors, and these are found not only across many parts of the brain but throughout the body, with a substantial representation in the digestive tract. Depending on the type of receptor in question and its location, serotonin is liable to make very different things happen—sometimes exciting a neuron to fire, other times inhibiting it. Think of it as a kind of word, the meaning or import of which can change radically depending on the context or even its placement in a sentence.

The group of tryptamines we call "the classic psychedelics" have a strong affinity with one particular type of serotonin receptor, called the 5-HT_{2A}. These receptors are found in large numbers in the human cortex, the outermost, and evolutionarily most recent, layer of the brain. Basically, the psychedelics resemble serotonin closely enough that they can attach themselves to this receptor site in such a way as to activate it to do various things.

Curiously, LSD has an even stronger affinity with the 5-HT_{2A} receptor—is "stickier"—than serotonin itself, making this an instance where the simulacrum is more convincing, chemically, than the original. This has led some scientists to speculate that the human body must produce some other, more bespoke chemical for the express purpose of activating the 5-HT_{2A} receptor—perhaps an endogenous psychedelic that is released under certain circumstances, perhaps when dreaming. One candidate for that chemical is the psychedelic molecule DMT, which has been found in trace amounts in the pineal gland of rats.

The science of serotonin and LSD has been closely intertwined since the 1950s; in fact, it was the discovery that LSD affected consciousness at such infinitesimal doses that helped to advance the new field of neurochemistry in the 1950s, leading to the development of the SSRI antidepressants. But it wasn't until 1998 that Franz Vollenweider, a Swiss researcher who is one of the pioneers of psychedelic neuroscience, demonstrated that psychedelics like LSD and psilocybin work on the human brain by binding with the 5-HT_{2A} receptors. He did this by giving subjects a drug called ketanserin that blocks the receptor; when he then administered psilocybin, nothing happened.

Yet Vollenweider's discovery, important as it was, is but a small step on the long (and winding) road from psychedelic chemistry to psychedelic consciousness. The 5-HT_{2A} receptor might be the lock on the door to the mind that those three molecules unlock, but how did that chemical opening lead, ultimately, to what I felt and experienced? To the dissolution of my ego, for example, and the collapse of any distinction between subject and object? Or to the morphing in my mind's eye of Mary into María Sabina? Put another way, what, if anything, can brain chemistry tell us about the "phenomenology" of the psychedelic experience?

All these questions concern the contents of consciousness, of course, which at least to this point has eluded the tools of neuroscience. By consciousness, I don't mean simply "being conscious"—the basic sensory awareness creatures have of changes in their environment, which is easy to measure experimentally. In this limited sense, even plants are "conscious," though it's doubtful they possess full-blown consciousness. What neuroscientists and philosophers and psychologists mean by consciousness is the unmistakable sense we have that we are, or possess, a self that has experiences.

Sigmund Freud wrote that "there is nothing of which we are more certain than the feeling of our self, our own ego." Yet it is difficult to be quite so certain that anyone else possesses consciousness, much less other creatures, because there is no outward physical evidence that consciousness as we experience it exists. The thing of which we are most certain is beyond the reach of our science, supposedly our surest way of knowing anything.

This dilemma has left ajar a door through which writers and philosophers have stepped. The classic thought experiment to determine whether another being is in possession of consciousness was proposed by Thomas Nagel, a philosopher, in a famous 1974 paper, "What Is It Like to Be a Bat?" He argued that if "there is something that it is like to be a bat"—if there is any subjective dimension to bat experience—then a bat possesses consciousness. He went on to suggest that this "what it is like" quality may not be reducible to material terms. Ever.

Whether or not Nagel's right about that is the biggest argument going in the field of consciousness studies. The question at its heart is often referred to as "the hard problem" or the "explanatory gap": How do you explain mind—the subjective quality of experience—in terms of meat, that is, in terms of the physical structures or chemistry of the brain? The question assumes, as most (but not all) scientists do, that consciousness is a product of brains and that it will eventually be explained as the epiphenomenon of material things like neurons and brain structures, chemicals and communications networks. That would certainly *seem* to be the most parsimonious hypothesis. Yet it is a long way from being proven, and a number of neuroscientists question whether it ever will be: whether something as elusive as subjective experience—what it feels like to be you—will *ever* yield to the reductions of science. These scientists and philosophers are sometimes called mysterians, which is not meant as a compliment. Some scientists have raised the possibility that consciousness may pervade the universe, suggesting we think of it the same way we do electromagnetism or gravity, as one of the fundamental building blocks of reality.

The idea that psychedelic drugs might shed some light on the problems of consciousness makes a certain sense. A psychedelic drug is powerful enough to disrupt the system we call normal waking consciousness in ways that may force some of its fundamental properties into view. True, anesthetics disrupt consciousness too, yet because such drugs shut it down, this kind of disturbance yields relatively little data. In contrast, someone on a psychedelic remains awake and able to report on what he or she is experiencing in real time. Nowadays, these subjective reports can be correlated with various measures of brain activity, using several different modes of imaging—tools unavailable to researchers during the first wave of psychedelic research in the 1950s and 1960s.

By deploying these technologies in combination with LSD and psilocybin, a handful of scientists working in both Europe and the United States are opening a new window onto consciousness, and what they are glimpsing through it promises to change our understanding of the links between our brains and our minds.

o   o   o

PERHAPS THE MOST AMBITIOUS neuroscientific expedition using psychedelics to map the terrain of human consciousness is taking place in a laboratory at the Centre for Psychiatry on the Hammersmith campus of Imperial College in West London. Recently completed, the campus consists of a futuristic but oddly depressing network of buildings, linked by glass-walled aerial walkways and glass doors that slide open silently at the detection of the proper identification. It is here in the lab of David Nutt, a prominent English psychopharmacologist, that a team led by a thirtysomething neuroscientist named Robin Carhart-Harris has been working since 2009 to identify the "neural correlates," or physical counterparts, of the psychedelic experience. By injecting volunteers with LSD and psilocybin and then using a variety of scanning technologies—including functional magnetic resonance imaging (fMRI) and magnetoencephalography (MEG)—to observe the changes in their brains, he and his team have given us our first glimpses of what something like ego dissolution, or a hallucination, actually looks like in the brain as it unfolds in the mind.

The fact that such an improbable and potentially controversial research project ever got off the ground owes to the convergence of three most unusual characters, and careers, in England in the year 2005: David Nutt, Robin Carhart-Harris, and Amanda Feilding, a.k.a. the Countess of Wemyss and March.

Robin Carhart-Harris's path to David Nutt's psychopharmacology lab was an eccentric one, having first passed through a graduate course in psychoanalysis. These days psychoanalysis is a theory few neuroscientists take seriously, regarding it less as a science than as a set of untestable beliefs. Carhart-Harris felt strongly otherwise. Steeped in the writings of Freud and Jung, he was fascinated by psychoanalytic theory while at the same time frustrated by its lack of scientific rigor, as well as by the limitations of its tools for exploring what it deemed most important about the mind: the unconscious.

"If the only way we can access the unconscious is via dreams and free association," he explained the first time we talked, "we aren't going to get anywhere. Surely there must be something else." One day he asked his seminar professor if that something else might be a drug. (I asked Robin if his hunch was based on personal experience or research, but he made clear this was not a subject he wished to discuss.) His professor sent him to read a book called *Realms of the Human Unconscious* by Stanislav Grof.

"I went to the library and read the book cover to cover. I was blown away. That set the course for the rest of my young life."

Carhart-Harris, who is a slender, intense young man in a hurry, with a neatly trimmed beard and large pale blue eyes that seldom blink, formulated a plan it would take him a few years to put into motion: he would use psychedelic drugs and modern brain-imaging technologies to build a foundation of hard science beneath the edifice of psychoanalysis. "Freud said dreams were the royal road to the unconscious," he reminded me. "Psychedelics could turn out to be the superhighway." Carhart-Harris's demeanor is modest, even humble, offering no clue to the audacity of his ambition. He likes to quote Grof's grand claim that what the telescope was for astronomy, or the microscope for biology, psychedelics will be for understanding the mind.

Carhart-Harris completed his master's in psychoanalysis in 2005 and began to plot his move into the neuroscience of psychedelics. He asked around and did some Internet research that eventually led him to David Nutt and Amanda Feilding as two people who might be interested in his project and in a position to help. He first approached Feilding, who in 1998 had established something called the Beckley Foundation to study the effects of psychoactive substances on the brain and to lobby for drug policy reform. The foundation is named for Beckley Park, the sprawling fourteenth-century Tudor manor where she grew up in Oxfordshire and where, in 2005, she invited Carhart-Harris to lunch. (On a recent visit of my own to Beckley, I counted two towers and three moats.)

Amanda Feilding, who was born in 1943, is an eccentric as only the English aristocracy can breed them. (She's descended from the house of Habsburg and two of Charles II's illegitimate children.) A student of comparative religion and mysticism, Feilding has had a long-standing interest in altered states of consciousness and, specifically, the role of blood flow to the brain, which in *Homo sapiens*, she believes, has been compromised ever since our species began standing upright. LSD, Feilding believes, enhances cognitive function and facilitates higher states of consciousness by increasing cerebral circulation. A second way to achieve a similar result is by means of the ancient practice of trepanation. This deserves a brief digression.

Trepanation involves drilling a shallow hole in the skull supposedly to improve cerebral blood circulation; in effect, it reverses the fusing of the cranial bones that happens in childhood. Trepanation was for centuries a common medical procedure, to judge by the number of ancient skulls that have turned up with neat holes in them. Convinced that trepanation would help facilitate higher states of consciousness, Feilding went looking for someone to perform the operation on her. When it became clear no professional would oblige, she trepanned herself in 1970, boring a small hole in the middle of her forehead with an electric drill. (She documented the procedure in a short but horrifying film called *Heartbeat in the Brain*.) Pleased with the results, Feilding went on to stand for election to Parliament, twice, on a platform of "Trepanation for the National Health."

But while Amanda Feilding may be eccentric, she is by no means feckless. Her work on both drug research and drug policy reform has been serious, strategic, and productive. In recent years, her focus has shifted from trepanation to the potential of psychedelics to improve brain function. In her own life, she has used LSD as a kind of "brain tonic," favoring a daily dose that hits "that sweet spot where creativity and enthusiasm is increased, but control is maintained." (She told me that there was a time when she put that tonic dose at 150 micrograms—far above a microdose and enough to send most people, myself included, on a full-fledged trip. But because frequent use of LSD can lead to tolerance, it's entirely possible that for some people 150 micrograms merely "adds a certain sparkle to consciousness.") I found Feilding to be disarmingly frank about the baggage she brings to the new conversation about psychedelic science: "I'm a druggie. I live in this big house. And I have a hole in my head. I guess that disqualifies me."

So, when an aspiring young scientist named Robin Carhart-Harris came for lunch at Beckley in 2005, sharing his ambition to combine research into LSD and Freud, Feilding immediately saw the potential, as well as an opportunity to put her theories about cerebral blood circulation to the test. Feilding indicated to Carhart-Harris that her foundation might be willing to fund such research and suggested that he contact David Nutt, then a professor at the University of Bristol and an ally of Feilding's in the campaign to reform drug policy.

In his own way, David Nutt is as notorious in England as Amanda Feilding. Nutt, who is a large, jolly fellow in his sixties with a mustache and a booming laugh, achieved his particular notoriety in 2009. That's when the home secretary fired him from the government's Advisory Council on the Misuse of Drugs, of which he had been chair. The committee is charged with advising the government on the classification of illicit drugs based on their risk to individuals and society. Nutt, who is an expert on addiction and on the class of drugs called benzodiazepines (such as Valium), had committed the fatal political error of quantifying empirically the risks of various psychoactive substances, both legal and illegal. He had concluded from his research, and would tell anyone who asked, that alcohol was more dangerous than cannabis and that using Ecstasy was safer than riding a horse.

"But the sentence that got me sacked," he told me when we met in his office at Imperial, "was when I went on live breakfast television.

I was asked, 'You're not seriously telling us that LSD is less harmful than alcohol, are you?' Of course I am!'*

Robin Carhart-Harris came to see David Nutt in 2005, hoping to study psychedelics and dreaming under him at Bristol; trying to be strategic, he mentioned the possibility of funding from Feilding. As Carhart-Harris recalls the interview, Nutt was blunt in his dismissal: "'The idea you want to do is incredibly far-fetched, you have no neuroscience experience, it's completely unrealistic.' But I told him I put all my eggs in this basket." Impressed by the young man's determination, Nutt made him an offer: "Come do a PhD with me. We'll start with something straightforward"—this turned out to be the effect of MDMA on the serotonin system—"and then maybe later on we can do psychedelics."

"Later on" came in 2009, when Carhart-Harris, armed with a PhD and working in Nutt's lab with funding from Amanda Feilding, received approval (from the National Health Service and the Home Office) to study the effect of psilocybin on the brain. (LSD would come a few years later.) Carhart-Harris put himself forward as the first volunteer. "If you're going to give this drug to people and put them in a scanner, I thought, the honest thing is to do it first to yourself." But, as he told Nutt, "I have an anxious disposition, and may not have been in the best place psychologically, so he dissuaded me; he also thought participating in the experiment might compromise my objectivity." In the end, a colleague became the first volunteer to receive an injection of psilocybin and then slide into an fMRI scanner to have his tripping brain imaged.

Carhart-Harris's working hypothesis was that their brains would

---

* In his 2012 book, *Drugs Without the Hot Air*, Nutt writes that "psychedelics overall are among the safest drugs we know of . . . It's virtually impossible to die from an overdose of them; they cause no physical harm; and if anything they're anti-addictive" (254).

exhibit increases in activity, particularly in the emotion centers. "I thought it would look like the dreaming brain," he told me. Employing a different scanning technology, Franz Vollenweider had published data indicating that psychedelics stimulated brain activity, especially in the frontal lobes. (An area responsible for executive and other higher cognitive functions.) But when the first set of data came in, Carhart-Harris got a surprise: "We were seeing *decreases* in blood flow"—blood flow being one of the proxies for brain activity that fMRI measures. "Had we made a mistake? It was a real headscratcher." But the initial data on blood flow was corroborated by a second measure that looks at changes in oxygen consumption to pinpoint areas of elevated brain activity. Carhart-Harris and his colleagues had discovered that psilocybin reduces brain activity, with the falloff concentrated in one particular brain network that at the time he knew little about: the default mode network.

Carhart-Harris began reading up on it. The default mode network, or DMN, was not known to brain science until 2001. That was when Marcus Raichle, a neurologist at Washington University, described it in a landmark paper published in the *Proceedings of the National Academy of Sciences*, or *PNAS*. The network forms a critical and centrally located hub of brain activity that links parts of the cerebral cortex to deeper (and older) structures involved in memory and emotion.[*]

The discovery of the default mode network was actually a scientific accident, a happy by-product of the use of brain-imaging tech-

---

[*] The key structures making up the default mode network are the medial prefrontal cortex, the posterior cingulate cortex, the inferior parietal lobule, the lateral temporal cortex, the dorsal medial prefrontal cortex, and the hippocampus formation. See Randy L. Buckner, Jessica R. Andrews-Hanna, and Daniel L. Schacter, "The Brain's Default Network," *Annals of the New York Academy of Sciences* 1124, no. 1 (2008). While neuroimaging indicates strong links between these structures, the concept of the default mode network remains new and is still not universally accepted.

nologies in brain research.* The typical fMRI experiment begins by establishing a "resting state" baseline for neural activity as the volunteer sits quietly in the scanner awaiting whatever tests the researcher has in store. Raichle had noticed that several areas in the brain exhibited heightened activity precisely when his subjects were doing nothing mentally. This was the brain's "default mode," the network of brain structures that light up with activity when there are no demands on our attention and we have no mental task to perform. Put another way, Raichle had discovered the place where our minds go to wander—to daydream, ruminate, travel in time, reflect on ourselves, and worry. It may be through these very structures that the stream of our consciousness flows.

The default network stands in a kind of seesaw relationship with the attentional networks that wake up whenever the outside world demands our attention; when one is active, the other goes quiet, and vice versa. But as any person can tell you, quite a lot happens in the mind when nothing much is going on outside us. (In fact, the DMN consumes a disproportionate share of the brain's energy.) Working at a remove from our sensory processing of the outside world, the default mode is most active when we are engaged in higher-level "metacognitive" processes such as self-reflection, mental time travel, mental constructions (such as the self or ego), moral reasoning, and "theory of mind"—the ability to attribute mental states to others, as when we try to imagine "what it is like" to be someone else. All these functions may belong

---

* It's important to keep in mind the limitations of fMRI and other neuroimaging technologies. Most of them measure not brain activity directly but proxies of it, such as blood flow and oxygen consumption. They also depend on complex software to translate faint signals into dramatic images, software the accuracy of which critics have recently questioned. In my experience, brain scientists who work with animals they can insert probes into are dismissive of fMRI, while brain scientists who work with humans accept it as the best tool available.

exclusively to humans, and specifically to adult humans, for the default mode network isn't operational until late in a child's development.

"The brain is a hierarchical system," Carhart-Harris explained in one of our interviews. "The highest-level parts"—those developed late in our evolution, typically located in the cortex—"exert an inhibitory influence on the lower-level [and older] parts, like emotion and memory." As a whole, the default mode network exerts a top-down influence on other parts of the brain, many of which communicate with one another through its centrally located hub. Robin has described the DMN variously as the brain's "orchestra conductor," "corporate executive," or "capital city," charged with managing and "holding the whole system together." And with keeping the brain's unrulier tendencies in check.

The brain consists of several different specialized systems—one for visual processing, for example, another to control motor activity—each doing its own thing. "Chaos is averted because all systems are not created equal," Marcus Raichle has written. "Electrical signaling from some brain areas takes precedence over others. At the top of this hierarchy resides the DMN, which acts as an uber-conductor to ensure that the cacophony of competing signals from one system do not interfere with those from another." The default mode network keeps order in a system so complex it might otherwise descend into the anarchy of mental illness.

As mentioned, the default mode network appears to play a role in the creation of mental constructs or projections, the most important of which is the construct we call the self, or ego.* This is why some

---

* I'm using the terms more or less interchangeably here. However, the ego, being closely associated with Freud's model of the mind, implies a construct that stands in a dynamic relationship to other parts of the mind, such as the unconscious, or id, acting on behalf of the self.

neuroscientists call it "the me network." If a researcher gives you a list of adjectives and asks you to consider how they apply to you, it is your default mode network that leaps into action. (It also lights up when we receive "likes" on our social media feeds.) Nodes in the default network are thought to be responsible for autobiographical memory, the material from which we compose the story of who we are, by linking our past experiences with what happens to us and with projections of our future goals.

The achievement of an individual self, a being with a unique past and a trajectory into the future, is one of the glories of human evolution, but it is not without its drawbacks and potential disorders. The price of the sense of an individual identity is a sense of separation from others and nature. Self-reflection can lead to great intellectual and artistic achievement but also to destructive forms of self-regard and many types of unhappiness. (In an often-cited paper titled "A Wandering Mind Is an Unhappy Mind," psychologists identified a strong correlation between unhappiness and time spent in mind wandering, a principal activity of the default mode network.) But, accepting the good with the bad, most of us take this self as an unshakable given, as real as anything we know, and as the foundation of our life as conscious human beings. Or at least I always took it that way, until my psychedelic experiences led me to wonder.

Perhaps the most striking discovery of Carhart-Harris's first experiment was that the steepest drops in default mode network activity correlated with his volunteers' subjective experience of "ego dissolution." ("I existed only as an idea or concept," one volunteer reported. Recalled another, "I didn't know where I ended and my surroundings began.") The more precipitous the drop-off in blood flow and oxygen consumption in the default network, the more likely a volunteer was to report the loss of a sense of self.[*]

Shortly after Carhart-Harris published his results in a 2012 paper in *PNAS* ("Neural Correlates of the Psychedelic State as Determined by fMRI Studies with Psilocybin"[†]), Judson Brewer, a researcher at Yale[‡] who was using fMRI to study the brains of experienced meditators, noticed that his scans and Robin's looked remarkably alike. The transcendence of self reported by expert meditators showed up on fMRIs as a quieting of the default mode network. It appears that when activity in the default mode network falls off precipitously, the ego temporarily vanishes, and the usual boundaries we experience between self and world, subject and object, all melt away.

This sense of merging into some larger totality is of course one of the hallmarks of the mystical experience; our sense of individuality and separateness hinges on a bounded self and a clear demarcation between subject and object. But all that may be a mental construction, a kind of illusion—just as the Buddhists have been trying to tell us. The psychedelic experience of "non-duality" suggests that consciousness survives the disappearance of the self, that it is not so indispensable as we—and it—like to think. Carhart-Harris suspects that the loss of a clear distinction between subject and object might help explain another feature of the mystical experience: the fact that the insights it sponsors are felt to be objectively true—revealed truths rather than plain old insights. It could be that in order to judge an insight as merely subjective, one person's opinion, you must

---

[*] It's worth noting that these findings seem to be at odds with Amanda Feilding's initial hypothesis that psychedelics work by *increasing* blood flow to the brain.

[†] David Nutt and Amanda Feilding are coauthors.

[‡] Brewer has since moved to the University of Massachusetts Medical School, where he's the director of research at the Center for Mindfulness.

first have a sense of subjectivity. Which is precisely what the mystic on psychedelics has lost.

The mystical experience may just be what it feels like when you deactivate the brain's default mode network. This can be achieved any number of ways: through psychedelics and meditation, as Robin Carhart-Harris and Judson Brewer have demonstrated, but perhaps also by means of certain breathing exercises (like Holotropic Breathwork), sensory deprivation, fasting, prayer, overwhelming experiences of awe, extreme sports, near-death experiences, and so on. What would scans of brains in the midst of those activities reveal? We can only speculate, but quite possibly we would see the same quieting of the default mode network Brewer and Carhart-Harris have found. This quieting might be accomplished by restricting blood flow to the network, or by stimulating the serotonin 2A receptors in the cortex, or by otherwise disturbing the oscillatory rhythms that normally organize the brain. But however it happens, taking this particular network off-line may give us access to extraordinary states of consciousness—moments of oneness or ecstasy that are no less wondrous for having a physical cause.

o   o   o

IF THE DEFAULT MODE network is the conductor of the symphony of brain activity, you would expect its temporary absence from the stage to lead to an increase in dissonance and mental disorder—as indeed appears to happen during the psychedelic journey. In a series of subsequent experiments using a variety of brain-imaging techniques, Carhart-Harris and his colleagues began to study what happens elsewhere in the neural orchestra when the default mode network puts down its baton.

Taken as a whole, the default mode network exerts an inhibitory influence on other parts of the brain, notably including the limbic regions involved in emotion and memory, in much the same way Freud conceived of the ego keeping the anarchic forces of the unconscious id in check. (David Nutt puts the matter bluntly, claiming that in the DMN "we've found the neural correlate for repression.") Carhart-Harris hypothesizes that these and other centers of mental activity are "let off the leash" when the default mode leaves the stage, and in fact brain scans show an increase in activity (as reflected by increases in blood flow and oxygen consumption) in several other brain regions, including the limbic regions, under the influence of psychedelics. This disinhibition might explain why material that is unavailable to us during normal waking consciousness now floats to the surface of our awareness, including emotions and memories and, sometimes, long-buried childhood traumas. It is for this reason that some scientists and psychotherapists believe psychedelics can be profitably used to surface and explore the contents of the unconscious mind.

But the default mode network doesn't only exert top-down control over material arising from within; it also helps regulate what is let into consciousness from the world outside. It operates as a kind of filter (or "reducing valve") charged with admitting only that "measly trickle" of information required for us to get through the day. If not for the brain's filtering mechanisms, the torrent of information the senses make available to our brains at any given moment might prove difficult to process—as indeed is sometimes the case during the psychedelic experience. "The question," as David Nutt puts it, "is why the brain is ordinarily so constrained rather than so open?" The answer may be as simple as "efficiency." Today most neuroscientists work under a paradigm of the brain as a prediction-making machine. To form a perception of something out in the world, the brain takes in as little sensory information as it needs to make an educated guess. We are forever cutting to the chase, basically, and leaping to conclusions, relying on prior experience to inform current perception.

The mask experiment I attempted to perform during my psilocybin journey is a powerful demonstration of this phenomenon. At least when it is working normally, the brain, presented with a few visual clues suggesting it is looking at a face, insists on seeing the face as a convex structure even when it is not, because that's the way faces usually are.

The philosophical implications of "predictive coding" are deep and strange. The model suggests that our perceptions of the world offer us not a literal transcription of reality but rather a seamless illusion woven from both the data of our senses and the models in our memories. Normal waking consciousness feels perfectly transparent, and yet it is less a window on reality than the product of our imaginations—a kind of controlled hallucination. This raises a question: How is normal waking consciousness any different from other, seemingly less faithful productions of our imagination—such as dreams or psychotic delusions or psychedelic trips? In fact, all these states of consciousness are "imagined": they're mental constructs that weave together some news of the world with priors of various kinds. But in the case of normal waking consciousness, the handshake between the data of our senses and our preconceptions is especially firm. That's because it is subject to a continual process of reality testing, as when you reach out to confirm the existence of the object in your visual field or, upon waking from a nightmare, consult your memory to see if you really did show up to teach a class without any clothes on. Unlike these other states of consciousness, ordinary waking consciousness has been optimized by natural selection to best facilitate our everyday survival.

Indeed, that feeling of transparency we associate with ordinary consciousness may owe more to familiarity and habit than it does to verisimilitude. As a psychonaut acquaintance put it to me, "If it were possible to temporarily experience another person's mental state, my guess is that it would feel more like a psychedelic state than a 'normal' state, because of its massive disparity with whatever mental state is habitual with you."

Another trippy thought experiment is to try to imagine the world as it appears to a creature with an entirely different sensory apparatus and way of life. You quickly realize there is no single reality out there waiting to be faithfully and comprehensively transcribed. Our senses have evolved for a much narrower purpose and take in only what serves our needs as animals of a particular kind. The bee perceives a substantially different spectrum of light than we do; to look at the world through its eyes is to perceive ultraviolet markings on the petals of flowers (evolved to guide their landings like runway lights) that don't exist for us. That example is at least a kind of seeing—a sense we happen to share with bees. But how do we even begin to conceive of the sense that allows bees to register (through the hairs on their legs) the electromagnetic fields that plants produce? (A weak charge indicates another bee has recently visited the flower; depleted of nectar, it's probably not worth a stop.) Then there is the world according to an octopus! Imagine how differently reality presents itself to a brain that has been so radically decentralized, its intelligence distributed across eight arms so that each of them can taste, touch, and even make its own "decisions" without consulting headquarters.

o   o   o

WHAT HAPPENS WHEN, under the influence of psychedelics, the usually firm handshake between brain and world breaks down? No one thing, as it turns out. I asked Carhart-Harris whether the tripping brain favors top-down predictions or bottom-up sensory data. "That's the classic dilemma," he suggested: whether the mind, unconstrained, will tend to favor its priors or the evidence of its senses. "You do often find a kind of impetuousness or overzealousness on the part of the priors, as when you see faces in the clouds." Eager to make sense of the data rushing in, the brain leaps to erroneous conclusions and, sometimes, a hallucination results. (The paranoid does much the same thing, ferociously imposing a false narrative on the stream of incoming information.) But in other cases, the reducing valve opens wide to admit lots more information, unedited and sometimes welcome.

People who are color-blind report being able to see certain colors for the first time when on psychedelics, and there is research to suggest that people hear music differently under the influence of these drugs. They process the timbre, or coloration, of music more acutely—a dimension of music that conveys emotion. When I listened to Bach's cello suite during my psilocybin journey, I was certain I heard more of it than I ever had, registering shadings and nuances and tones that I hadn't been able to hear before and haven't heard since.

Carhart-Harris thinks that psychedelics render the brain's usual handshake of perception less stable and more slippery. The tripping brain may "slip back and forth" between imposing its priors and admitting the raw evidence of its senses. He suspects that there are moments during the psychedelic experience when confidence in our usual top-down concepts of reality collapses, opening the way for more bottom-up information to get through the filter. But when all that sensory information threatens to overwhelm us, the mind furiously generates new concepts (crazy or brilliant, it hardly matters) to make sense of it all—"and so you might see faces coming out of the rain.

"That's the brain doing what the brain does"—that is, working to reduce uncertainty by, in effect, telling itself stories.

THE HUMAN BRAIN is an inconceivably complex system—perhaps the most complex system ever to exist—in which an order has emerged, the highest expression of which is the sovereign self and our normal waking consciousness. By adulthood, the brain has gotten very good at observing and testing reality and developing reliable predictions about it that optimize our investments of energy (mental and otherwise) and therefore our chances of survival. Uncertainty is a complex brain's biggest challenge, and predictive coding evolved to help us reduce it. In general, the kind of precooked or conventionalized thinking this adaptation produces serves us well. But only up to a point.

Precisely where that point lies is a question Robin Carhart-Harris and his colleagues have explored in an ambitious and provocative paper titled "The Entropic Brain: A Theory of Conscious States Informed by Neuroimaging Research with Psychedelic Drugs," published in *Frontiers in Human Neuroscience* in 2014. Here, Carhart-Harris attempts to lay out his grand synthesis of psychoanalysis and cognitive brain science. The question at its heart is, do we pay a price for the achievement of order and selfhood in the adult human mind? The paper concludes that we do. While suppressing entropy (in this context, a synonym for uncertainty) in the brain "serves to promote realism, foresight, careful reflection and an ability to recognize and overcome wishful and paranoid fantasies," at the same time this achievement tends to "constrain cognition" and exert "a limiting or narrowing influence on consciousness."

After a series of Skype interviews, Robin Carhart-Harris and I were meeting for the first time, in his fifth-floor walk-up in an unposh section of Notting Hill, a few months after the publication of the entropy paper. In person, I was struck by Robin's youthfulness and intensity. For all his ambition, his affect is strikingly self-effacing and does little to prepare you for his willingness to venture out onto intellectual limbs that would scare off less intrepid scientists.

The entropy paper asks us to conceive of the mind as an uncertainty-reducing machine with a few serious bugs in it. The sheer complexity of the human brain and the greater number of different mental states in its repertoire (as compared with other animals) make the maintenance of order a top priority, lest the system descend into chaos.

Once upon a time, Carhart-Harris writes, the human or protohuman brain exhibited a much more anarchic form of "primary consciousness," characterized by "magical thinking"—beliefs about the world that have been shaped by wishes and fears and supernatural interpretation. (In primary consciousness, Carhart-Harris writes, "cognition is less meticulous in its sampling of the external world and is instead easily biased by emotion, e.g., wishes and anxieties.") Magical thinking is one way for human minds to reduce their uncertainty about the world, but it is less than optimal for the success of the species.

A better way to suppress uncertainty and entropy in the human brain emerged with the evolution of the default mode network, Carhart-Harris contends, a brain-regulating system that is absent or undeveloped in lower animals and young children. Along with the default mode network, "a coherent sense of self or 'ego' emerges" and, with that, the human capacity for self-reflection and reason. Magical thinking gives way to "a more reality-bound style of thinking, governed by the ego." Borrowing from Freud, he calls this more highly evolved mode of cognition "secondary consciousness." Secondary consciousness "pays deference to reality and diligently seeks to represent the world as precisely as possible" in order to minimize "surprise and uncertainty (i.e. entropy)."

The article offers an intriguing graphic depicting a "spectrum of cognitive states," ranging from high-entropy mental states to low ones. At the high-entropy end of the spectrum, he lists psychedelic states; infant consciousness; early psychosis; magical thinking; and divergent or creative thinking. At the low-entropy end of the spectrum, he lists narrow or rigid thinking; addiction; obsessive-compulsive disorder; depression; anesthesia; and, finally, coma.

Carhart-Harris suggests that the psychological "disorders" at the low-entropy end of the spectrum are not the result of a lack of order in the brain but rather stem from an *excess* of order. When the grooves of self-reflective thinking deepen and harden, the ego becomes over-bearing. This is perhaps most clearly evident in depression, when the ego turns on itself and uncontrollable introspection gradually shades out reality. Carhart-Harris cites research indicating that this debili-tating state of mind (sometimes called heavy self-consciousness or depressive realism) may be the result of a hyperactive default mode network, which can trap us in repetitive and destructive loops of rumination that eventually close us off from the world outside. Hux-ley's reducing valve contracts to zero. Carhart-Harris believes that people suffering from a whole range of disorders characterized by excessively rigid patterns of thought—including addiction, obses-sions, and eating disorders as well as depression—stand to benefit from "the ability of psychedelics to disrupt stereotyped patterns of thought and behavior by disintegrating the patterns of [neural] activ-ity upon which they rest."

So it may be that some brains could stand to have a little more entropy, not less. This is where psychedelics come in. By quieting the default mode network, these compounds can loosen the ego's grip on the machinery of the mind, "lubricating" cognition where before it had been rusted stuck. "Psychedelics alter consciousness by *disorganizing* brain activity," Carhart-Harris writes. They increase the amount of entropy in the brain, with the result that the system reverts to a less constrained mode of cognition.*

"It's not just that one system drops away," he says, "but that an older system reemerges." That older system is primary consciousness, a mode of thinking in which the ego temporarily loses its dominion and the unconscious, now unregulated, "is brought into an observable space." This, for Carhart-Harris, is the heuristic value of psychedelics to the study of the mind, though he sees therapeutic value as well.

It's worth noting that Carhart-Harris does not romanticize psychedelics and has little patience for the sort of "magical thinking" and "metaphysics" that they nourish in their acolytes—such as the idea that consciousness is "transpersonal," a property of the universe rather than the human brain. In his view, the forms of consciousness that psychedelics unleash are regressions to a "more primitive" mode of cognition. With Freud, he believes that the loss of self, and the sense of oneness, characteristic of the mystical experience—whether occasioned by chemistry or religion—return us to the psychological condition of the infant on its mother's breast, a stage when it has yet to develop a sense of itself as a separate and bounded individual. For

---

* Exactly how psychedelics accomplish this, neurochemically, is still uncertain, but some of Carhart-Harris's research points to a plausible mechanism. Because of their affinity with the serotonin 2A receptors, psychedelic compounds cause a set of neurons in the cortex ("layer five pyramidal neurons," to be exact) that are rich in these receptors to fire in such a way as to desynchronize the usual oscillations of the brain. Carhart-Harris likens these oscillations, which help to organize brain activity, to the synchronized clapping of an audience. When a few wayward individuals clap out of order, the applause becomes less rhythmic and more chaotic. Similarly, the excitation of these cortical neurons appears to disrupt oscillations in a particular frequency—the alpha waves—that have been correlated with activity in the default mode network and, specifically, in self-reflection.

Carhart-Harris, the pinnacle of human development is the achievement of this differentiated self, or ego, and its imposition of order on
the anarchy of a primitive mind buffeted by fears and wishes and given
to various forms of magical thinking. While he holds with Aldous
Huxley that psychedelics throw open the doors of perception, he does
not agree that everything that comes through that opening—including the "Mind at Large" that Huxley glimpsed—is necessarily real.
"The psychedelic experience can yield a lot of fool's gold," he told me.

Yet Carhart-Harris also believes there is genuine gold in the psychedelic experience. When we met, he offered examples of scientists
whose own experiences with LSD had supplied them with insights
into the workings of the brain. Too much entropy in the human
brain may lead to atavistic thinking and, at the far end, madness, yet
too little can cripple us as well. The grip of an overbearing ego can
enforce a rigidity in our thinking that is psychologically destructive.
It may be socially and politically destructive too, in that it closes the
mind to information and alternative points of view.

In one of our conversations, Robin speculated that a class of drugs
with the power to overturn hierarchies in the mind and sponsor un-
conventional thinking has the potential to reshape users' attitudes
toward authority of all kinds; that is, the compounds may have a
political effect. Many believe LSD played precisely that role in the
political upheaval of the 1960s.

"Was it that hippies gravitated to psychedelics, or do psychedelics
create hippies? Nixon thought it was the latter. He may have been
right!" Robin believes that psychedelics may also subtly shift people's
attitudes toward nature, which also underwent a sea change in the
1960s. When the influence of the DMN declines, so does our sense
of separateness from our environment. His team at Imperial College
has tested volunteers on a standard psychological scale that measures
"nature relatedness" (respondents rate their agreement with state-

ments like "I am not separate from nature, but a part of nature"). A psychedelic experience elevated people's scores.[*]

o   o   o

So what does a high-entropy brain look like? The various scanning technologies that the Imperial College lab has used to map the tripping brain show that the specialized neural networks of the brain—such as the default mode network and the visual processing system—each become disintegrated, while the brain as a whole becomes *more* integrated as new connections spring up among regions that ordinarily kept mainly to themselves or were linked only via the central hub of the DMN. Put another way, the various networks of the brain became less specialized.

"Distinct networks became less distinct under the drug," Carhart-Harris and his colleagues wrote, "implying that they communicate more openly," with other brain networks. "The brain operates with greater flexibility and interconnectedness under hallucinogens."

In a 2014 paper published in the *Journal of the Royal Society Interface*, the Imperial College team demonstrated how the usual lines of communications within the brain are radically reorganized when the default mode network goes off-line and the tide of entropy is allowed to rise. Using a scanning technique called magnetoencephalography, which maps electrical activity in the brain, the authors produced a map of the brain's internal communications during normal waking consciousness and after an injection of psilocybin (shown

---

[*] This research was published in 2017: Matthew M. Nour et al., "Psychedelics, Personality, and Political Perspectives," *Journal of Psychoactive Drugs*. "Ego dissolution experienced during a participant's 'most intense' psychedelic experience positively predicted liberal political views, openness and nature relatedness, and negatively predicted authoritarian political views."

on the following pages). In its normal state, shown on the left, the brain's various networks (here depicted lining the circle, each represented by a different shude) talk mostly to themselves, with a relatively few heavily trafficked pathways among them.

But when the brain operates under the influence of psilocybin, as shown on the right, thousands of new connections form, linking far-flung brain regions that during normal waking consciousness don't exchange much information. In effect, traffic is rerouted from a relatively small number of interstate highways onto myriad smaller roads linking a great many more destinations. The brain appears to become less specialized and more globally interconnected, with considerably more intercourse, or "cross talk," among its various neighborhoods.

There are several ways this temporary rewiring of the brain may affect mental experience. When the memory and emotion centers are allowed to communicate directly with the visual processing centers, it's possible our wishes and fears, prejudices and emotions, begin to inform what we see—a hallmark of primary consciousness and a recipe for magical thinking. Likewise, the establishment of new linkages across brain systems can give rise to synesthesia, as when sense information gets cross-wired so that colors become sounds or sounds become tactile. Or the new links give rise to hallucination, as when the contents of my memory transformed my visual perception of Mary into María Sabina, or the image of my face in the mirror into a vision of my grandfather. The forming of still other kinds of novel connections could manifest in mental experience as a new idea, a fresh perspective, a creative insight, or the ascribing of new meanings to familiar things—or any number of the bizarre mental phenomena people on psychedelics report. The increase in entropy allows a thousand mental states to bloom, many of them bizarre and senseless, but some number of them revelatory, imaginative, and, at least potentially, transformative.

One way to think about this blooming of mental states is that it temporarily boosts the sheer amount of diversity in our mental life. If problem solving is anything like evolutionary adaptation, the more possibilities the mind has at its disposal, the more creative its solutions will be. In this sense, entropy in the brain is a bit like variation in evolution: it supplies the diversity of raw materials on which selection can then operate to solve problems and bring novelty into the

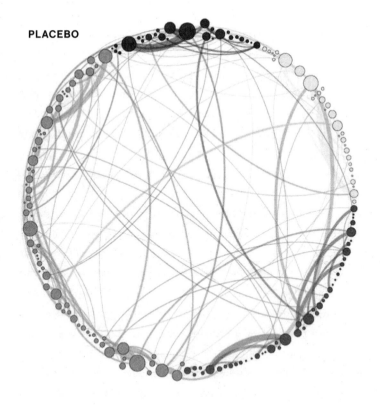

PLACEBO

world. If, as so many artists and scientists have testified, the psychedelic experience is an aid to creativity—to thinking "outside the box"—this model might help explain why that is the case. Maybe the problem with "the box" is that it is singular.

A key question that the science of psychedelics has not even begun to answer is whether the new neural connections that psychedelics make possible endure in any way, or if the brain's wiring returns to the status quo ante once the drug wears off. The finding by Roland Griffiths's lab that the psychedelic experience leads to long-term

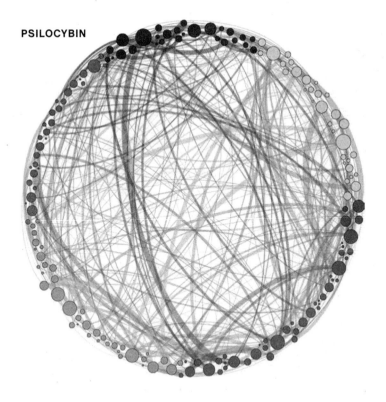

PSILOCYBIN

changes in the personality trait of openness raises the possibility that some kind of learning takes place while the brain is rewired and that it might in some way persist. Learning entails the establishment of new neural circuits; these get stronger the more exercise they get. The long-term fate of the novel connections formed during the psychedelic experience—whether they prove durable or evanescent— might depend on whether we recall and, in effect, exercise them after the experience ends. (This could be as simple as recollecting what we experienced, reinforcing it during the integration process, or using meditation to reenact the altered state of consciousness.) Franz Vollenweider has suggested that the psychedelic experience may facilitate "neuroplasticity": it opens a window in which patterns of thought and behavior become more plastic and so easier to change. His model sounds like a chemically mediated form of cognitive behavioral therapy. But so far this is all highly speculative; as yet there has been little mapping of the brain before and after psychedelics to determine what, if anything, the experience changes in a lasting way.

Carhart-Harris argues in the entropy paper that even a temporary rewiring of the brain is potentially valuable, especially for people suffering from disorders characterized by mental rigidity. A high-dose psychedelic experience has the power to "shake the snow globe," he says, disrupting unhealthy patterns of thought and creating a space of flexibility—entropy—in which more salubrious patterns and narratives have an opportunity to coalesce as the snow slowly resettles.

o   o   o

THE IDEA that increasing the amount of entropy in the human brain might actually be good for us is surely counterintuitive. Most of us bring a negative connotation to the term: entropy suggests the gradual deterioration of a hard-won order, the disintegration of a system over time. Certainly getting older *feels* like an entropic process—a gradual running down and disordering of the mind and body. But maybe that's the wrong way to think about it. Robin Carhart-Harris's paper got me wondering if, at least for the mind, aging is really a process of *declining* entropy, the fading over time of what we should regard as a positive attribute of mental life.

Certainly by middle age, the sway of habitual thinking over the operations of the mind is nearly absolute. By now, I can count on past experience to propose quick and usually serviceable answers to just about any question reality poses, whether it's about how to soothe a child or mollify a spouse, repair a sentence, accept a compliment, answer the next question, or make sense of whatever's happening in the world. With experience and time, it gets easier to cut to the chase and leap to conclusions—clichés that imply a kind of agility but that in fact may signify precisely the opposite: a petrifaction of thought. Think of it as predictive coding on the scale of life; the priors—and by now I've got millions of them—usually have my back, can be relied on to give me a decent enough answer, even if it isn't a particularly fresh or imaginative one. A flattering term for this regime of good enough predictions is "wisdom."

Reading Robin's paper helped me better understand what I was looking for when I decided to explore psychedelics: to give my own snow globe a vigorous shaking, see if I could renovate my everyday mental life by introducing a greater measure of entropy, and uncertainty, into it. Getting older might render the world more predictable (in every sense), yet it also lightens the burden of responsibility, creating a new space for experiment. Mine had been to see if it wasn't too late to skip out of some of the deeper grooves of habit that the been-theres and done-thats of long experience had inscribed on my mind.

o   o   o

IN BOTH PHYSICS and information theory, entropy is often associated with expansion—as in the expansion of a gas when it is heated
or freed from the constraints of a container. As the gas's molecules
diffuse in space, it becomes harder to predict the location of any
given one; the uncertainty of the system thus increases. In a throw-
away line at the end of his entropy paper, Carhart-Harris reminds us
that in the 1960s the psychedelic experience was usually described as
"consciousness-expansion"; knowingly or not, Timothy Leary and
his colleagues had hit on exactly the right metaphor for the entropic
brain. This expansion metaphor also chimes with Huxley's reducing
valve, implying as it does that consciousness exists in a state of opening or contraction.

As a matter of experience, a quality as abstract as entropy is almost
impossible for us to perceive, but expansion, perhaps, is not. Judson
Brewer, the neuroscientist who studies meditation, has found that a
felt sense of expansion in consciousness correlates with a drop in
activity in one particular node of the default mode network—the
posterior cingulate cortex (PCC), which is associated with self-
referential processing. One of the most interesting things about a
psychedelic experience is that it sharpens one's sensitivity to one's
own mental states, especially in the days immediately following.
The usual seamlessness of consciousness is disturbed in such a way
as to make any given state—mind wandering, focused attention,
rumination—both more salient and somewhat easier to manipulate.
In the wake of my psychedelic experiences (and, perhaps, in the wake
of interviewing Judson Brewer), I found that when I put my mind to
it, I could locate my own state of consciousness on a spectrum ranging from contraction to expansion.

When, for example, I'm feeling especially generous or grateful, open to feelings and people and nature, I register a sense of expansion. This feeling is often accompanied by a diminution of ego, as well as a falloff in the attention paid to past and future on which the ego feasts. (And depends.) By the same token, there is a pronounced sense of contraction when I'm obsessing about things or feeling fearful, defensive, rushed, worried, and regretful. (These last two feelings don't exist without time travel.) At such times, I feel altogether more *me*, and not in a good way. If the neuroscientists are right, what I'm observing in my mind has a physical correlate in the brain: the default mode network is either online or off; entropy is either high or low. What exactly *to do* with this information I'm not yet sure.

o   o   o

BY NOW, it may be lost to memory, but all of us, even the psychedelically naive, have had direct personal experience of an entropic brain and the novel type of consciousness it sponsors—as a young child. Baby consciousness is so different from adult consciousness as to constitute a mental country of its own, one from which we are expelled sometime early in adolescence. Is there a way back in? The closest we can come to visiting that foreign land as adults may be during the psychedelic journey. This at least is the startling hypothesis of Alison Gopnik, a developmental psychologist and philosopher who happens to be a colleague of mine at Berkeley.

Alison Gopnik and Robin Carhart-Harris come at the problem of consciousness from what seem like completely different directions and disciplines, but soon after they learned of each other's work (I had e-mailed a PDF of Robin's entropy paper to Alison and told him about her superb book, *The Philosophical Baby*), they struck up a conversation that has proven to be remarkably illuminating, at least for me. In April 2016, their conversation wound up on a stage at a conference on consciousness in Tucson, Arizona, where the two met for the first time and shared a panel.*

In much the same way psychedelics have given Carhart-Harris an oblique angle from which to approach the phenomena of normal consciousness by exploring an altered state of it, Gopnik proposes we regard the mind of the young child as another kind of "altered state," and in a number of respects it is a strikingly similar one. She cautions that our thinking about the subject is usually constrained by our own restricted experience of consciousness, which we naturally take to be the whole of it. In this case, most of the theories and generalizations about consciousness have been made by people who share a fairly limited subtype of it she calls "professor consciousness," which she defines as "the phenomenology of your average middle-aged professor."

"As academics, either we're incredibly focused on a particular problem," Gopnik told the audience of philosophers and neuroscientists in Tucson, "or we're sitting there saying to ourselves, 'Why can't I focus on this problem I'm supposed to be focused on, and why instead am I daydreaming?'" Gopnik herself looks the part of a Berkeley professor in her early sixties, with her colorful scarves, flowing skirts, and sensible shoes. A child of the 1960s who is now a grandmother, she has a speaking style that is at once lighthearted and learned, studded with references indicating a mind as much at home in the humanities as the sciences.

"If you thought, as people often have thought, that this was all there was to consciousness . . . you might very well find yourself thinking that young children were actually less conscious than we were," because both focused attention and self-reflection are absent in young children. Gopnik asks us to think about child conscious-

---

* The panel was recorded and is available on YouTube: https://www.youtube.com/watch?v=v2VzRMevUXg.

ness in terms of not what's missing from it or undeveloped but rather what is uniquely and wonderfully present—qualities that she believes psychedelics can help us to better appreciate and, possibly, reexperience.

In *The Philosophical Baby*, Gopnik draws a useful distinction between the "spotlight consciousness" of adults and the "lantern consciousness" of young children. The first mode gives adults the ability to narrowly focus attention on a goal. (In his own remarks, Carhart-Harris called this "ego consciousness" or "consciousness with a point.") In the second mode—lantern consciousness—attention is more widely diffused, allowing the child to take in information from virtually anywhere in her field of awareness, which is quite wide, wider than that of most adults. (By this measure, children are more conscious than adults, rather than less.) While children seldom exhibit sustained periods of spotlight consciousness, adults occasionally experience that "vivid panoramic illumination of the everyday" that lantern consciousness affords us. To borrow Judson Brewer's terms, lantern consciousness is expansive, spotlight consciousness narrow, or contracted.

The adult brain directs the spotlight of its attention where it will and then relies on predictive coding to make sense of what it perceives. This is not at all the child's approach, Gopnik has discovered. Being inexperienced in the way of the world, the mind of the young child has comparatively few priors, or preconceptions, to guide her perceptions down the predictable tracks. Instead, the child approaches reality with the astonishment of an adult on psychedelics.

What this means for cognition and learning can be best understood by looking at machine learning, or artificial intelligence, Gopnik suggests. In teaching computers how to learn and solve problems, AI designers speak in terms of "high temperature" and "low temperature" searches for the answers to questions. A low-temperature search (so-called because it requires less energy) involves reaching for the most probable or nearest-to-hand answer, like the one that worked for a similar problem in the past. Low-temperature searches succeed more often than not. A high-temperature search requires more energy because it involves reaching for less likely but possibly more ingenious and creative answers—those found outside the box of preconception. Drawing on its wealth of experience, the adult mind performs low-temperature searches most of the time.

Gopnik believes that both the young child (five and under) and the adult on a psychedelic have a stronger predilection for the high-temperature search; in their quest to make sense of things, their minds explore not just the nearby and most likely but "the entire space of possibilities." These high-temperature searches might be inefficient, incurring a higher rate of error and requiring more time and mental energy to perform. High-temperature searches can yield answers that are more magical than realistic. Yet there are times when hot searches are the only way to solve a problem, and occasionally they return answers of surpassing beauty and originality. $E=mc^2$ was the product of a high-temperature search.

Gopnik has tested this hypothesis on children in her lab and has found that there are learning problems that four-year-olds are better at solving than adults. These are precisely the kinds of problems that require thinking outside the box, those times when experience hobbles rather than greases the gears of problem solving, often because the problem is so novel. In one experiment, she presented children with a toy box that lights up and plays music when a certain kind of block is placed on top of it. Normally, this "blicket detector" is set to respond to a single block of a certain color or shape, but when the experimenter reprograms the machine so that it responds only when *two* blocks are placed on it, four-year-olds figure it out much faster than adults do.

"Their thinking is less constrained by experience, so they will try even the most unlikely possibilities"; that is, they'll conduct lots of high-temperature searches, testing the most far-out hypotheses. "Children are better learners than adults in many cases when the solutions are nonobvious" or, as she puts it, "further out in the space of possibilities," a realm where they are more at home than we are. Far out, indeed.

"We have the longest childhood of any species," Gopnik says. "This extended period of learning and exploration is what's distinctive about us. I think of childhood as the R&D stage of the species, concerned exclusively with learning and exploring. We adults are production and marketing." Later I asked her if she meant to say that children perform R&D for the individual, not the species, but in fact she meant exactly what she said.

"Each generation of children confronts a new environment," she explained, "and their brains are particularly good at learning and thriving in that environment. Think of the children of immigrants, or four-year-olds confronted with an iPhone. Children don't invent these new tools, they don't create the new environment, but in every generation they build the kind of brain that can best thrive in it. Childhood is the species' ways of injecting noise into the system of cultural evolution." "Noise," of course, is in this context another word for "entropy."

"The child's brain is extremely plastic, good for learning, not accomplishing"—better for "exploring rather than exploiting." It also has a great many more neural connections than the adult brain. (During the panel, Carhart-Harris showed his map of the mind on psilocybin, with its dense forest of lines connecting every region to every other.) But as we reach adolescence, most of those connections get pruned, so that the "human brain becomes a lean, mean acting machine." A key element of that developmental process is the suppression of entropy, with all of its implications, both good and bad. The system cools, and hot searches become the exception rather than the rule. The default mode network comes online.

"Consciousness narrows as we get older," Gopnik says. "Adults have congealed in their beliefs and are hard to shift," she has written, whereas "children are more fluid and consequently more willing to entertain new ideas.

"If you want to understand what an expanded consciousness looks like, all you have to do is have tea with a four-year-old."

Or drop a tab of LSD. Gopnik told me she has been struck by the similarities between the phenomenology of the LSD experience and her understanding of the consciousness of children: hotter searches, diffused attention, more mental noise (or entropy), magical thinking, and little sense of a self that is continuous over time.

"The short summary is, babies and children are basically tripping all the time."

o o o

Surely this insight is interesting, but is it useful? Both Gopnik and Carhart-Harris believe it is, believe that the psychedelic experience, as they conceptualize it, has the potential to help people who are sick and people who are not. For the well, psychedelics, by introducing more noise or entropy into the brain, might shake people out of their usual patterns of thought—"lubricate cognition," in Carhart-Harris's words—in ways that might enhance well-being, make us more open and boost creativity. In Gopnik's terms, the drugs could help adults achieve the kind of fluid thinking that is second nature to kids, expanding the space of creative possibility. If, as Gopnik hypothesizes, "childhood is a way of injecting noise—and novelty—into the system of cultural evolution," psychedelics might do the same thing for the system of the adult mind.

As for the unwell, the patients who stand to gain the most are probably those suffering from the kinds of mental disorders characterized by mental rigidity: addiction, depression, obsession.

"There are a range of difficulties and pathologies in adults, like depression, that are connected with the phenomenology of rumination and an excessively narrow, ego-based focus," Gopnik says. "You get stuck on the same thing, you can't escape, you become obsessive, perhaps addicted. It seems plausible to me that the psychedelic experience could help us get out of those states, create an opportunity in which the old stories of who we are might be rewritten." The experience could work as a kind of reset—as when you "introduce a burst of noise into a system" that has gotten locked into a rigid pattern. Quieting the default mode network and loosening the grip of the ego—which she suggests may be illusory anyway—might also be helpful to such people. Gopnik's idea of a brain reboot sounded very much like Carhart-Harris's notion of shaking the snow globe: a way to boost entropy, or heat, in a system that has gotten frozen stuck.

Soon after publishing his entropy paper, Carhart-Harris resolved to put some of his theories into practice by testing them on patients. For the first time, the lab expanded its focus from pure research to a clinical application of that work. David Nutt secured a grant from the U.K. government for the lab to conduct a small pilot study looking at the potential of psilocybin to relieve the symptoms of "treatment-resistant depression"—patients who hadn't responded to the usual therapeutic protocols and drugs.

Doing clinical work was definitely outside Carhart-Harris's experience and comfort zone, as well as the lab's. One unfortunate early episode pointed up the inherent tensions between the roles of the clinician, devoted solely to the patient's welfare, and the scientist, intent on gathering data as well. After being injected with LSD in a trial Carhart-Harris was running (not a clinical trial, it should be pointed out), a volunteer in his late thirties named Toby Slater began feeling anxious in the fMRI scanner and asked to get out. After taking a break, Slater, perhaps hoping to please the researchers, volunteered to get back in the machine so they could complete the experiment. ("I'm afraid he could see my disappointment," Carhart-Harris recalls, ruefully.) But Slater's anxiety returned: "I felt like a lab rat," he told me. He asked to get out again and tried to leave the lab. The researchers had to persuade him to stay and let them administer a sedative.

Carhart-Harris describes the episode—one of the very few adverse events seen in the Imperial research—as "a learning experience" and, by all accounts, he has since shown himself to be a compassionate and effective clinician as well as an original scientist—surely a rare combination. The response of most patients in the depression trial, as we will see in the following chapter, has been remarkably positive, at least in the short term. Over dinner at a restaurant in West London, Robin told me about one severely depressed woman in the trial whom over the course of several meetings he had never once seen smile. As he sat with her during her psilocybin journey, "she smiled for the very first time.

"'It's nice to smile,' she said.

"After it was over, she told me she had been visited by a guardian angel. She described a presence of some kind, a voice that was entirely supportive and wanted her to be well. It would say things like 'Darling, you need to smile more, hold your head up high, stop looking down at the ground. Then it reached over and pushed up my cheeks,' she said, 'lifting the corners of my mouth.'

"That must have been what was happening in her mind when I observed her smiling," Robin said, now smiling himself, broadly if a bit sheepishly. In the aftermath of her experience, the woman's depression score dropped from thirty-six to four.

"I have to say, that was a very nice feeling."

CHAPTER SIX

THE TRIP TREATMENT
Psychedelics in Psychotherapy

One: Dying

AT NEW YORK UNIVERSITY, psilocybin trips take place in a treatment room carefully decorated to look more like a cozy den than a hospital suite. The effect almost works, but not entirely, for the stainless steel and plastic fittings of modern medicine peek through the domestic scrim here and there, chilly reminders that the room you are tripping in is still in the belly of a big city hospital complex.

Against one wall is a comfortable couch long enough for a patient to stretch out on during a session. An abstract painting—or is it a cubist landscape?—hangs on the opposite wall, and on the book-shelves large-format books about art and mythology share space with native craft items and spiritual knickknacks—a large glazed ceramic mushroom, a Buddha, a crystal. This could be the apartment of a well-traveled shrink of a certain age, one with an interest in Eastern religions and the art of what used to be called primitive cultures. Yet the illusion crumbles as soon as you lift your gaze to the ceiling, where the tracks that would ordinarily support the curtains dividing one hospital bed from another traverse the white acoustic tiles. And then there is the supersized bathroom, ablaze with fluorescent light and outfitted with the requisite grab bars and pedals.

It was here in this room that I first heard the story of Patrick Mettes, a volunteer in NYU's psilocybin cancer trial who, in the course of a turbulent six-hour psilocybin journey on the couch where I now sat, had a life-changing—or perhaps I should say death-changing—experience. I had come to interview Tony Bossis, the palliative care psychologist who guided Mettes that day, and his colleague Stephen Ross, the Bellevue psychiatrist who directed the trial, which sought to determine whether a single high dose of psilocybin could alleviate the anxiety and depression that often follow a life-threatening cancer diagnosis.

While Bossis, hirsute and bearish, looks the part of a fifty-something Manhattan shrink with an interest in alternative therapies, Ross, who is in his forties, comes across as more of a straight arrow; neatly trimmed in a suit and tie, he could pass for a Wall Street banker. A bookish teenager growing up in L.A., Ross says he had no personal experience of psychedelics and knew next to nothing about them before a colleague happened to mention that LSD had been used successfully to treat alcoholics in the 1950s and 1960s. This being his psychiatric specialty, Ross did some research and was astonished to discover a "completely buried body of knowledge." By the 1990s, when he began his residency in psychiatry at Columbia and the New York State Psychiatric Institute, the history of psychedelic therapy had been erased from the field, never to be mentioned.

The trial at NYU, along with a sister study conducted in Roland Griffiths's lab at Johns Hopkins, represents one of a handful of efforts to pick up the thread of inquiry that got dropped in the 1970s when sanctioned psychedelic therapy ended. While the NYU and Hopkins trials are assessing the potential of psychedelics to help the dying, other trials now under way are exploring the possibility that psychedelics (usually psilocybin rather than LSD, because, as Ross explained, it "carries none of the political baggage of those three letters") could be used to lift depression and break addictions—to alcohol, cocaine, and tobacco.

None of this work is exactly new: to delve into the history of clinical research with psychedelics is to realize that most of this ground has already been tilled. Charles Grob, the UCLA psychiatrist whose 2011 pilot study of psilocybin for cancer anxiety cleared the path for the NYU and Hopkins trials, acknowledges that "in a lot of ways we are simply picking up the torch from earlier generations of researchers who had to put it down because of cultural pressures." But if psychedelics are ever to find acceptance in modern medicine, all this buried knowledge will need to be excavated and the experiments that produced it reprised according to the prevailing scientific standards.

Yet even as psychedelic therapies are being tested by modern science, the very strangeness of these molecules and their actions upon the mind is at the same time testing whether Western medicine can deal with the implicit challenges they pose. To cite one obvious example, conventional drug trials of psychedelics are difficult if not impossible to blind: most participants can tell whether they've received psilocybin or a placebo, and so can their guides. Also, in testing these drugs, how can researchers hope to tease out the chemical's effect from the critical influence of set and setting? Western science and modern drug testing depend on the ability to isolate a single variable, but it isn't clear that the effects of a psychedelic drug can *ever* be isolated, whether from the context in which it is administered, the presence of the therapists involved, or the volunteer's expectations. Any of these factors can muddy the waters of causality.

And how is Western medicine to evaluate a psychiatric drug that appears to work not by means of any strictly pharmacological effect but by administering a certain kind of experience in the minds of the people who take it?

Add to this the fact that the kind of experience these drugs sponsor often goes under the heading of "spiritual," and you have, with psychedelic therapy, a very large pill for modern medicine to swallow. Charles Grob well appreciates the challenge but is also refreshingly unapologetic about it: he describes psychedelic therapy as a form of "applied mysticism." This is surely an odd phrase to hear on the lips of a scientist, and to many ears it sounds dangerously unscientific.

"For me that is not a medical concept," Franz Vollenweider, the pioneering psychedelic researcher, told *Science* magazine, when asked to comment on the role of mysticism in psychedelic therapy. "It's more like an interesting shamanic concept." But other researchers working on psychedelics don't run from the idea that elements of shamanism might have a role to play in psychedelic therapy—as indeed it has probably done for several thousand years before there was such a thing as science. "If we are to develop optimal research designs for evaluating the therapeutic utility of hallucinogens," Grob has written, "it will not be sufficient to adhere to strict standards of scientific methodology alone. We must also pay heed to the examples provided us by such successful applications of the shamanic paradigm." Under that paradigm, the shaman/therapist carefully orchestrates "extrapharmacological variables" such as set and setting in order to put the "hyper-suggestible properties" of these medicines to best use. This is precisely where psychedelic therapy seems to be operating: on a frontier between spirituality and science that is as provocative as it is uncomfortable.

Yet the new research into psychedelics comes along at a time when mental health treatment in this country is so "broken"—to use the word of Tom Insel, who until 2015 was director of the National Institute of Mental Health—that the field's willingness to entertain radical new approaches is perhaps greater than it has been in a generation. The pharmacological toolbox for treating depression—which afflicts nearly a tenth of all Americans and, worldwide, is the leading cause of disability—has little in it today, with antidepressants losing their effectiveness* and the pipeline for new psychiatric drugs drying up. Pharmaceutical companies are no longer investing in the development of so-called CNS drugs—medicines targeted at the central nervous system. The mental health system reaches only a fraction of the people suffering from mental disorders, most of whom are discouraged from seeking treatment by its cost, social stigma, or ineffectiveness. There are almost forty-three thousand suicides every year in America (more than the number of deaths from either breast cancer or auto accidents), yet only about half of the people who take their lives have ever received mental health treatment. "Broken" does not seem too harsh a characterization of such a system.

Jeffrey Guss, a Manhattan psychiatrist and a coinvestigator on the NYU trial, thinks the moment could be ripe for psychotherapy to entertain a completely new paradigm. Guss points out that for many years now "we've had this conflict between the biologically based treatments and psychodynamic treatments. They've been fighting one another for legitimacy and resources. Is mental illness a disorder of chemistry, or is it a loss of meaning in one's life? Psychedelic therapy is the wedding of those two approaches."

In recent years, "psychiatry has gone from being brainless to

* As in the case of many drugs, the SSRI antidepressants introduced in the 1980s were much more effective when they were new, probably owing to the placebo effect. Today, they perform only slightly better than a placebo.

being mindless," as one psychoanalyst has put it. If psychedelic therapy proves successful, it will be because it succeeds in rejoining the brain and the mind in the practice of psychotherapy. At least that's the promise.

For the therapists working with people approaching the end of life, these questions are of more than academic interest. As I chatted with Stephen Ross and Tony Bossis in the NYU treatment room, I was struck by their excitement, verging on giddiness, at the results they were observing in their cancer patients—after a single guided psilocybin session. At first, Ross couldn't believe what he was seeing: "I thought the first ten or twenty people were plants—that they must be faking it. They were saying things like 'I understand love is the most powerful force on the planet' or 'I had an encounter with my cancer, this black cloud of smoke.' People were journeying to early parts of their lives and coming back with a profound new sense of things, new priorities. People who had been palpably scared of death—they lost their fear. The fact that a drug given once could have such an effect for so long is an unprecedented finding. We have never had anything like that in the psychiatric field."

This is when Tony Bossis first told me about his experience sitting with Patrick Mettes as he journeyed to a place in his mind that, somehow, lifted the siege of his terror.

"You're in this room, but you're in the presence of something large. I remember how, after two hours of silence, Patrick began to cry softly and say, twice, 'Birth and death is a lot of work.' It's humbling to sit there. It's the most rewarding day of your career."

As a palliative care specialist, Bossis spends a lot of his time with the dying. "People don't realize how few tools we have in psychiatry to address existential distress." Existential distress is what psychologists call the complex of depression, anxiety, and fear common in people confronting a terminal diagnosis. "Xanax isn't the answer." If there is an answer, Bossis believes, it is going to be more spiritual in nature than pharmacological.

"So how do we not explore this," he asks, "if it can recalibrate how we die?"

o   o   o

IT WAS ON AN APRIL MONDAY in 2010 that Patrick Mettes, a fifty-three-year-old television news director being treated for a cancer of his bile ducts, read the article on the front page of the *New York Times* that would change his death. His diagnosis had come three years earlier, shortly after his wife, Lisa Callaghan, noticed that the whites of his eyes had suddenly turned yellow. By 2010, the cancer had spread to Patrick's lungs, and he was buckling under the weight of an especially debilitating chemotherapy regime and the dawning realization that he might not survive. The article, headlined "Hallucinogens Have Doctors Tuning In Again," briefly mentioned research at NYU, where psilocybin was being tested to relieve existential distress in cancer patients. According to Lisa, Patrick had no experience with psychedelics, but he immediately determined to call NYU and volunteer.

Lisa was against the idea. "I didn't want there to be an easy way out," she told me. "I wanted him to fight."

Patrick placed the call anyway and, after filling out some forms and answering a long list of questions, was accepted into the trial. He was assigned to Tony Bossis. Tony was roughly the same age as Patrick; he is also a soulful man of uncommon warmth and compassion, and the two immediately hit it off.

At their first meeting, Bossis told Patrick what to expect. After three or four preparatory sessions of talking therapy, Patrick would be scheduled for two dosings—one of them an "active placebo" (in this case a high dose of niacin, which produces a tingling sensation), and the other a capsule containing twenty-five milligrams of psilo-cybin. Both sessions would take place in the treatment room where I met Bossis and Ross. During each session, which would last the better part of a day, Patrick would lie on the couch wearing eyeshades and listening through headphones to a playlist of carefully curated music—Brian Eno, Philip Glass, Pat Metheny, and Ravi Shankar, as well as some classical and New Age compositions. Two sitters—one of them male (Bossis) and the other female (Krystallia Kalliontzi)—would be in attendance for the duration, saying very little but available to help should he run into any trouble. In preparation, the two shared with Patrick the set of "flight instructions" written by the Hopkins researcher Bill Richards.

Bossis suggested that Patrick use the phrase "Trust and let go" as a kind of mantra for his journey. Go wherever it takes you, he advised: "Climb staircases, open doors, explore paths, fly over landscapes." But the most important advice for the journey he offered is always to move toward, rather than try to flee, anything truly threatening or monstrous you encounter—look it straight in the eyes. "Dig in your heels and ask, 'What are you doing in my mind?' Or, 'What can I learn from you?'"

o   o   o

THE IDEA OF GIVING a psychedelic drug to the dying was first broached not by a therapist or scientist but by Aldous Huxley in a letter to Humphry Osmond, proposing a research project involving "the administration of LSD to terminal cancer cases, in the hope that it would make dying a more spiritual, less strictly physiological process." Huxley himself had his wife, Laura, give him an injection of LSD when he was on his own deathbed, on November 22, 1963.

By then, Huxley's idea had been tested on a number of cancer patients in North America. In 1965, Sidney Cohen wrote an essay for *Harper's* ("LSD and the Anguish of Dying") exploring the potential of psychedelics to "alter[] the experience of dying." He described treatment with LSD as "therapy by self-transcendence." The premise behind the approach was that our fear of death is a function of our egos, which burden us with a sense of separateness that can become unbearable as we approach death. "We are born into an egoless world," Cohen wrote, "but we live and die imprisoned within ourselves."

The idea was to use psychedelics to escape the prison of self. "We wanted to provide a brief, lucid interval of complete egolessness to demonstrate that personal intactness was not absolutely necessary, and that perhaps there was something 'out there'"—something greater than our individual selves that might survive our demise. Cohen quoted a patient, a woman dying of ovarian cancer, describing the shift in her perspective following an LSD session:

> My extinction is not of great consequence at this moment, not even for me. It's just another turn in the swing of existence and non-existence. I feel it has little to do with the church or talk of death. I suppose that I'm detached—that's it—away from myself and my pain and my decaying. I could die nicely now—if it should be so. I do not invite it, nor do I put it off.

In 1972, Stanislav Grof and Bill Richards, who were working together at Spring Grove, wrote that LSD gave patients an experience "of cosmic unity" such that death, "instead of being seen as the absolute end of everything and a step into nothingness, appears suddenly as a transition into another type of existence . . . The idea of possible continuity of consciousness beyond physical death becomes much more plausible than the opposite."

VOLUNTEERS IN THE NYU psilocybin trial are required to write an account of their journey soon after its completion, and Patrick Mettes, who worked in journalism, took the assignment seriously. His wife, Lisa, said that after his Friday session Patrick labored all weekend to make sense of the experience and write it down. Lisa agreed to share his account with me and also gave Patrick's therapist, Tony Bossis, permission to show me the notes he took during the session, as well as his notes from several follow-up psychotherapy sessions.

Lisa, who at the time worked as a marketing executive for a cookware company, had an important meeting on that January morning in 2011, so Patrick came by himself to the treatment room in the NYU dental school on First Avenue and Twenty-fourth Street, taking the subway from their apartment in Brooklyn. (The treatment room was in the dental college because, at the time, both Bellevue and NYU's cancer center wanted to keep their distance from a trial involving psychedelics.) Tony Bossis and Krystallia Kalliontzi, his guides, greeted him, reviewed the day's plans, and then at 9:00 a.m. presented Patrick with a chalice containing the pill; whether it contained psilocybin or the placebo, none of them would know for at least thirty minutes. Patrick was asked to state his intention, which he said was to learn to cope better with the anxiety and depression he felt about his cancer and to work on what he called his "regret in life." He placed a few photographs around the room, of himself and Lisa on their wedding day and of their dog, Arlo.

At 9:30, Patrick lay down on the couch, put on the headphones and eyeshades, and fell quiet. In his account, Patrick likened the start of the journey to the launch of a space shuttle: "a physically violent and rather clunky liftoff which eventually gave way to the blissful serenity of weightlessness."

Many of the volunteers I interviewed reported initial episodes of intense fear and anxiety before giving themselves up to the experience, as the guides encourage them to do. This is where the flight instructions come in. Their promise is that if you surrender to whatever happens ("trust, let go, and be open" or "relax and float downstream"), whatever at first might seem terrifying will soon morph into something else, and likely something pleasant, even blissful.

Early in his journey, Patrick encountered his brother's wife, who died of cancer more than twenty years earlier, at forty-three. "Ruth acted as my tour guide," he wrote, and "didn't seem surprised to see me. She 'wore' her translucent body so I would know her . . . This period of my journey seemed to be about the feminine." Michelle Obama made an appearance. "The considerable feminine energy all around me made clear the idea that a mother, any mother, regardless of her shortcomings . . . could never NOT love her offspring. This was very powerful. I knew I was crying . . . it was here that I felt as if I was coming out of the womb . . . being birthed again. My rebirth was smooth . . . comforting."

Outwardly, however, what was happening to Patrick appeared to be anything but smooth. He was crying, Bossis noted, and breathing heavily. This is when he first said, "Birth and death is a lot of work," and seemed to be convulsing. Then Patrick reached out and clutched Kalliontzi's hand while pulling up his knees and pushing, as if he were delivering a baby. From Bossis's notes:

> 11:15 "Oh God."
> 11:25 "It's really so simple."
> 11:47 "Who knew a man could give birth?" And then,
>   "I gave birth, to what I don't know."

> 12:10 "It's just too amazing." Patrick is alternately laugh-
> ing and crying at this point. "Oh God, it all makes
> sense now, so simple and beautiful."

Now Patrick asked to take a break. "It was getting too intense," he wrote. He removed the headphones and eyeshades. "I sat up and spoke with Tony and Krystallia. I mentioned that everyone deserved to have this experience . . . that if everyone did, no one could ever do harm to another again . . . wars would be impossible to wage. The room and everything in it was beautiful. Tony and Krystallia, sitting on [their] pillows, were radiant!" They helped him to the bathroom. "Even the germs (if there were any present) were beautiful, as was everything in our world and universe."

Afterward, he voiced some reluctance to "go back in."

"The work was considerable but I loved the sense of adventure." Eventually, he put his eyeshades and headphones on and lay back down.

"From here on, love was the only consideration . . . It was and is the only purpose. Love seemed to emanate from a single point of light . . . and it vibrated . . . I could feel my physical body trying to vibrate in unity with the cosmos . . . and, frustratingly, I felt like a guy who couldn't dance . . . but the universe accepted it. The sheer joy . . . the bliss . . . the nirvana . . . was indescribable. And in fact there are no words to accurately capture my experience . . . my state . . . this place. I know I've had no earthly pleasure that's ever come close to this feeling . . . no sensation, no image of beauty, noth-ing during my time on earth has felt as pure and joyful and glorious as the height of this journey." Aloud, he said, "Never had an orgasm of the soul before." The music loomed large in the experience: "I was learning a song and the song was simple . . . it was one note . . . C . . .

it was the vibration of the universe . . . a collection of everything that ever existed . . . all together equaling God."

Patrick then described an epiphany having to do with simplicity. He was thinking about politics and food, music and architecture, and—his field—television news, which he realized was, like so much else, "over-produced. We put too many notes in a song . . . too many ingredients in our recipes . . . too many flourishes in the clothes we wear, the houses we live in . . . it all seemed so pointless when really all we needed to do was focus on the love." Just then he saw Derek Jeter, then the Yankee shortstop, "making yet another balletic turn to first base."

"I was convinced in that moment I had figured it all out . . . It was right there in front of me . . . love . . . the only thing that mattered. This was now to be my life's cause."

Then he said something that Bossis jotted down at 12:15: "Ok, I get it! You can all punch out now. Our work is done."

But it wasn't done, not yet. Now "I took a tour of my lungs . . . I remember breathing deeply to help facilitate the 'seeing.'" Bossis noted that at 2:30 Patrick had said, "I went into my lungs and saw two spots. They were no big deal.

"I was being told (without words) not to worry about the cancer . . . it's minor in the scheme of things . . . simply an imperfection of your humanity and that the more important matter . . . the real work to be done is before you. Again, love."

Now Patrick experienced what he called "a brief death."

"I approached what appeared to be a very sharp, pointed piece of stainless steel. It had a razor blade quality to it. I continued up to the apex of this shiny metal object and as I arrived, I had a choice, to look or not look, over the edge and into the infinite abyss . . . the vastness of the universe . . . the eye of everything . . . [and] of nothing. I was hesitant but not frightened. I wanted to go all in but felt that if I did, I would possibly leave my body permanently . . . death from this life. But it was not a difficult decision . . . I knew there was much more for me here." Telling his guides about his choice, Patrick explained that he "was not ready to jump off and leave Lisa."

Then, rather suddenly around 3:00 p.m., it was over. "The transition from a state where I had no sense of time or space to the relative dullness of now, happened quickly. I had a headache."

When Lisa arrived to take him home, Patrick "looked like he had run a race," she recalled. "The color in his face was not good, he looked tired and sweaty, but he was on fire. He was lit up with all the things he wanted to tell me and all the things he couldn't." He told her he "had touched the face of God."

o   o   o

EVERY PSYCHEDELIC JOURNEY is different, yet a few common themes seem to recur in the journeys of those struggling with cancer. Many of the cancer patients I interviewed described an experience of either giving birth or being reborn, though none quite as intense as Patrick's. Many also described an encounter with their cancer (or their fear of it) that had the effect of shrinking its power over them. I mentioned earlier the experience of Dinah Bazer, a petite and mild New Yorker in her sixties, a figure-skating instructor, who was diagnosed with ovarian cancer in 2010. When we met in the NYU treatment room, Dinah, who has auburn curls and wore large hoop earrings, told me that even after a successful course of chemotherapy she was paralyzed by the fear of a recurrence and wasted her days "waiting for the other shoe to drop."

She too worked with Tony Bossis and in the difficult first moments of her session imagined herself trapped in the hold of a ship, rocking back and forth, consumed by fear. "I stuck my hand out from under the blanket and said, 'I am so scared.' Tony took my hand and told me to just go with it. His hand became my anchor.

"I saw my fear. Almost as in a dream, my fear was located under my rib cage on the left side; it was not my tumor, but it was this black thing in my body. And it made me immensely angry; I was enraged by my fear. I screamed, *'Get the fuck out! I won't be eaten alive.'* And you know what? It was gone! It went away. I drove it away with my anger." Dinah reports that years later it hasn't returned. "The cancer is something completely out of my control, but the fear, I realized, is not."

Dinah's epiphany gave way to feelings of "overwhelming love" as her thoughts turned from her fear to her children. She told me she was and remains a "solid atheist," and yet "the phrase that I used—which I hate to use but it's the only way to describe it—is that I felt 'bathed in God's love.'" Paradox is a hallmark of the mystical experience, and the contradiction between the divine love Dinah felt and "not having a shred of belief" didn't seem to faze her. When I pointed this out, she shrugged and then smiled: "What other way is there to express it?"

Not surprisingly, visions of death loom large in the journeys taken by the cancer patients I interviewed at NYU and Hopkins. A breast cancer survivor in her sixties (who asked to remain anonymous) described zipping merrily through space as if in a video game until she arrived smack at the wall of a crematorium and realized, with a fright, "I've died and now I'm going to be cremated. (But I didn't have the experience of burning—how could I? I was dead!) The next thing I know, I'm belowground in this gorgeous forest, deep woods, loamy and brown. There are roots all around me and I'm seeing the trees growing, and I'm part of them. I had died but I was there in the ground with all these roots and it didn't feel sad or happy, just natural, contented, peaceful. I wasn't gone. I was part of the earth."

Several cancer patients described edging up to the precipice of death and looking over to the other side before drawing back. Tammy Burgess, diagnosed with ovarian cancer at fifty-five, found herself peering across "the great plane of consciousness. It was very serene and beautiful. I felt alone, but I could reach out and touch anyone I'd ever known.

"When my time came, that's where my life would go once it left me, and that was okay."

The uncanny authority of the psychedelic experience might help explain why so many cancer patients in the trials reported that their fear of death had lifted or at least abated: they had stared directly at death and come to know something about it, in a kind of dress rehearsal. "A high-dose psychedelic experience is death practice," says Katherine MacLean, the former Hopkins psychologist. "You're losing everything you know to be real, letting go of your ego and your body, and that process can feel like dying." And yet the experience brings the comforting news that there is *something* on the other side of that death—whether it is the "great plane of consciousness" or one's ashes underground being taken up by the roots of trees—and some abiding, disembodied intelligence to somehow know it. "Now I am aware that there is a whole other 'reality,'" one NYU volunteer told a researcher a few months after her journey. "Compared to other people, it is like I know another language."

At a follow-up session with Tony Bossis a few weeks after his journey, Patrick Mettes—whom his wife, Lisa, describes as "an earthy, connected person, a doer"—discussed the idea of an afterlife. Bossis's notes indicate that Patrick interpreted his journey as "pretty clearly a window . . . [on] a kind of afterlife, something beyond this physical body." He spoke of "the plane of existence of love" as "infinite." In subsequent sessions, Patrick talked about his body and cancer "as [a] type of illusion." It also became clear that, psychologically at least, Patriek was doing remarkably well in the aftermath of his session. He was meditating regularly, felt he had become better able to live in the present, and "described loving [his] wife even more." In a session in March, two months out from his journey, Bossis noted that Patrick, though slowly dying of cancer, "feels the happiest in his life."

"I am the luckiest man on earth."

o   o   o

HOW MUCH SHOULD THE AUTHENTICITY of these experiences concern us? Most of the therapists involved in the research take a scrupulously pragmatic view of the question. They're fixed on relieving their patients' suffering and exhibit scant interest in metaphysical theories or questions of truth. "That's above my pay grade," Tony Bossis said with a shrug when I asked him whether he thought the experiences of cosmic consciousness described by his patients were fictive or real. Asked the same question, Bill Richards cited William James, who suggested we judge the mystical experience not by its veracity, which is unknowable, but by "its fruits": Does it turn someone's life in a positive direction?

Many researchers acknowledge that a strong placebo effect may be at work when a drug as suggestible as psilocybin is administered by medical professionals with legal and institutional sanction: under such conditions, the expectations of the therapist are much more likely to be fulfilled by the patient. (And bad trips are much less likely to occur.) Here we bump into one of the richer paradoxes of the psilocybin trials: while it succeeds in no small part because it has the sanction and authority of science, its effectiveness seems to depend on a mystical experience that leaves people convinced there is more to this world than science can explain. Science is being used to validate an experience that would appear to undermine the scientific perspective in what might be called White-Coat Shamanism.

Are questions of truth important, if the therapy helps people who are suffering? I had difficulty finding anyone involved in the research who was troubled by such questions. David Nichols, the retired Purdue University chemist and pharmacologist who founded the Heffter Research Institute in 1993 to support psychedelic research (including the trials at Hopkins, for which he synthesized the psilocybin), puts the pragmatic case most baldly. In a 2014 interview with *Science* magazine, he said, "If it gives them peace, if it helps people to die peacefully with their friends and their family at their side, I don't care if it's real or an illusion."

For his part, Roland Griffiths acknowledges that "authenticity is a scientific question not yet answered. All we have to go by is the phenomenology"—that is, what people tell us about their internal experiences. That's when he began querying me about my own spiritual development, which I confessed was still fairly rudimentary; I told him my worldview has always been staunchly materialist.

"Okay, then, but what about the miracle that we are conscious? Just think about that for a second, that we are aware *and* that we are aware that we are aware! How unlikely is *that*?" How can we be certain, he was suggesting, that our experience of consciousness is "authentic"? The answer is we can't; it is beyond the reach of our science, and yet who doubts its reality? In fact, the evidence for the existence of consciousness is much like the evidence for the reality of the mystical experience: we believe it exists not because science can independently verify it but because a great many people have been convinced of its reality; here, too, all we have to go on is the phenomenology. Griffiths was suggesting that insofar as I was on board for one "miracle" well beyond the reach of materialist science—"the marvel of consciousness," as Vladimir Nabokov once called it, "that sudden window swinging open on a sunlit landscape amidst the night of non-being"—maybe I needed to keep a more open mind to the possibility of others.

In December 2016, a front-page story in the *New York Times* reported on the dramatic results of the Johns Hopkins and NYU psilocybin cancer studies, which were published together in a special issue of the *Journal of Psychopharmacology*, along with nearly a dozen commentaries from prominent voices in the mental health establishment—including two past presidents of the American Psychiatric Association—hailing the findings.

In both the NYU and the Hopkins trials, some 80 percent of cancer patients showed clinically significant reductions in standard measures of anxiety and depression, an effect that endured for at least six months after their psilocybin session. In both trials, the intensity of the mystical experience volunteers reported closely correlated with the degree to which their symptoms subsided. Few if any psychiatric interventions of any kind have demonstrated such dramatic and sustained results.*

The trials were small—eighty subjects in all—and will have to be repeated on a larger scale before the government will consider re-

---

* The statistical "effect size" of these results—at or above 1.0 for most of the outcome measures used in both trials—is remarkable for a psychiatric treatment. As a comparison, when the SSRI antidepressants had their first clinical trials, the effect size was only 0.3—which was good enough for them to be approved.

scheduling psilocybin and approving the treatment.* But the results were encouraging enough to win the attention and cautious support of the mental health community, which has called for more research. Dozens of medical schools have asked to participate in future trials, and funders have stepped forward to underwrite those trials. After decades in the shadows, psychedelic therapy is suddenly respectable again, or nearly so. New York University, which proudly promoted the results of a trial it had once only tolerated somewhat grudgingly, invited Stephen Ross to move his treatment room from the dental college into the main hospital. Even the NYU cancer center, which had initially been reluctant to refer patients to the psilocybin trial, asked Ross to set up a treatment room on its premises for an upcoming trial.

The papers offered little in the way of a theory to explain the effects of psilocybin, except to point out that the patients with the best outcomes were the ones who had the most complete mystical experience. But exactly why should that experience translate into relief from anxiety and depression? Is it the intimation of some kind of immortality that accounts for the effect? This seems too simple and fails to account for the variety of experiences people had, many of which did not dwell on an afterlife. And some of the ones that did conceived of what happens after death in naturalistic terms, as when the anonymous volunteer imagined herself as "part of the earth," molecules of matter being taken up by the roots of trees. This really happens.

---

* A few critical voices were heard. In a pair of blog posts on *PLOS*, James Coyne raised several methodological objections having to do with the size and composition of the patient group, the reliability of the diagnoses, the placebo control, the blinding, and the theoretical assumptions: "Since when are existential/spiritual well-being issues psychiatric?" http://blogs.plos.org/mindthebrain/2016/12/14/psilocybin-as-a-treatment-for-cancer-patients-who-are-not-depressed-the-nyu-study/.

Of course the mystical experience consists of several components, most of which don't require a supernatural explanation. The dissolution of the sense of self, for example, can be understood in either psychological or neurobiological terms (as possibly the disintegration of the default mode network) and may explain many of the benefits people experienced during their journeys without resort to any spiritual conception of "oneness." Likewise, the sense of "sacredness" that classically accompanies the mystical experience can be understood in more secular terms as simply a heightened sense of meaning or purpose. It's still early days in our understanding of consciousness, and no single one of our vocabularies for approaching the subject—the biological, the psychological, the philosophical, or the spiritual—has yet earned the right to claim it has the final word. It may be that by layering these different perspectives one upon the other, we can gain the richest picture of what might be going on.

In a follow-up study to the NYU trial, "Patient Experiences of Psilocybin-Assisted Psychotherapy," published in the *Journal of Humanistic Psychology* in 2017, Alexander Belser, a member of the NYU team, interviewed volunteers to better understand the psychological mechanisms underlying the transformations they experienced. I read the study as a subtle attempt to move beyond the mystical experience paradigm to a more humanistic one and at the same time to underscore the importance of the psychotherapist in the psychedelic experience. (Note the use of the term "psilocybin-assisted psychotherapy" in the title; neither of the papers in *Psychopharmacology* mentioned psychotherapy in its title, only the drug.)

A few key themes emerged. All of the patients interviewed described powerful feelings of connection to loved ones ("relational embeddedness" is the term the authors used) and, more generally, a shift "from feelings of separateness to interconnectedness." In most cases, this shift was accompanied by a repertoire of powerful emotions, including "exalted feelings of joy, bliss, and love." Difficult passages during the journey were typically followed by positive feelings of surrender and acceptance (even of their cancers) as people's fears fell away.

Jeffrey Guss, a coauthor on the paper and a psychiatrist, interprets what happens during the session in terms of the psilocybin's "egolytic" effects—the drug's ability to either silence or at least muffle the voice of the ego. In his view, which is informed by his psychoanalytic training, the ego is a mental construct that performs certain functions on behalf of the self. Chief among these are maintaining the boundary between the conscious and the unconscious realms of the mind and the boundary between self and other, or subject and object. It is only when these boundaries fade or disappear, as they seem to do under the influence of psychedelics, that we can "let go of rigid patterns of thought, allowing us to perceive new meanings with less fear."

The whole question of meaning is central to the approach of the NYU therapists,* and is perhaps especially helpful in understanding the experience of the cancer patients on psilocybin. For many of these patients, a diagnosis of terminal cancer constitutes, among other things, a crisis of meaning. *Why me? Why have I been singled out for this fate? Is there any sense to life and the universe?* Under the weight of this existential crisis, one's horizon shrinks, one's emotional repertoire contracts, and one's focus narrows as the mind turns in on

---

* Several of the NYU therapists referred me to the writing of Viktor E. Frankl, the Viennese psychoanalyst and the author of *Man's Search for Meaning.* Frankl, who survived both Auschwitz and Dachau, believed that the crucial human drive is not for pleasure, as his teacher Freud maintained, or power, as Alfred Adler maintained, but meaning. Frankl concurs with Nietzsche, who wrote, "He who has a Why to live for can bear almost any How."

itself, shutting out the world. Loops of rumination and worry come to occupy more of one's mental time and space, reinforcing habits of thought it becomes ever more difficult to escape.

Existential distress at the end of life bears many of the hallmarks of a hyperactive default network, including obsessive self-reflection and an inability to jump the deepening grooves of negative thinking. The ego, faced with the prospect of its own extinction, turns inward and becomes hypervigilant, withdrawing its investment in the world and other people. The cancer patients I interviewed spoke of feeling closed off from loved ones, from the world, and from the full range of emotions; they felt, as one put it, "existentially alone."

By temporarily disabling the ego, psilocybin seems to open a new field of psychological possibility, symbolized by the death and rebirth reported by many of the patients I interviewed. At first, the falling away of the self feels threatening, but if one can let go and surrender, powerful and usually positive emotions flow in—along with formerly inaccessible memories and sense impressions and meanings. No longer defended by the ego, the gate between self and other—Huxley's reducing valve—is thrown wide open. And what comes through that opening for many people, in a great flood, is love. Love for specific individuals, yes, but also, as Patrick Mettes came to feel (to *know!*), love for everyone and everything—love as the meaning and purpose of life, the key to the universe, and the ultimate truth.

So it may be that the loss of self leads to a gain in meaning. Can this be explained biologically? Probably not yet, but recent neuroscience offers a few intriguing clues. Recall that the Imperial College team found that when the default mode network disintegrates (taking with it the sense of self), the brain's overall connectivity increases, allowing brain regions that don't ordinarily communicate to form new lines of connection. Is it possible that some of these new connections in the brain manifest in the mind as new meanings or perspectives? The connecting of formerly far-flung dots?

It may also be that psychedelics can directly imbue otherwise irrelevant sensory information with meaning. A recent paper in *Current Biology** described an experiment in which pieces of music that held no personal relevance for volunteers were played for them while on LSD. Under the influence of the psychedelic, however, volunteers attributed marked and lasting personal meaning to the same songs. These medicines may help us construct meaning, if not discover it.

No doubt the suggestibility of the mind on psychedelics and the guiding presence of psychotherapists also play a role in attributing meaning to the experience. In preparing volunteers for their journeys, Jeffrey Guss speaks explicitly about the acquisition of meaning, telling his patients "that the medicine will show you hidden or unknown shadow parts of yourself; that you will gain insight into yourself, and come to learn about the meaning of life and existence." (He also tells them they may have a mystical or transcendent experience but carefully refrains from defining it.) "As a result of this molecule being in your body, you'll understand more about yourself and life and the universe." And more often than not this happens. Replace the science-y word "molecule" with "sacred mushroom" or "plant teacher," and you have the incantations of a shaman at the start of a ceremonial healing.

But however it works, and whatever vocabulary we use to explain

---

* Katrin H. Preller et al., "The Fabric of Meaning and Subjective Effects in LSD-Induced States Depend on Serotonin 2A Receptor Activation," *Current Biology* 27, no. 3 (2017): 451–57. The work was done in Franz Vollenweider's lab. When the serotonin 5-HT_{2A} receptors were blocked with a drug (ketanserin), "the LSD-induced attribution of personal relevance to previously meaningless stimuli" was also blocked, leading the authors to conclude that these receptors play a role in the generation and attribution of personal meaning.

it, this seems to me the great gift of the psychedelic journey, especially to the dying: its power to imbue everything in our field of experience with a heightened sense of purpose and consequence. Depending on one's orientation, this can be understood either in humanistic or in spiritual terms—for what is the Sacred but a capitalized version of significance? Even for atheists like Dinah Bazer—*like me!*—psychedelics can charge a world from which the gods long ago departed with the pulse of meaning, the immanence with which they once infused it. The sense of a cold and arbitrary universe governed purely by chance is banished. Especially in the absence of faith, these medicines, in the right hands, may offer powerful antidotes for the existential terrors that afflict not only the dying.

To believe that life has any meaning at all is of course a large presumption, requiring in some a leap of faith, but surely it is a helpful one, and never more so than at the approach of death. To situate the self in a larger context of meaning, whatever it is—a sense of oneness with nature or universal love—can make extinction of the self somewhat easier to contemplate. Religion has always understood this wager, but why should religion enjoy a monopoly? Bertrand Russell wrote that the best way to overcome one's fear of death "is to make your interests gradually wider and more impersonal, until bit by bit the walls of the ego recede, and your life becomes increasingly merged in the universal life." He goes on:

> An individual human existence should be like a river: small at first, narrowly contained within its banks, and rushing passionately past rocks and over waterfalls. Gradually, the river grows wider, the banks recede, the waters flow more quietly, and in the end, without any visible break, they become merged in the sea, and painlessly lose their individual being.

o   o   o

PATRICK METTES lived seventeen months after his psilocybin session, and according to Lisa those months were filled with a great many unexpected satisfactions, alongside Patrick's dawning acceptance that he was going to die.

Lisa had initially been wary of the NYU trial, interpreting Patrick's desire to participate as a sign he'd given up the fight. In the event, he came away convinced he still had much to do in this life—much love to give and receive—and wasn't yet ready to leave it and, especially, his wife. Patrick's psychedelic journey had shifted his perspective, from a narrow lens trained on the prospect of dying to a renewed focus on how best to live the time left to him. "He had a new resolve. That there was a point to his life, that he got it, and was moving with it.

"We still had our arguments," Lisa recalled, "and we had a very trying summer" as they endured a calamitous apartment renovation in Brooklyn. "That was hell on earth," Lisa recalled, but Patrick "had changed. He had a sense of patience he had never had before, and with me he had real joy about things. It was as if he had been relieved of the duty of caring about the details of life, and he could let all that go. Now it was about being with people, enjoying his sandwich and the walk on the promenade. It was as if we lived a lifetime in a year."

After the psilocybin session, Lisa somehow convinced herself that Patrick was not going to die after all. He continued with his chemo and his spirits improved, but she now thinks all this time "he knew very well he wasn't going to make it." Lisa continued to work, and Patrick spent his good days walking the city. "He would walk everywhere, try every restaurant for lunch, and tell me about all the great places he discovered. But his good days got fewer and fewer." Then, in March 2012, he told her he wanted to stop chemo.

"He didn't want to die," Lisa says, "but I think he just decided that this is not how he wanted to live."

That fall his lungs began to fail, and Patrick wound up in the hospital. "He gathered everyone together and said good-bye and explained that this is how he wanted to die. He had a very conscious death." Patrick's seeming equanimity in the face of death exerted a powerful influence on everyone around him, Lisa said, and his room in the palliative care unit at Mount Sinai became a center of gravity in the hospital. "Everyone, the nurses and the doctors, wanted to hang out in our room; they just didn't want to leave. Patrick would talk and talk. It was like he was a yogi. He put out so much love." When Tony Bossis visited Patrick a week before he died, he was struck by the mood in the room and by Patrick's serenity.

"He was consoling *me*. He said his biggest sadness was leaving his wife. But he was not afraid."

Lisa e-mailed me a photograph of Patrick she had taken a few days before he died, and when the image popped open on my screen, it momentarily took my breath away. Here was an emaciated man in a hospital gown, an oxygen clip in his nose, but with bright, shining blue eyes and a broad smile. On the eve of death, the man was beaming.

Lisa stayed with Patrick in his hospital room night after night, the two of them often talking into the wee hours. "I feel like I have one foot in this world and one in the next," he told her at one point. "One of the last nights we were together, he said, 'Honey, don't push me. I'm finding my way.'" At the same time, he sought to comfort her. "This is simply the wheel of life," she recalls him saying. "'You feel like you're being ground down by it now, but the wheel is going to turn and you'll be on top again.'"

Lisa hadn't had a shower in days, and her brother finally persuaded her to go home for a few hours. Minutes before she returned to his bedside, Patrick slipped away. "I went home to shower and he died." We were speaking on the phone, and I could hear her crying softly. "He wasn't going to die as long as I was there. My brother had told me, 'You need to let him go.'"

Patrick was gone by the time she got back to the hospital. "He had died seconds before. It was like something had evaporated from him. I sat with him for three hours. It's a long time before the soul is out of the room."

"It was a good death," Lisa told me, a fact she credits to the people at NYU and to Patrick's psilocybin journey. "I feel indebted to them for what they allowed him to experience—the deep resources they allowed him to tap into. These were his own deep resources. That, I think, is what these mind-altering drugs do."

"Patrick was far more spiritual than I was to begin with," Lisa told me the last time we spoke. It was clear his journey had changed her too. "It was an affirmation of a world I knew nothing about. But there are more dimensions to this world than I ever knew existed."

Two: Addiction

The dozen or so Apollo astronauts who have escaped Earth's orbit and traveled to the moon had the privilege of seeing the planet from a perspective never before available to our species, and several of them reported that the experience changed them in profound and enduring ways. The sight of that "pale blue dot" hanging in the infinite black void of space erased the national borders on our maps and rendered Earth small, vulnerable, exceptional, and precious.

Edgar Mitchell, returning from the moon on Apollo 14, had what he has described as a mystical experience, specifically a *savikalpa samadhi*, in which the ego vanishes when confronted with the immensity of the universe during the course of a meditation on an object—in this case, planet Earth.

"The biggest joy was on the way home," he recalled. "In my cockpit window, every two minutes: the earth, the moon, the sun, and the whole panorama of the heavens. That was a powerful, overwhelming experience.

"And suddenly I realized that the molecules of my body, and the molecules of my spacecraft, the molecules in the body of my partners, were prototyped, manufactured in some ancient generation of stars. [I felt] an overwhelming sense of oneness, of connectedness . . . It wasn't 'Them and Us,' it was 'That's me! That's all of it, it's one thing.' And it was accompanied by an ecstasy, a sense of 'Oh my God, wow, yes'—an insight, an epiphany."*

It was the power of this novel perspective—the same perspective that Stewart Brand, after his 1966 LSD trip on a North Beach rooftop, worked so hard to disseminate to the culture—that helped to inspire the modern environmental movement as well as the Gaia hypothesis, the idea that Earth and its atmosphere together constitute a single living organism.

I thought about this so-called overview effect during my conversations with volunteers in the psilocybin trials, and especially with those who had overcome their addictions after a psychedelic journey—to inner space, if you will. Several volunteers described achieving a new distance on their own lives, a vantage from which matters that had once seemed daunting now seemed smaller and more

---

* The experience would shape his post-NASA work: the former engineer established the Institute of Noetic Sciences to study consciousness and paranormal phenomena.

manageable, including their addictions. It sounded as though the psychedelic experience had given many of them an overview effect on the scenes of their own lives, making possible a shift in worldview and priorities that allowed them to let go of old habits, sometimes with remarkable ease. As one lifetime smoker put it to me in terms so simple I found it hard to believe, "Smoking became irrelevant, so I stopped."

The smoking cessation pilot study in which this man took part—his name is Charles Bessant, and he has been abstinent now for six years—was directed by Matthew Johnson, a protégé of Roland Griffiths's at Johns Hopkins, where the study took place. Johnson is a psychologist in his early forties who, like Griffiths, trained as a behaviorist, studying things like "operant conditioning" in rats. Tall, slender, and angular, Johnson wears a scrupulously trimmed black beard and oversized retro-nerd black glasses that make him look a little like Ira Glass. His interest in psychedelics goes back to his college days, when he read Ram Dass and learned about the Harvard Psilocybin Project, but never did he dare to imagine he would someday have a job working with them in a laboratory.

"I had it in the back of my mind that someday I wanted to do research with the psychedelic compounds," he told me when we first met in his Hopkins office, "but I figured that was a long way off in the future." Yet soon after Johnson arrived at Johns Hopkins to do a pharmacology postdoc in 2004, "I found out that Roland had this super hush-hush project with psilocybin. Everything lined up perfectly."

Johnson worked on the lab's early psilocybin studies, serving as a guide for several dozen sessions and helping to crunch the data, before launching a study of his own in 2009. The smoking study gave fifteen volunteer smokers who were trying to quit several sessions of cognitive behavioral therapy followed by two or three doses of psilocybin. A so-called open-label study, there was no placebo, so they all knew they were getting the drug. Volunteers had to stop smoking before their psilocybin session; they had their carbon-monoxide levels measured at several intervals to ensure compliance and confirm they remained abstinent.

The study was tiny and not randomized, but the results were nevertheless striking, especially when you consider that smoking is one of the most difficult addictions to break—harder, some say, than heroin. Six months after their psychedelic sessions, 80 percent of the volunteers were confirmed as abstinent; at the one-year mark, that figure had fallen to 67 percent, which is still a better rate of success than the best treatment now available. (A much larger randomized study, comparing the effectiveness of psilocybin therapy with the nicotine patch, is currently under way.) As in the cancer-anxiety studies, the volunteers who had the most complete mystical experiences had the best outcomes; they were, like Charles Bessant, able to quit smoking.

After interviewing cancer patients confronted with the prospect of death, people who had had epic journeys in which they confronted their cancers and traveled to the underworld, I wondered how the experience would compare when the stakes were lower: What kinds of journeys would ordinary people simply hoping to break a bad habit have, and what kinds of insights would they return with?

Surprisingly banal, it turns out. Not that their journeys were banal—psilocybin transported them all over the world and through history and to outer space—but the insights they brought back with them were mundane in the extreme. Alice O'Donnell, a sixtyish book editor born in Ireland, reveled "in the freedom to go everywhere" in the course of her journey. She grew feathers that allowed her to travel back in time to various scenes of European history, died three times, watched her "soul move from her body to a funeral pyre floating on the Ganges," and found herself "standing on the edge of the universe, witnessing the dawn of creation." She had the "humbling" realization that "everything in the universe is of equal importance, including yourself.

"Instead of being so narrowly focused, moving through this little tunnel of adult life," she found that the journey "returned me to the child's wider sense of wonder—to the world of Wordsworth. A part of my brain that had gone to sleep was awakened.

"The universe was so great and there were so many things you could do and see in it that killing yourself seemed like a dumb idea. It put smoking in a whole new context. Smoking seemed very unimportant; it seemed kind of stupid, to be honest."

Alice imagined herself throwing out lots of junk from her house, emptying the attic and the basement: "I had an image of tossing everything over the ledge, all the stuff I didn't need anymore. It's amazing how you can whittle things down to the few really important things that are necessary for survival. And the most important thing of all is the breath. When that stops, you're dead." She emerged from her journey with the conviction "that you should cherish your breath." She has not had a cigarette since her psilocybin journey. Whenever she feels a craving, she goes back in memory to her session "and thinks of all the wonderful things I experienced, and how it felt to be on that much higher plane."

Charles Bessant had his epiphany while on a similarly "higher plane." Bessant, a museum exhibit designer in his sixties, found himself standing on a mountaintop in the Alps, "the German states stretching out before me all the way to the Baltic." (Wagner was playing in his headphones.) "My ego had dissolved, yet I'm telling you this. It was terrifying." He sounded like a nineteenth-century Romantic describing an encounter with the sublime, at once terrible and awe inspiring.

"People use words like 'oneness,' 'connectivity,' 'unity'—I get it! I was part of something so much larger than anything I had ever imagined." We were speaking by phone on a Saturday morning, and at one point Bessant paused in his account to describe the scene before him.

"Right now, I'm standing here in my garden, and the light is coming through the canopy of leaves. For me to be able to stand here in the beauty of this light, talking to you, it's only because my eyes are open to see it. If you don't stop to look, you'll never see it. It's the statement of an obvious thing, I know, but to feel it, to look and be amazed by this light" is a gift he attributes to his session, which gave him "a feeling of connectedness to everything."

Bessant followed up on our conversation by e-mail with a series of clarifications and elaborations, striving to find the words equal to the immensity of the experience. It was in the face of this immensity that smoking suddenly seemed pitifully small. "Why quit smoking? Because I found it irrelevant. Because other things had become so much more important."

Some volunteers marveled themselves at the simultaneous power and banality of their insights. Savannah Miller is a single mom in her thirties who works as a bookkeeper for her father's company in Maryland. Possibly because she spent her twenties tangled in an abusive relationship with a man she describes as "a psychopath," her trip was painful but ultimately cathartic; she remembers crying uncontrollably and producing tremendous amounts of snot (something her guides confirmed really happened). Savannah gave little thought to her habit during the journey, except toward the end when she pictured herself as a smoking gargoyle.

"You know how gargoyles look, crouched down with their shoulders hunched? That's how I felt and saw myself, a little golem creature smoking, pulling in the smoke and not letting it out, until my chest hurts and I'm choking. It was powerful and disgusting. I can still see it now, that hideous coughing gargoyle, whenever I picture

364 HOW TO CHANGE YOUR MIND

myself as a smoker." Months later, she says the image is still helpful when the inevitable cravings arise.

In the middle of her session, Savannah suddenly sat up and announced she had discovered something important, an "epiphany" that her guides needed to write down so it wouldn't be lost to posterity: "Eat right. Exercise. Stretch."

Matt Johnson refers to these realizations as "duh moments" and says they are common among his volunteers and not at all insignificant. Smokers know perfectly well that their habit is unhealthy, disgusting, expensive, and unnecessary, but under the influence of psilocybin that knowing acquires a new weight, becomes "something they feel in the gut and the heart. Insights like this become more compelling, stickier, and harder to avoid thinking about. These sessions deprive people of the luxury of mindlessness"—our default state, and one in which addictions like smoking can flourish.

Johnson believes the value of psilocybin for the addict is in the new perspective—at once obvious and profound—that it opens onto one's life and its habits. "Addiction is a story we get stuck in, a story that gets reinforced every time we try and fail to quit: 'I'm a smoker and I'm powerless to stop.' The journey allows them to get some distance and see the bigger picture and to see the short-term pleasures of smoking in the larger, longer-term context of their lives."

Of course, this re-contextualization of an old habit doesn't just happen; countless people have taken psilocybin and continued to smoke. If it does happen, it's because breaking the habit is the avowed intention of the session, strongly reinforced by the therapist in the preparatory meetings and the integration afterward. The "set" of the psychedelic journey is carefully orchestrated by the therapist in much the same way a shaman would use his authority and stagecraft to maximize the medicine's deep powers of suggestion. This is why it is important to understand that "psychedelic therapy" is not simply treatment with a psychedelic drug but rather a form of "psychedelic-assisted therapy," as many of the researchers take pains to emphasize.

Yet what accounts for the unusual authority of the rather ordinary insights volunteers brought back from their journeys? "You don't get that on any other drug," Roland Griffiths points out. Indeed, after most drug experiences, we're fully aware of, and often embarrassed by, the *inauthenticity* of what we thought and felt while under the influence. Though neither Griffiths nor Johnson mentioned it, the connection between seeing and believing might explain this sense of authenticity. Very often on psychedelics our thoughts become visible. These are not hallucinations, exactly, because the subject is often fully aware that what she is seeing is not really before her, yet these thoughts made visible are nevertheless remarkably concrete, vivid, and therefore memorable.

This is a curious phenomenon, as yet unexplained by neuroscience, though some interesting hypotheses have recently been proposed. When neuroscientists who study vision use fMRIs to image brain activity, they find that the same regions in the visual cortex light up whether one is seeing an object live—"online"—or merely recalling or imagining it, off-line. This suggests that the ability to visualize our thoughts should be the rule rather than the exception. Some neuroscientists suspect that during normal waking hours something in the brain inhibits the visual cortex from presenting to consciousness a visual image of whatever it is we're thinking about. It's not hard to see why such an inhibition might be adaptive: cluttering the mind with vivid images would complicate reasoning and abstract thought, not to mention everyday activities like walking or driving a car. But when we *are* able to visualize our thoughts—such as the thought of ourselves as a smoker looking like a coughing gargoyle—those thoughts take on added weight, feel more real to us. Seeing is believing.

Perhaps this is one of the things psychedelics do: relax the brain's inhibition on visualizing our thoughts, thereby rendering them more authoritative, memorable, and sticky. The overview effect reported by the astronauts didn't add anything to our intellectual understanding of this "pale blue dot" in the vast sea of space, but seeing it made it real in a way it had never been before. Perhaps the equally vivid overview effect on the scenes of their lives that psychedelics afford some people is what makes it possible for them to change their behavior.

Matt Johnson believes that psychedelics can be used to change all sorts of behaviors, not just addiction. The key, in his view, is their power to occasion a sufficiently dramatic experience to "dope-slap people out of their story. It's literally a reboot of the system—a biological control-alt-delete. Psychedelics open a window of mental flexibility in which people can let go of the mental models we use to organize reality."

In his view, the most important such model is the self, or ego, which a high-dose psychedelic experience temporarily dissolves. He speaks of "our addiction to a pattern of thinking with the self at the center of it." This underlying addiction to a pattern of thinking, or cognitive style, links the addict to the depressive and to the cancer patient obsessed with death or recurrence.

"So much of human suffering stems from having this self that needs to be psychologically defended at all costs. We're trapped in a story that sees ourselves as independent, isolated agents acting in the world. But that self is an illusion. It can be a *useful* illusion, when you're swinging through the trees or escaping from a cheetah or trying to do your taxes. But at the systems level, there is no truth to it. You can take any number of more accurate perspectives: that we're a swarm of genes, vehicles for passing on DNA; that we're social creatures through and through, unable to survive alone; that we're organisms in an ecosystem, linked together on this planet floating in the middle of nowhere. Wherever you look, you see that the level of interconnectedness is truly amazing, and yet we insist on thinking of ourselves as individual agents." Albert Einstein called the modern human's sense of separateness "a kind of optical delusion of his consciousness."*

"Psychedelics knock the legs out from under that model. That can be dangerous in the wrong circumstances, leading to bad trips and worse." Johnson brought up the case of Charles Manson, who reportedly used LSD to break down and brainwash his followers, a theory of the case he deems plausible. "But in the right setting, where your safety is assured, it may be a good intervention for dealing with some of the problems of the self"—of which addiction is only one. Dying, depression, obsession, eating disorders—all are exacerbated by the tyranny of an ego and the fixed narratives it constructs about our relationship to the world. By temporarily overturning that tyranny and throwing our minds into an unusually plastic state (Robin Carhart-Harris would call it a state of heightened entropy), psychedelics, with the help of a good therapist, give us an opportunity to propose some new, more constructive stories about the self and its relationship to the world, stories that just might stick.

This is a very different kind of therapy than we are accustomed to in the West, because it is neither purely chemical nor purely psychodynamic—neither mindless nor brainless. Whether Western medicine is ready to accommodate such a radically novel—and

---

* "A human being is a part of the whole called by us 'Universe,' a part limited in time and space. He experiences himself, his thoughts and feeling as something separated from the rest—a kind of optical delusion of his consciousness. This delusion is a kind of prison for us, restricting us to our personal desires and to affection for a few persons nearest to us. Our task must be to free ourselves from this prison by widening our circle of compassion to embrace all living creatures and the whole of nature in its beauty." (Walter Sullivan, "The Einstein Papers: A Man of Many Parts," *The New York Times*, March 29, 1972.)

ancient—model for mental transformation is an open question. In taking people safely through the liminal state psychedelics occasion, with its radical suggestibility, Johnson acknowledges that the doctors and researchers "play the same role as shamans or elders.

"Whatever we're delving into here, it's in the same realm as the placebo. But a placebo on rocket boosters."

o   o   o

THE WHOLE IDEA of using a psychedelic drug to treat addiction is not new. Native Americans have long used peyote as both a sacrament and a treatment for alcoholism, a scourge of the indigenous community since the arrival of the white man. Speaking at a meeting of the American Psychiatric Association in 1971, the psychiatrist Karl Menninger said that "peyote is not harmful to these people . . . It is a better antidote to alcohol than anything the missionaries, the white man, the American Medical Association, and the public health services have come up with."*

Thousands of alcoholics were treated with LSD and other psychedelics in the 1950s and 1960s, though until recently it's been hard to say anything definitive about the results. For a time, the therapy was deemed effective enough to become a standard treatment for alcoholism in Saskatchewan. Clinical reports were enthusiastic, yet most of the formal studies conducted were poorly designed and badly controlled, if at all. Results were notably impressive when the studies were performed by sympathetic therapists (and especially by therapists who themselves had taken LSD) and notably dismal when conducted by inexperienced investigators who gave mammoth doses to patients with no attention to set or setting.

---

* Quoted in Charles S. Grob, "Psychiatric Research with Hallucinogens: What Have We Learned?," *Heffter Review of Psychedelic Research* 1 (1998).

The record was a complete muddle until 2012, when a meta-analysis that combined data from the six best randomized controlled studies done in the 1960s and 1970s (involving more than five hundred patients in all) found that indeed there had been a statistically robust and clinically "significant beneficial effect on alcohol misuse" from a single dose of LSD, an effect that lasted up to six months. "Given the evidence for a beneficial effect of LSD on alcoholism," the authors concluded, "it is puzzling why this treatment has been largely overlooked."

Since then, psychedelic therapy for alcohol and other addictions has undergone a modest and so far encouraging revival, both in university studies and in various underground settings.* In a 2015 pilot study conducted at the University of New Mexico ten alcoholics received psilocybin, combined with "motivational enhancement therapy," a type of cognitive behavioral therapy designed expressly to treat addiction. By itself, the psychotherapy had little effect on drinking behavior, but after the psilocybin session drinking decreased significantly, and these changes were sustained during the thirty-six weeks of follow-up. Michael Bogenschutz, the lead investigator, reported a strong correlation between the "strength of the experience and the effect" on drinking behavior. The New Mexico results were encouraging enough to warrant a much larger phase 2 trial, involving 180 volunteers, which Bogenschutz is now conducting at NYU in collaboration with Stephen Ross and Jeffrey Guss.

"Alcoholism can be understood as a spiritual disorder," Ross told me the first time we met, in the treatment room at NYU. "Over time you lose your connection to everything but this compound.

---

* Ibogaine, a psychedelic derived from the root of an African shrub, is being used underground as well as in clinics in Mexico to treat opiate addiction; ayahuasca has also been reported to be helpful breaking addictions.

Life loses all meaning. At the end, nothing is more important than that bottle, not even your wife and your kids. Eventually, there is nothing you won't sacrifice for it."

It was Ross who first told me the story of Bill W., the founder of AA, how he got sober after a mystical experience on belladonna and in the 1950s sought to introduce LSD into the fellowship. To use a drug to promote sobriety might sound counterintuitive, even crazy, yet it makes a certain sense when you consider how reliably psychedelics can sponsor spiritual breakthroughs as well as the conviction, central to the AA philosophy, that before she can hope to recover, the alcoholic must first acknowledge her "powerlessness." AA takes a dim view of the human ego and, like psychedelic therapy, attempts to shift the addict's attention from the self to a "higher power" as well as to the consolations of fellowship—the sense of interconnectedness.

Michael Bogenschutz put me in touch with a woman I'll call Terry McDaniels, a volunteer in his alcoholism pilot study in New Mexico—a surprising introduction, I came to think, because hers wasn't the kind of unqualified success story researchers like to give journalists. I spoke to McDaniels by phone from her trailer park outside Albuquerque, where she lives on disability a few trailers down from her daughter. She hasn't been able to work since 1997, when "my ex-husband beat my head in with a cast-iron skillet. Since that occurred, I've had a real problem with my memory."

McDaniels, who was born in 1954, has had a tough life, going back to her childhood, when her parents left her for long periods in the indifferent care of older siblings. "Even to this day I have a hard time laughing." She told me she spends many of her days mired in feelings of regret, anger, envy, self-loathing, and, especially, a deep sense of guilt toward her children. "I feel very bad I haven't given them the life I could have if I had stayed away from drink. I think about that other life I might have had all the time."

When I asked McDaniels how long she had been sober, she surprised me: she wasn't. She'd actually been on a bender just a few weeks earlier, after her daughter "hurt my feelings by asking for money I owed her." But the binge lasted only a day, and she had only had beer and wine to drink; in the years before her psychedelic session, she would binge on hard liquor for two weeks at a time, the drinking interrupted only when she blacked out. For McDaniels, a one-day binge now and again represents progress.

McDaniels read about the psilocybin trial in the local alternative weekly. She had never before used a psychedelic but felt desperate and willing to try something new. She had made many attempts to get sober, had been in rehab, therapy, and AA, but always fell back on the bottle. She worried that her head injury might disqualify her from the trial, but she was accepted and in the event had a powerful spiritual experience.

The first part of the trip was unbearably dark: "I saw my children and I was bawling and bawling, for the life they never had." But eventually it turned into something awe inspiring.

"I saw Jesus on the cross," she recalled. "It was just his head and shoulders, and it was like I was a little kid in a tiny helicopter circling around his head. But he was on the cross. And he just sort of gathered me up in his hands, you know, the way you would comfort a small child. I felt such a great weight lift from my shoulders, felt very much at peace. It was a beautiful experience."

The teaching of the experience, she felt, was self-acceptance. "I spend less time thinking about people who have a better life than me. I realize I'm not a bad person; I'm a person who's had a lot of bad things happen. Jesus might have been trying to tell me it was okay, that these things happen. He was trying to comfort me." Now, McDaniels says, "I read my Bible every day and keep a conscious contact with God."

By her own lights, McDaniels is doing, if not well exactly, then somewhat better. The experience has helped her begin to rethink the story of her life she tells herself: "I don't take everything so personally, like I used to. I have more self-acceptance, and that is a gift, because for a lot of years, I did not like myself. But I am not a bad person."

That one's perspective could shift in such a way in the absence of any change in circumstance strikes me as both hopeful and poignant. I was reminded of an experiment that several of the addiction researchers I interviewed had told me about—the so-called rat park experiment. It's well known in the field of drug abuse research that rats in a cage given access to drugs of various kinds will quickly addict themselves, pressing the little levers for the drug on offer in preference to food, often to the point of death. Much less well known, however, is the fact that if the cage is "enriched" with opportunities for play, interaction with other rats, and exposure to nature, the same rats will utterly ignore the drugs and so never become addicted. The rat park experiments lend support to the idea that the propensity to addiction might have less to do with genes or chemistry than with one's personal history and environment.

Now comes a class of chemicals that may have the power to change how we *experience* our personal history and environment, no matter how impoverished or painful they may be. "Do you see the world as a prison or a playground?" is the key question Matt Johnson takes away from the rat park experiment. If addiction represents a radical narrowing of one's perspective and behavior and emotional repertoire, the psychedelic journey has the potential to reverse that constriction, open people up to the possibility of change by disrupting and enriching their interior environment.

"People come out of these experiences seeing the world a little more like a playground."

ONE GOOD WORD to describe the experiences of both the Apollo astronauts and the volunteers on their psilocybin journeys is "awe," a human emotion that can perhaps help weave together the disparate strands of psychological interpretation proposed by the psychedelic researchers with whom I spoke. It was Peter Hendricks, a young psychologist at the University of Alabama conducting a trial using psilocybin to treat cocaine addicts, who first suggested to me that the experience of awe might offer the psychological key to explain the power of psychedelics to alter deeply rooted patterns of behavior.

"People who are addicted know they're harming themselves—their health, their careers, their social well-being—but they often fail to see the damage their behavior is doing to others." Addiction is, among other things, a radical form of selfishness. One of the challenges of treating the addict is getting him to broaden his perspective beyond a consuming self-interest in his addiction, the behavior that has come to define his identity and organize his days. Awe, Hendricks believes, has the power to do this.

Hendricks mentioned the research of Dacher Keltner, a psychologist at Berkeley who happens to be a close friend. "Keltner believes that awe is a fundamental human emotion, one that evolved in us because it promotes altruistic behavior. We are descendants of those who found the experience of awe blissful, because it's advantageous for the species to have an emotion that makes us feel part of something much larger than ourselves." This larger entity could be the social collective, nature as a whole, or a spirit world, but it is something sufficiently overpowering to dwarf us and our narrow self-interest. "Awe promotes a sense of the 'small self' that directs our attention away from the individual to the group and the greater good."

Keltner's lab at Berkeley has done a clever series of experiments demonstrating that after people have had even a relatively modest experience of awe, such as looking at soaring trees, they're more likely to come to the assistance of others. (In this experiment, conducted in a eucalyptus grove on the Berkeley campus, volunteers spent a minute looking either at the trees or at the façade of a nearby building. Then a confederate walked toward the participants and stumbled, scattering pens on the ground. Bystanders who had looked at the trees proved more likely to come to her aid than those who had looked at the building.) In another experiment, Keltner's lab found that if you ask people to draw themselves before and after viewing awe-inspiring images of nature, the after-awe self-portraits will take up considerably less space on the page. An experience of awe appears to be an excellent antidote for egotism.

"We now have a pharmacological intervention that can occasion truly profound experiences of awe," Hendricks pointed out. Awe in a pill. For the self-obsessed addict, "it can be blissful to feel a part of something larger and greater than themselves, to feel reconnected to other people"—to the weave of social and family relations that addiction reliably frays. "Very often they come to recognize the harm they're doing not only to themselves but to loved ones. That's where the motivation to change often comes from—a renewed sense of connection and responsibility, as well as the positive feeling of being a small self in the presence of something greater."

The concept of awe, I realized, could help connect several of the dots I'd been collecting in the course of my journey through the landscape of psychedelic therapy. Whether awe is a cause or an effect of the mental changes psychedelics sponsor isn't entirely clear. But either way, awe figures in much of the phenomenology of psychedelic consciousness, including the mystical experience, the overview effect, self-transcendence, the enrichment of our inner environment, and even the generation of new meanings. As Keltner has written, the overwhelming force and the mystery of awe are such that the experience can't readily be interpreted according to our accustomed frames of thought. By rocking those conceptual frameworks, awe has the power to change our minds.

Three: Depression

Something unexpected happened when, early in 2017, Roland Griffiths and Stephen Ross brought the results of their clinical trials to the FDA, hoping to win approval for a larger, phase 3 trial of psilocybin for cancer patients. Impressed by their data—and seemingly undeterred by the unique challenges posed by psychedelic research, such as the problem of blinding, the combining of therapy and medicine, and the fact that the drug in question is still illegal—the FDA staff surprised the researchers by asking them to expand their focus and ambition: to test whether psilocybin could be used to treat the much larger and more pressing problem of depression in the general population. As the regulators saw it, the data contained a strong enough "signal" that psilocybin could relieve depression; it would be a shame not to test the proposition, given the enormity of the need and the limitations of the therapies now available. Ross and Griffiths had focused on cancer patients because they thought it would be easier to win approval to study a controlled substance in people who were already seriously ill or dying. Now the government was telling them to raise their sights. "It was surreal," Ross told me, twice, as he recounted the meeting, still somewhat stunned at the response and outcome. (The FDA declined to confirm or deny this account of the meeting, explaining that it doesn't comment on drugs in development or under regulatory review.)

Much the same thing happened in Europe, when, in 2016, researchers approached the European Medicines Agency (EMA)—the European Union's drug-regulating body—seeking approval to use psilocybin in the treatment of anxiety and depression in patients with life-changing diagnoses. "Existential distress" is not an official *DSM* diagnosis, the regulators pointed out, so the national health services won't cover it. But there's a signal here that psilocybin could be useful in treating depression, so why don't you do a big, multisite trial for that?

The EMA was responding not only to the Hopkins and NYU data but also to the small "feasibility study" of the potential of using psilocybin to treat depression that Robin Carhart-Harris had directed in David Nutt's lab at Imperial College. In the study, the initial results of which appeared in *Lancet Psychiatry* in 2016, researchers gave psilocybin to six men and six women suffering from "treatment-resistant depression"—meaning they had already tried at least two treatments without success. There was no control group, so everyone knew he or she was getting psilocybin.

After a week, all of the volunteers showed improvement in their symptoms, and two-thirds of them were depression-free, in some cases for the first time in years. Seven of the twelve volunteers still showed substantial benefit after three months. The study was expanded to include a total of twenty volunteers; after six months, six remained in remission, while the others had relapsed to one degree or another, suggesting the treatment might need to be repeated. The study was modest in scale and not randomized, but it demonstrated that psilocybin was well tolerated in this population, with no adverse events, and most of the subjects had seen benefits that were marked and rapid.* The EMA was sufficiently impressed with the data to suggest a much larger trial for treatment-resistant depression, which afflicts more than 800,000 people in Europe. (This is out of a total of some 40 million Europeans with depressive disorders, according to the World Health Organization.)

Rosalind Watts was a young clinical psychologist working for the National Health Service when she read an article about psychedelic therapy in the *New Yorker*.† The idea that you might actually be able to cure mental illness rather than just manage its symptoms inspired her to write to Robin Carhart-Harris, who hired her to help out with the depression study, the lab's first foray into clinical research. Watts guided several sessions and then conducted qualitative interviews with all of the volunteers six months after their treatments, hoping to understand exactly how the psychedelic session had affected them.

Watts's interviews uncovered two "master" themes. The first was that the volunteers depicted their depression foremost as a state of "disconnection," whether from other people, their earlier selves, their senses and feelings, their core beliefs and spiritual values, or nature. Several referred to living in "a mental prison," others to being "stuck" in endless circles of rumination they likened to mental "grid-

---

* As for the three volunteers who received no benefit, they had mild or unremarkable sessions. This might be because they were still on SSRIs, which may block the effects of psychedelics, or because some fraction of the population simply doesn't respond to the drugs. The Hopkins team, too, has occasionally seen cases of "dud trips" that leave people unaffected.

† By me, as it happened. "The Trip Treatment," *New Yorker*, Feb. 9, 2015.

lock." I was reminded of Carhart-Harris's hypothesis that depression might be the result of an overactive default mode network—the site in the brain where rumination appears to take place.

The Imperial depressives also felt disconnected from their senses. "I would look at orchids," one told Watts, "and intellectually understand that there was beauty, but not experience it."

For most of the volunteers, the psilocybin experience had sprung them from their mental jails, if only temporarily. One woman in the study told me that the month following her session was the first time she had been free from depression since 1991. Others described similar experiences:

"It was like a holiday away from the prison of my brain. I felt free, carefree, reenergized."

"It was like the light switch being turned on in a dark house."

"You're not immersed in thought patterns; the concrete coat has come off."

"It was like when you defrag the hard drive on your computer . . . I thought, 'My brain is being defragged, how brilliant is that!'"

For many of the volunteers, these changes in the experience of their own minds persisted:

"My mind works differently. I ruminate much less, and my thoughts feel ordered, contextualized."

Several reported reconnecting to their senses:

"A veil dropped from my eyes, things were suddenly clear, glowing, bright. I looked at plants and felt their beauty. I can still look at my orchids and feel that: that is one thing that has really lasted."

Some reconnected to themselves:

"I had an experience of tenderness toward myself."

"At its most basic, I feel like I used to before the depression."

Others reconnected to other people:

"I was talking to strangers. I had these full long conversations with everybody I came into contact with."

"I would look at people on the street and think, 'How interesting we are'—I felt connected to them all."

And to nature:

"Before, I enjoyed nature; now I feel part of it. Before I was looking at it as a thing, like TV or painting. You're part of it, there's no separation or distinction, you are it."

"I was everybody, unity, one life with 6 billion faces. I was the one asking for love and giving love, I was swimming in the sea, and the sea was me."

The second master theme was a new access to difficult emotions, emotions that depression often blunts or closes down completely. Watts hypothesizes that the depressed patient's incessant rumination constricts his or her emotional repertoire. In other cases, the depressive keeps emotions at bay because it is too painful to experience them.

This is especially true in cases of childhood trauma. Watts put me in touch with a thirty-nine-year-old man in the study, a music journalist named Ian Roullier, who, along with his older sister, had been abused by his father as a child. As adults, the siblings brought charges against their father that put him in jail for several years, but this hadn't relieved the depression that has trailed Ian for most of his life.

"I can remember the moment when the horrible cloud first came over me. It was in the family room of a pub called the Fighting Cocks in St. Albans. I was ten." Antidepressants helped for a while, but "putting the plaster over the wound doesn't heal anything." On psilocybin, he was able for the first time to confront his lifelong pain—and his father.

"Normally, when Dad comes up in my head, I just push the thought away. But this time I went the other way." His guide had told him he should "go in and through" any frightening material that arose during his journey.

"So this time I looked him in the eye. That was a really big thing for me, to literally face the demon. And there he was. But he was a horse! A military horse standing on its hind legs, dressed in a military outfit with a helmet, and holding a gun. It was terrifying, and I wanted to push the image aside, but I didn't. *In and through*: Instead, I looked the horse in the eyes—and promptly started to laugh, it was so ridiculous.

"That's when what had been a bad trip really turned. Now I had every sort of emotion, positive, negative, it didn't matter. I thought about the [Syrian] refugees in Calais and started crying for them, and I saw that every emotion is as valid as any other. You don't cherry-pick happiness and enjoyment, the so-called good emotions; it was okay to have negative thoughts. That's life. For me, trying to resist emotions just amplified them. Once I was in this state, it was beautiful—a feeling of deep contentment. I had this overwhelming feeling—it wasn't even a thought—that everything and everyone needs to be approached with love, including myself."

Ian enjoyed several months of relief from his depression as well as a new perspective on his life—something no antidepressant had ever given him. "Like Google Earth, I had zoomed out," he told Watts in his six-month interview. For several weeks after his session, "I was absolutely connected to myself, to every living thing, to the universe." Eventually, Ian's overview effect faded, however, and he ended up back on Zoloft.

"The sheen and shine that life and existence had regained immediately after the trial and for several weeks after gradually faded," he wrote one year later. "The insights I gained during the trial have never left and will never leave me. But they now feel more like ideas," he says. He says he's doing better than before and has been able to hold down a job, but his depression has returned. He told me he wishes he could have another psilocybin session at Imperial. Because that's currently not an option, he'll sometimes meditate and listen to the playlist from his session. "That really does help put me back in that place."

More than half of the Imperial volunteers saw the clouds of their depression eventually return, so it seems likely that psychedelic therapy for depression, should it prove useful and be approved, will not be a onetime intervention. But even the temporary respite the volunteers regarded as precious, because it reminded them there was another way to be that was worth working to recapture. Like electroconvulsive therapy for depression, which it in some ways resembles, psychedelic therapy is a shock to the system—a "reboot" or "defragging"—that may need to be repeated every so often. (Assuming the treatment works as well when repeated.) But the potential of the therapy has regulators and researchers and much of the mental health community feeling hopeful.

"I believe this could revolutionize mental health care," Watts told me. Her conviction is shared by every other psychedelic researcher I interviewed.

o   o   o

"If many remedies are prescribed for an illness," wrote Anton Chekhov, who was a physician as well as a writer, "you may be certain that the illness has no cure." But what about the reverse of Chekhov's statement? What are we to make of a single remedy being prescribed for a great many illnesses? How could it be that psychedelic therapy might be helpful for disorders as different as depression, addiction, the anxiety of the cancer patient, not to mention obsessive-compulsive disorder (about which there has been one encouraging study) and eating disorders (which Hopkins now plans to study)?

We shouldn't forget that irrational exuberance has afflicted psychedelic research since the beginning, and the belief that these molecules are a panacea for whatever ails us is at least as old as Timothy Leary. It could well be that the current enthusiasm will eventually give way to a more modest assessment of their potential. New treatments always look shiniest and most promising at the beginning. In early studies with small samples, the researchers, who are usually biased in favor of finding an effect, have the luxury of selecting the volunteers most likely to respond. Because their number is so small, these volunteers benefit from the care and attention of exceptionally well-trained and dedicated therapists, who are also biased in favor of success. Also, the placebo effect is usually strongest in a new medicine and tends to fade over time, as observed in the case of antidepressants; they don't work nearly as well today as they did upon their introduction in the 1980s. None of these psychedelic therapies have yet proven themselves to work in large populations; what successes have been reported should be taken as promising signals standing out from the noise of data, rather than as definitive proofs of cure.

Yet the fact that psychedelics have produced such a signal across a range of indications can be interpreted in a more positive light. When a single remedy is prescribed for a great many illnesses, to paraphrase Chekhov, it could mean those illnesses are more alike than we're accustomed to think. If a therapy contains an implicit theory of the disorder it purports to remedy, what might the fact that psychedelic therapy seems to address so many indications have to tell us about what those disorders might have in common? And about mental illness in general?

I put this question to Tom Insel, the former head of the National Institute of Mental Health. "It doesn't surprise me at all" that the same treatment should show promise for so many indications. He points out that the *DSM*—the *Diagnostic and Statistical Manual of Mental Disorders*, now in its fifth edition—draws somewhat arbitrary lines between mental disorders, lines that shift with each new edition.

"The *DSM* categories we have don't reflect reality," Insel said; they exist for the convenience of the insurance industry as much as anything else. "There's much more of a continuum between these disorders than the *DSM* recognizes." He points to the fact that SSRIs, when they work, are useful for treating a range of conditions besides depression, including anxiety and obsessive-compulsive disorder, suggesting the existence of some common underlying mechanism.

Andrew Solomon, in his book *The Noonday Demon: An Atlas of Depression*, traces the links between addiction and depression, which frequently co-occur, as well as the intimate relationship between depression and anxiety. He quotes an expert on anxiety who suggests we should think of the two disorders as "fraternal twins": "Depression is a response to past loss, and anxiety is a response to future loss." Both reflect a mind mired in rumination, one dwelling on the past, the other worrying about the future. What mainly distinguishes the two disorders is their tense.

A handful of researchers in the mental health field seem to be groping toward a grand unified theory of mental illness, though they would not be so arrogant as to call it that. David Kessler, the physician and former head of the FDA, recently published a book called *Capture: Unraveling the Mystery of Mental Suffering* that makes the case for such an approach. "Capture" is his term for the common mechanism underlying addiction, depression, anxiety, mania, and obsession; in his view, all these disorders involve learned habits of negative thinking and behavior that hijack our attention and trap us in loops of self-reflection. "What started as a pleasure becomes a need; what was once a bad mood becomes continuous self-indictment; what was once an annoyance becomes persecution," in a process he describes as a form of "inverse learning." "Every time we respond [to a stimulus], we strengthen the neural circuitry that prompts us to repeat" the same destructive thoughts or behaviors.

Could it be that the science of psychedelics has a contribution to make to the development of a grand unified theory of mental illness—or at least of some mental illnesses? Most of the researchers in the field—from Robin Carhart-Harris to Roland Griffiths, Matthew Johnson, and Jeffrey Guss—have become convinced that psychedelics operate on some higher-order mechanisms in the brain and mind, mechanisms that may underlie, and help explain, a wide variety of mental and behavioral disorders, as well as, perhaps, garden-variety unhappiness.

It could be as straightforward as the notion of a "mental reboot"—Matt Johnson's biological control-alt-delete key—that jolts the brain out of destructive patterns (such as Kessler's "capture"), affording an opportunity for new patterns to take root. It could be that, as Franz Vollenweider has hypothesized, psychedelics enhance neuroplasticity. The myriad new connections that spring up in the brain during the psychedelic experience, as mapped by the neuroimaging done at Imperial College, and the disintegration of well-traveled old connections, may serve simply to "shake the snow globe," in Robin Carhart-Harris's phrase, a predicate for establishing new pathways.

Mendel Kaelen, a Dutch postdoc in the Imperial lab, proposes a more extended snow metaphor: "Think of the brain as a hill covered in snow, and thoughts as sleds gliding down that hill. As one sled after another goes down the hill, a small number of main trails will appear in the snow. And every time a new sled goes down, it will be drawn into the preexisting trails, almost like a magnet." Those main trails represent the most well-traveled neural connections in your brain, many of them passing through the default mode network. "In time, it becomes more and more difficult to glide down the hill on any other path or in a different direction.

"Think of psychedelics as temporarily flattening the snow. The deeply worn trails disappear, and suddenly the sled can go in other directions, exploring new landscapes and, literally, creating new pathways." When the snow is freshest, the mind is most impressionable, and the slightest nudge—whether from a song or an intention or a therapist's suggestion—can powerfully influence its future course.

Robin Carhart-Harris's theory of the entropic brain represents a promising elaboration on this general idea, and a first stab at a unified theory of mental illness that helps explain all three of the disorders we've examined in these pages. A happy brain is a supple and flexible brain, he believes; depression, anxiety, obsession, and the cravings of addiction are how it feels to have a brain that has become excessively rigid or fixed in its pathways and linkages—a brain with more order than is good for it. On the spectrum he lays out (in his entropic brain article) ranging from excessive order to excessive entropy, depression, addiction, and disorders of obsession all fall on the too-much-order end. (Psychosis is on the entropy end of the spectrum, which is why it probably doesn't respond to psychedelic therapy.)

The therapeutic value of psychedelics, in Carhart-Harris's view, lies in their ability to temporarily elevate entropy in the inflexible brain, jolting the system out of its default patterns. Carhart-Harris uses the metaphor of annealing from metallurgy: psychedelics introduce energy into the system, giving it the flexibility necessary for it to bend and so change. The Hopkins researchers use a similar metaphor to make the same point: psychedelic therapy creates an interval of maximum plasticity in which, with proper guidance, new patterns of thought and behavior can be learned.

All these metaphors for brain activity are just that—metaphors—and not the thing itself. Yet the neuroimaging of tripping brains that's been done at Imperial College (and that has since been replicated in several other labs using not only psilocybin but also LSD and ayahuasca) has identified measurable changes in the brain that lend credence to these metaphors. In particular, the changes in activity and connectivity in the default mode network on psychedelics suggest it may be possible to link the felt experience of certain types of mental suffering with something observable—and alterable—in the brain. If the default mode network does what neuroscientists think it does, then an intervention that targets that network has the potential to help relieve several forms of mental illness, including the handful of disorders psychedelic researchers have trialed so far.

So many of the volunteers I spoke to, whether among the dying, the addicted, or the depressed, described feeling mentally "stuck," captured in ruminative loops they felt powerless to break. They talked about "prisons of the self," spirals of obsessive introspection that wall them off from other people, nature, their earlier selves, and the present moment. All these thoughts and feelings may be the products of an overactive default mode network, that tightly linked set of brain structures implicated in rumination, self-referential thought, and metacognition—thinking about thinking. It stands to reason that by quieting the brain network responsible for thinking about ourselves, and thinking about thinking about ourselves, we might be able to jump that track, or erase it from the snow.

The default mode network appears to be the seat not only of the ego, or self, but of the mental faculty of time travel as well. The two are of course closely related: without the ability to remember our past and imagine a future, the notion of a coherent self could hardly be said to exist; we define ourselves with reference to our personal history and future objectives. (As meditators eventually discover, if we can manage to stop thinking about the past or future and sink into the present, the self seems to disappear.) Mental time travel is constantly taking us off the frontier of the present moment. This can be highly adaptive; it allows us to learn from the past and plan for the future. But when time travel turns obsessive, it fosters the backward-looking gaze of depression and the forward pitch of anxiety. Addiction, too, seems to involve uncontrollable time travel. The addict uses his habit to organize time: *When was the last hit, and when can I get the next?*

To say the default mode network is the seat of the self is not a simple proposition, especially when you consider that the self may not be exactly real. Yet we can say there is a set of mental operations, time travel among them, that are associated with the self. Think of it simply as the locus of this particular set of mental activities, many of which appear to have their home in the structures of the default mode network.

Another type of mental activity that neuroimaging has located in the DMN (and specifically in the posterior cingulate cortex) is the work performed by the so-called autobiographical or experiential self: the mental operation responsible for the narratives that link our first person to the world, and so help define us. *"This is who I am." "I don't deserve to be loved." "I'm the kind of person without the willpower to break this addiction."* Getting overly attached to these narratives, taking them as fixed truths about ourselves rather than as stories subject to revision, contributes mightily to addiction, depression, and anxiety. Psychedelic therapy seems to weaken the grip of these narratives, perhaps by temporarily disintegrating the parts of the default mode network where they operate.

And then there is the ego, perhaps the most formidable creation of the default mode network, which strives to defend us from threats both internal and external. When all is working as it should be, the ego keeps the organism on track, helping it to realize its goals and provide for its needs, notably for survival and reproduction. It gets the job done. But it is also fundamentally conservative. "The ego keeps us in our grooves," as Matt Johnson puts it. For better and, sometimes, for worse. For occasionally the ego can become tyrannical and turn its formidable powers on the rest of us.* Perhaps this is the link between the various forms of mental illness that psychedelic therapy seems to help most: all involve a disordered ego—overbearing, punishing, or misdirected.†

In a college commencement address he delivered three years before his suicide, David Foster Wallace asked his audience to "think of the old cliché about 'the mind being an excellent servant but a terrible master.' This, like many clichés, so lame and unexciting on the surface, actually expresses a great and terrible truth," he said.

"It is not the least bit coincidental that adults who commit suicide with firearms almost always shoot themselves in the head. They shoot the terrible master."

---

* This is how Freud understood depression, which he called melancholia: after the loss of an object of desire, the ego splits in two, with one part punishing the other, which has taken the place of the lost love in our attentions. In his view, depression is a misplaced form of revenge for a loss, retribution that has been misdirected at the self.

† Tom Insel, who after leaving the NIMH went to work for Google's life science subsidiary, Verily, before joining a mental health start-up called Mindstrong Health, told me that there are now algorithms that can reliably diagnose depression based on the frequency and context of one's use of the first-person pronoun.

o   o   o

OF ALL THE PHENOMENOLOGICAL EFFECTS that people on psychedelics report, the dissolution of the ego seems to me by far the most important and the most therapeutic. I found little consensus on terminology among the researchers I interviewed, but when I unpack their metaphors and vocabularies—whether spiritual, humanistic, psychoanalytic, or neurological—it is finally the loss of ego or self (what Jung called "psychic death") they're suggesting is the key psychological driver of the experience. It is this that gives us the mystical experience, the death rehearsal process, the overview effect, the notion of a mental reboot, the making of new meanings, and the experience of awe.

Consider the case of the mystical experience: the sense of transcendence, sacredness, unitive consciousness, infinitude, and blissfulness people report can all be explained as what it can feel like to a mind when its sense of being, or having, a separate self is suddenly no more.

Is it any wonder we would feel one with the universe when the boundaries between self and world that the ego patrols suddenly fall away? Because we are meaning-making creatures, our minds strive to come up with new stories to explain what is happening to them during the experience. Some of these stories are bound to be supernatural or "spiritual," if only because the phenomena are so extraordinary they can't be easily explained in terms of our usual conceptual categories. The predictive brain is getting so many error signals that it is forced to develop extravagant new interpretations of an experience that transcends its capacity for understanding.

Whether the most magnificent of these stories represent a regression to magical thinking, as Freud believed, or access to transpersonal realms such as the "Mind at Large," as Huxley believed, is itself a matter of interpretation. Who can say for certain? Yet it seems to me very likely that losing or shrinking the self would make anyone feel more "spiritual," however you choose to define the word, and that this is apt to make one feel better.

The usual antonym for the word "spiritual" is "material." That at least is what I believed when I began this inquiry—that the whole issue with spirituality turned on a question of metaphysics. Now I'm inclined to think a much better and certainly more useful antonym for "spiritual" might be "egotistical." Self and Spirit define the opposite ends of a spectrum, but that spectrum needn't reach clear to the heavens to have meaning for us. It can stay right here on earth. When the ego dissolves, so does a bounded conception not only of our self but of our self-interest. What emerges in its place is invariably a broader, more openhearted and altruistic—that is, more spiritual— idea of what matters in life. One in which a new sense of connection, or love, however defined, seems to figure prominently.

"The psychedelic journey may not give you what you want," as more than one guide memorably warned me, "but it will give you what you need." I guess that's been true for me. It might have been nothing like the one I signed up for, but I can see now that the journey has been a spiritual education after all.

Coda: Going to Meet My Default Mode Network

I got the opportunity—a non-pharmacological opportunity—to peer into my own default mode network soon after I interviewed Judson Brewer, the psychiatrist and neuroscientist who studies the brains of meditators. It was Brewer, you'll recall, who discovered that the brains of experienced meditators look much like the brains of people on psilocybin: the practice and the medicine both dramatically reduce activity in the default mode network.

Brewer invited me to visit his lab at the Center for Mindfulness at the University of Massachusetts medical school in Worcester to run some experiments on my own default mode network. His lab has developed a neural feedback tool that allows researchers (and their volunteers) to observe in real time the activity in one of the key brain structures in the default mode network: the posterior cingulate cortex.

Until now I have tried to spare you the names and functions of specific parts of brain anatomy, but I do need to describe this one in a bit more detail. The posterior cingulate cortex is a centrally located node within the default mode network involved in self-referential mental processes. Situated in the middle of the brain, it links the prefrontal cortex—site of our executive function, where we plan and exercise will—with the centers of memory and emotion in the hippocampus. The PCC is believed to be the locus of the experiential or narrative self; it appears to generate the narratives that link what happens to us to our abiding sense of who we are. Brewer believes that this particular operation, when it goes awry, is at the root of several forms of mental suffering, including addiction.

As Brewer explains it, activity in the PCC is correlated not so much with our thoughts and feelings as with "how we *relate* to our thoughts and feelings." It is where we get "caught up in the push and pull of our experience." (This has particular relevance for the addict: "It's one thing to have cravings," as Brewer points out, "but quite another to get caught up in your cravings.") When we take something that happens to us personally? That's the PCC doing its (egotistical) thing. To hear Brewer describe it is to suspect neuroscience might have at last found the address for the "But enough about you" center of the brain.

Buddhists believe that attachment is at the root of all forms of mental suffering; if the neuroscience is right, a lot of these attachments have their mooring in the PCC, where they are nurtured and sustained. Brewer thinks that by diminishing its activity, whether by means of meditation or psychedelics, we can learn "to be with our thoughts and cravings without getting caught up in them." Achieving such a detachment from our thoughts, feelings, and desires is what Buddhism (along with several other wisdom traditions) teaches is the surest path out of human suffering.

Brewer took me into a small, darkened room where a comfortable chair faced a computer monitor. One of his laboratory assistants brought in the contraption: a red rubber bathing cap with 128 sensors arrayed in a dense grid across every centimeter of its surface. Each of the sensors was linked to a cable. After the assistant carefully fitted the cap onto my skull, she squirted a dab of conductive gel beneath each of the 128 electrodes to ensure the faint electrical signals emanating from deep within my brain could readily traverse my scalp. Brewer took a picture of me on my phone: I had sprouted a goofy tangle of high-tech dreadlocks.

To calibrate a baseline level of activity for my PCC, Brewer projected a series of adjectives on the screen—"courageous," "cheap," "patriotic," "impulsive," and so on. Simply reading the list does nothing to activate the PCC, which is why he told me now to think about how these adjectives either applied or didn't apply to me. *Take it personally*, in other words. This is precisely the thought process that the PCC exists to perform, relating thoughts and experiences to our sense of who we are.

Once he had established a baseline, Brewer, from another room, led me through a series of exercises to see if I could alter the activity of my PCC by thinking different kinds of thoughts. At the completion of each "run"—lasting a few minutes—he would project a bar graph on the screen in front of me; the length of each bar indicates to what extent the activity in my PCC had exceeded or dropped below baseline, in ten-second increments. I could also follow the ups and downs of my PCC activity by listening to rising and falling tones on a monitor, but I found that too distracting.

I began by trying to meditate, something I'd gotten into the habit of doing early in my foray into the science and practice of psychedelic consciousness. A brief daily meditation had become a way for me to stay in touch with the kind of thinking I'd done on psychedelics. I discovered my trips had made it easier for me to drop into a mentally quiet place, something that in the past had always eluded me. So I closed my eyes and began to follow my breath. I had never tried to meditate in front of other people, and it felt awkward, but when Brewer put the graph up on the screen, I could see that I had succeeded in quieting my PCC—not by a lot, but most of the bars dipped below baseline. Yet the graph was somewhat jagged, with several bars leaping above baseline. Brewer explained that this is what happens when you're trying too hard to meditate and become conscious of the effort. There it was in black and white: the graph of my effortfulness and self-criticism.

Next Brewer asked me to do a "loving-kindness" meditation. This is one where you're supposed to close your eyes and think warm and charitable thoughts about people: first yourself, then those closest to you, and finally people you don't know—humanity at large. The bars dropped smartly below baseline, deeper than before: *I was good at this!* (A self-congratulatory thought that no doubt shot a bar skyward.)

For the next and last run, I told Brewer I had an idea for a mental exercise I wanted to try but didn't want to tell him what it was until afterward. I closed my eyes and tried to summon scenes from my psychedelic journeys. The one that came to mind first was an image of a pastoral landscape, a gently rolling quilt of field and forest and pond, directly above which hovered some kind of gigantic rectangular frame made of steel. The structure, which was a few stories tall but hollow, resembled a pylon for electrical transmission lines or something a kid might build from an Erector set—a favorite toy of my childhood. Anyway, by the odd logic of psychedelic experience, it was clear to me even in the moment that this structure represented my ego, and the landscape above which it loomed was, I presumed, the rest of me.

The description makes it sound as though the structure were menacing, hovering overhead like a UFO, but in fact the emotional tone of the image was mostly benign. The structure had revealed itself as empty and superfluous and had lost its purchase on the ground—on me. The scene had given me a kind of overview effect: *behold your ego*, sturdy, gray, empty, and floating free, like an untethered pylon. Consider how much more beautiful the scene would be were it not in the way. The phrase "child's play" looped in my mind: the structure was nothing more than a toy that a child could assemble and disassemble at will. During the trip the structure continued to loom, casting an intricate shadow over the scene, but now in my recollection I could picture it drifting off, leaving me . . . to be.

Who knows what kinds of electrical signals were leaking from my default mode network during this reverie, or for that matter what the image symbolized. You've read this chapter: obviously, I've been giving a lot of thought to the ego and its discontents. Here was some of that thinking rendered starkly visible. I had succeeded in detaching myself from my ego, at least imaginatively, something I would never have thought possible before psychedelics. Aren't we identical with our ego? What's left of us without it? The lesson of both psychedelics and meditation is the same: *No!* on the first count, and *More than enough* on the second. Including this lovely landscape of the mind, which became lovelier still when I let that ridiculous steel structure float away, taking its shadow with it.

A beep indicated the run was over. Brewer's voice came on the loudspeaker: "What in the world were you *thinking*?" Apparently, I'd dropped way below baseline. I told him, in general terms. He sounded excited by the idea that the mere recollection of a psychedelic experience might somehow replicate what happens in the brain during the real thing. Maybe that's what was going on. Or maybe it was the specific content of the image, and the mere thought of bidding adieu to my ego, watching it float away like a hot-air balloon, that had the power to silence my default mode network.

Brewer started spouting hypotheses. Which is really all that science can offer us at this point: hunches, theories, so many more experiments to try. We have plenty of clues, and more now than before the renaissance of psychedelic science, but we remain a long way from understanding exactly what happens to consciousness when we alter it, either with a molecule or with meditation. Yet gazing at the bars on the graph before me, these crude hieroglyphs of psychedelic thought, I felt as if I were standing on the edge of a wide-open frontier, squinting to make out something wondrous.

In Praise of Neural Diversity

IN APRIL 2017, the international psychedelic community gathered in the Oakland Convention Center for Psychedelic Science, an every-few-years-or-so event organized by MAPS, the Multidisciplinary Association for Psychedelic Studies, the nonprofit established by Rick Doblin in 1986 with the improbable goal of returning psychedelics to scientific and cultural respectability. In 2016, Doblin himself seemed stunned at how far and fast things had come and how close to hand victory now seemed. Earlier in the year, the FDA had approved phase 3 trials of MDMA, and psilocybin was not far behind. If the results of these trials come anywhere near those of phase 2, the government will presumably have to reschedule the two drugs, and then doctors will be able to prescribe them. "We are not the counterculture," Doblin told a reporter during the conference. "We are the culture."

What had been as recently as 2010 a modest gathering of psychonauts and a handful of renegade researchers was now a six-day convention-cum-conference that had drawn more than three thousand people from all over the world to hear researchers from twenty-five countries present their findings. Not that there weren't also plenty of psychonauts and legions of the psychedelically curious. Between the lectures and panels and plenaries, they browsed a sprawling marketplace offering psychedelic books, psychedelic artwork, and psychedelic music.

For me, the event turned out to be a kind of reunion, bringing together most of the characters in my story under one roof. I was able to catch up with virtually all the scientists I'd interviewed (though Robin Carhart-Harris, with a baby on the way, had to skip), as well as several of the underground guides with whom I'd worked.

Everyone, it seemed, was here, scientists rubbing shoulders with guides and shamans, veteran psychonauts, a large contingent of therapists eager to add psychedelics to their practice, plus funders and filmmakers and even a smattering of entrepreneurs sniffing out business opportunities. And although I picked up snippets of concern about the new attorney general's efforts to rekindle the drug war, on the whole the mood was unmistakably celebratory.

When I asked conferencegoers which session they deemed most memorable, almost invariably they mentioned the plenary panel called "Future of Psychedelic Psychiatry." What was most noteworthy about this panel was the identity of the panelists, which, at a psychedelic convention, was cause for cognitive dissonance. Here was Paul Summergrad, MD, the former head of the American Psychiatric Association, seated next to Tom Insel, MD, the former head of the National Institute of Mental Health. The panel was organized and moderated by George Goldsmith, an American entrepreneur and health industry consultant based in London. In the last several years, he and his wife, Ekaterina Malievskaia, a Russian-born physician, have devoted their considerable energy and resources to winning approval for psilocybin-assisted therapy in the European Union.

It was clear to everyone in the standing-room crowd exactly what the three men on the panel represented: the recognition of psychedelic therapy by the mental health establishment. Insel spoke of how poorly the record of mental health care stacks up against the achievements of the rest of medicine. He pointed out that it has failed to lower mortality from serious psychiatric disorders and spoke of the promise of new models of mental health treatment such as psychedelic therapy. "I'm really impressed by the approach here," he told the group. "People don't say, simply, we're gonna give psychedelics. They talk about 'psychedelic-assisted psychotherapy.' . . . I think it's a really novel approach." Insel tempered his enthusiasm, however, by noting that such a novel paradigm may bedevil regulators accustomed to evaluating new drugs in isolation.

George Goldsmith asked both men what advice they would give to the researchers in the room, men and women who have been working diligently for years to bring psychedelic therapy to patients. Without hesitating, Insel turned to the audience and said, "Don't screw it up!"

"There may be lots of promise here," Insel said, "but it's really easy to forget about issues related to safety, issues related to rigor, issues related to reputational risks." He suggested that psychedelics would probably need to be rebranded in the public mind and that it would be essential to steer clear of anything that smacked of "recreational use." He and Summergrad both warned that a single sloppy researcher, or a patient with a disastrous experience, could poison the well for everybody. Nobody needed to mention the name Timothy Leary.

o o o

How CLOSE ARE WE to a world in which psychedelic therapy is sanctioned and routine, and what would such a world look like? Bob Jesse was in the audience when the former head of NIMH took his swipe against "recreational use," and though I didn't see it, I can picture his grimace. *And what exactly is wrong with re-creating ourselves?* Bob Jesse worries that the "medicalization" of psychedelics these men were advocating as the one true path would be a mistake.

Not that medicalization will be easy. Several steep regulatory hurdles will first need to be overcome. Phase 3 trials involve multiple sites and hundreds of volunteers; they can cost tens of millions of dollars. Normally Big Pharma foots the bill for such trials, but thus far the pharmaceutical companies have shown scant interest in psychedelics. For one thing, this class of drugs offers them little if any intellectual property: psilocybin is a product of nature, and the patent on LSD expired decades ago. For another, Big Pharma mostly invests in drugs for chronic conditions, the pills you have to take every day. Why would it invest in a pill patients might only need to take once in a lifetime?

Psychiatry faces a similar dilemma: it too is wedded to interminable therapies, whether that means the daily antidepressant or the weekly psychotherapy session. It is true that a psychedelic session lasts several hours and usually requires two therapists be present for the duration, but if the therapy works as it's supposed to, there won't be a lot of repeat business. It's not at all clear what the business model might be. Yet.

Several of the researchers and therapists I've interviewed nevertheless look forward to a time, not far off, when psychedelic therapy is routine and widely available, in the form of a novel hybrid of pharmacology and psychotherapy. George Goldsmith envisions a network of psychedelic treatment centers, facilities in attractive natural settings where patients will go for their guided sessions. He has formed a company called Compass Pathways to build these centers in the belief they can offer a treatment for a range of mental illnesses sufficiently effective and economical that Europe's national health services will reimburse for them. Goldsmith has so far raised three million pounds to fund and organize psilocybin trials (starting with treatment-resistant depression) at multiple sites in Europe. Already he is working with designers at IDEO, the international design firm, to redesign the entire experience of psychedelic therapy. Paul Summergrad and Tom Insel have both joined his advisory board.

Kathcrine MacLean, the former Hopkins researcher who wrote the landmark paper on openness, hopes someday to establish a "psychedelic hospice," a retreat center somewhere out in nature where not only the dying but their loved ones can use psychedelics to help them let go—the patient and the loved ones both.

"If we limit psychedelics just to the patient," she explains, "we're sticking to the old medical model. But psychedelics are more radical than that. I get nervous when people say they should only be prescribed by a doctor. I imagine a broader application."

In MacLean's words it's easy to hear echoes of the 1960s experience with psychedelics—the excitement about their potential to help not only the sick but everyone else too. This kind of thinking—or talking—makes some of her mainstream colleagues nervous. It's exactly the kind of talk that Insel and Summergrad were warning the community against. Good luck with that.

"The betterment of well people" is very much on the minds of most of the researchers I interviewed, even if some of them were more reluctant to discuss it on the record than institutional outsiders like Bob Jesse and Rick Doblin and Katherine MacLean. For them, medical acceptance is a first step toward a much broader cultural acceptance—outright legalization, in Doblin's view, or something more carefully controlled in MacLean's and Jesse's. Jesse would like to see the drugs administered by trained guides working in what he calls "longitudinal multigenerational contexts," which, as he describes them, sound a lot like churches. (Think of the churches that use ayahuasca in a ritual context, administered by experienced elders in a group setting.) Others envision a time when people seeking a psychedelic experience—whether for reasons of mental health or spiritual seeking or simple curiosity—could go, very occasionally, to something like a "mental health club," as Julie Holland, a psychiatrist who used to work with Stephen Ross at Bellevue, described it. "Sort of like a cross between a spa/retreat and a gym, where people can experience psychedelics in a safe, supportive environment."*

Everyone speaks of the importance of well-trained psychedelic guides—"board certified"—and the need to help people afterward integrate the powerful experiences they have had in order to make sense of them and render them truly useful. Tony Bossis paraphrases the religious scholar (and Good Friday Experiment volunteer) Huston Smith on this point: "A spiritual experience does not by itself make a spiritual life." Integration is essential to making sense of the experience, whether in or out of the medical context. Or else it remains just a drug experience.

As for the guides themselves, they are already being trained and certified: late in 2016, the California Institute of Integral Studies graduated its first class of forty-two psychedelic therapists. (This is a development that worries some in the underground, who fear being left behind when psychedelic therapy is legitimized. Yet it's hard to

---

* Or at least people who can afford it. One advantage of medicalizing psychedelic therapy is that it would presumably be accessible to everyone with health insurance.

imagine such experienced and highly skilled practitioners won't continue to find clients, especially among the well.)

When I asked Rick Doblin if he worries about another backlash, he pointed out that our culture has come a long way from the 1960s and has shown a remarkable ability to digest a great many of the cultural novelties first cooked up during that era.

"That was a very different time. People wouldn't even talk about cancer or death then. Women were tranquilized to give birth; men weren't allowed in the delivery room! Yoga and meditation were totally weird. Now mindfulness is mainstream and everyone does yoga, and there are birthing centers and hospices all over. We've integrated all these things into our culture. And now I think we're ready to integrate psychedelics."

Doblin points out that many of the people now in charge of our institutions are of a generation well acquainted with these molecules. This, he suggests, is the true legacy of Timothy Leary. It's all well and good for today's researchers to disdain his "antics" and blame him for derailing the first wave of research, and yet, as Doblin points out with a smile, "there would be no second wave if Leary hadn't turned on a whole generation." Indeed. Consider the case of Paul Summergrad, who has spoken publicly of his own youthful use of psychedelics. In a videotaped interview with Ram Dass that was shown at the 2015 meeting of the American Psychiatric Association, he told his colleagues that an acid trip he took in college had been formative in his intellectual development. (Jeffrey Lieberman, another past president of the American Psychiatric Association, has also written of the insights gleaned from his youthful experiments with LSD.*)

---

* He recounts these experiences in his book *Shrinks: The Untold Story of Psychiatry* (New York: Little, Brown, 2015), 190–93.

And yet, and yet . . . As much as I want to believe Doblin's sunny forecast, it's not hard to imagine things easily going off the rails. Tony Bossis agrees, as much as he hopes that psychedelics will someday be routine in palliative care.

"We don't die well in America. Ask people where do you want to die, and they will tell you, at home with their loved ones. But most of us die in an ICU. The biggest taboo in America is the conversation about death. Sure, it's gotten better; now we have hospices, which didn't exist not so long ago. But to a doctor, it's still an insult to let a patient go." In his view, psychedelics have the potential not only to open up that difficult conversation but to change the experience of dying itself. *If* the medical community will embrace them.

"This culture has a fear of death, a fear of transcendence, and a fear of the unknown, all of which are embodied in this work." Psychedelics may by their very nature be too disruptive for our institutions ever to embrace them. Institutions generally like to mediate the individual's access to authority of whatever kind—whether medical or spiritual—whereas the psychedelic experience offers something akin to direct revelation, making it inherently antinomian. And yet some cultures have successfully devised ritual forms to contain and harness the Dionysian energies of psychedelics; think of the Eleusinian mysteries of ancient Greece or the shamanic ceremonies surrounding peyote or ayahuasca in the Americas today. It is not impossible.

The first time I raised Jesse's idea of the betterment of well people with Roland Griffiths, he seemed to squirm a bit in his chair and then chose his words with care. "Culturally right now, that is a dangerous idea to promote." And yet, as we've talked, now over the course of three years, it's become clear that he too feels that many of us, and not just those dealing with cancer or depression or addiction, stand to benefit from these remarkable molecules and, even more,

from the spiritual experiences to which he believes—indeed, his research has demonstrated—they can open a door.

"We're all dealing with death," as he told me the first time we
met. "This is far too valuable to limit to sick people." A careful man,
mindful of the political land mines that may yet lie ahead, Griffiths
amended that last sentence just slightly, recast it in the future tense:
"This will be far too valuable to limit to sick people."

o   o   o

I, FOR ONE, sincerely hope that the kinds of experiences I've had
on psychedelics will *not* be limited to sick people and will someday
become more widely available. Does that mean I think these drugs
should simply be legalized? Not exactly. It is true I had a very positive experience using psilocybin "recreationally"—on my own, that
is, without the support of a guide—and for some people this might
be fine. But sooner or later, it seems, everyone has a trip for which
"bad" is far too pallid a modifier. I would hate to be alone when that
happens. For me, working one-on-one with an experienced guide in
a safe place removed from my everyday life turned out to be the ideal
way to explore psychedelics. Yet there are other ways to structure the
psychedelic journey—to provide a safe container for its potentially
overwhelming energies. Ayahuasca and peyote are typically used in a
group, with the leader, often but not necessarily a shaman, acting in
a supervisory role and helping people to navigate and interpret their
experiences. But whether individually or in a group, the presence of
someone with training and experience who can "hold the space"—to
use that hoary New Age locution—is more meaningful and comforting than I would have imagined.

Not only did my guides create a setting in which I felt safe enough
to surrender to the psychedelic experience, but they also helped me
to make sense of it afterward. Just as important, they helped me to

see there was something here worth making sense of. This is by no means self-evident. It is all too easy to dismiss what unfolds in our minds during a psychedelic journey as simply a "drug experience," and that is precisely what our culture encourages us to do. Matt Johnson made this point the first time we spoke: "Let's say you have some nineteen-year-olds taking mushrooms at a party. One of them has a profound experience. He's come to understand what God is, or his connection to the universe. What do his friends say? 'Oh, man, you had too much last night! No more mushrooms for you!'

"'Were you drinking or on drugs?' is what our culture says when you have a powerful experience."

Yet even a moment's reflection tells you that attributing the content of the psychedelic experience to "drugs" explains virtually nothing about it. The images and the narratives and the insights don't come from nowhere, and they certainly don't come from a chemical. They come from inside our minds,* and at the very least have something to tell us about *that*. If dreams and fantasies and free associations are worth interpreting, then surely so is the more vivid and detailed material with which the psychedelic journey presents us. It opens a new door on one's mind.

And about that my psychedelic journeys have taught me a great many interesting things. Many of these were the kinds of things one might learn in the course of psychotherapy: insights into important relationships; the outlines of fears and desires ordinarily kept out of view; repressed memories and emotions; and, perhaps most interesting and useful, a new perspective on how one's mind works.

This, I think, is the great value of exploring non-ordinary states of consciousness: the light they reflect back on the ordinary ones,

---

* I don't dismiss the possibility they may come from somewhere else, but will confine myself here to the more parsimonious explanation.

which no longer seem quite so transparent or so ordinary. To realize, as William James concluded, that normal waking consciousness is but one of many potential forms of consciousness—ways of perceiving or constructing the world—separated from it by merely "the filmiest of screens," is to recognize that our account of reality, whether inward or outward, is incomplete at best. Normal waking consciousness might seem to offer a faithful map to the territory of reality, and it is good for many things, but it is only a map—and not the only map. As to why these other modes of consciousness exist, we can only speculate. Most of the time, it is normal waking consciousness that best serves the interests of survival—and is most adaptive. But there are moments in the life of an individual or a community when the imaginative novelties proposed by altered states of consciousness introduce exactly the sort of variation that can send a life, or a culture, down a new path.

For me, the moment I recognized the tenuousness and relativity of my own default consciousness came that afternoon on Fritz's mountaintop, when he taught me how to enter a trance state by means of nothing more than a pattern of rapid breathing and the sounds of rhythmic drumming. *Where in the world has* that *been all my life?* This is nothing Freud or any number of psychologists and behavioral economists haven't told us, but the idea that "normal" consciousness is but the tip of a large and largely uncharted psychic iceberg is now for me something more than a theory; the hidden vastness of the mind is a felt reality.

I don't mean to suggest I have achieved this state of ego-transcending awareness, only tasted it. These experiences don't last, or at least they didn't for me. After each of my psychedelic sessions came a period of several weeks in which I felt noticeably different—more present to the moment, much less inclined to dwell on what's next. I was also notably more emotional and surprised myself on several occasions by how little it took to make me tear up or smile. I found myself thinking about things like death and time and infinity, but less in angst than in wonder. (I spent an unreasonable amount of time reflecting on how improbable and fortunate it is to be living here and now at the frontier of two eternities of nonexistence.) All at once and unexpectedly, waves of compassion or wonder or pity would wash over me.

This was a way of being I treasured, but, alas, every time it eventually faded. It's difficult not to slip back into the familiar grooves of mental habit; they are so well worn; the tidal pull of what the Buddhists call our "habit energies" is difficult to withstand. Add to this the expectations of other people, which subtly enforce a certain way of being yourself, no matter how much you might want to attempt another. After a month or so, it was pretty much back to baseline.

But not quite, not completely. For much like the depressed patients I interviewed in London, who described being nourished and even inspired by their furloughs from the cage of depression, the experience of some other way of being in the world survives in memory, as a possibility and a destination.

For me, the psychedelic experience opened a door to a specific mode of consciousness that I can now occasionally recapture in meditation. I'm speaking of a certain cognitive space that opens up late in a trip or in the midst of a mild one, a space where you can entertain all sorts of thoughts and scenarios without reaching for any kind of resolution. It somewhat resembles hypnagogic consciousness, that liminal state perched on the edge of sleep when all kinds of images and scraps of story briefly surface before floating away. But this is sustained, and what comes up can be clearly recalled. And though the images and ideas that appear are not under your direct control, but rather seem to be arriving and departing of their own accord, you *can* launch a topic or change it, like a channel. The ego is not entirely absent—you haven't been blasted into particles, or have returned from that particular state—but the stream of consciousness is taking its own desultory course, and you are bobbing and drifting along with it, looking neither forward nor back, immersed in the currents of being rather than doing. And yet a certain kind of mental work *is* getting done, and occasionally I have emerged from the state with usable ideas, images, or metaphors.*

My psychedelic adventures familiarized me with this mental territory, and, sometimes, not always, I find I can return to it during my daily meditation. I don't know if this is exactly where I'm supposed to be when I'm meditating, but I'm always happy to find myself floating in this particular mental stream. I would never have found it if not for psychedelics. This strikes me as one of the great gifts of the experience they afford: the expansion of one's repertoire of conscious states.

Just because the psychedelic journey takes place entirely in one's mind doesn't mean it isn't real. It is an experience and, for some of us, one of the most profound a person can have. As such, it takes its place as a feature in the landscape of a life. It can serve as a reference point, a guidepost, a wellspring, and, for some, a kind of spiritual sign or shrine. For me, the experiences have become landmarks to circle around and interrogate for meaning—meanings about myself, obviously, but also about the world. Several of the images that appeared in the course of my trips I think about all the time, hoping to unwrap

---

* In a 1969 essay in the *Harvard Theological Review*, Walter Pahnke described several distinct modes of psychedelic consciousness, including one he termed "the cognitive psychedelic experience." This is "characterized by astonishingly lucid thought. Problems can be seen from a novel perspective, and the inner relationships of many levels or dimensions can be seen all at once. The creative experience may have something in common with this kind of psychedelic experience, but such a possibility must await the result of future investigation."

what feels like a gift of meaning—from where or what or whom, I cannot say. There was that steel pylon hovering over the landscape of self. Or the image of my grandfather's skull staring back at me in Mary's mirror. The majestic but now hollowed-out trees in which my parents appeared to me, liable to topple in the next windstorm. Or the inky well of Yo-Yo Ma's cello, resonating with Bach's warm embrace of death. But there is one other image I haven't shared that I keep thinking must contain some important teaching, even as it continues to mystify me.

My last psychedelic journey was on ayahuasca. I was invited to join a circle of women who gather every three or four months to work with a legendary guide, a woman in her eighties who had trained under Leo Zeff. (She in turn had trained Mary, the woman who guided my psilocybin journey.) This journey was different from the others in that it took place in the company of a dozen other travelers, all of them strangers to me. Befitting this particular psychedelic, which is a tea brewed from two Amazonian plants (one a vine, the other a leaf), there was a considerable amount of ceremony in the shamanic mode: the singing of traditional *icaros*, prayers and invocations to "the grandmother" (a.k.a. the "plant teacher" or ayahuasca), bells and rattles and *shakapas*, and the blowing on us of various scents and smokes. All of which contributed to a mood of deep mystery and a suspension of disbelief that was especially welcome, inasmuch as we were in a yoga studio a long way from any jungle.

As has been the case with all of my journeys, the night before had been sleepless, as part of me worked to convince the rest of me not to do this crazy thing. That part was of course my ego, which before every trip has fought the threat to its integrity with ferocity and ingenuity, planting doubts and scenarios of disaster I had trouble batting away. *What about your heart, pal? You could die! What if you lose*

*your lunch or, even worse, your shit?! And what if "the grandmother"
dredges up some childhood trauma? Do you really want to lose it among
these strangers? These women?* (Part of the power of the ego flows from
its command of one's rational faculties.) By the time I arrived for the
circle, I was a nervous wreck, assailed by second and third thoughts
as to the wisdom of what I was about to do.

But, as has happened every time, as soon as I swallowed the medicine and slipped past the point of no return, the voice of doubt went
quiet and I surrendered to whatever was in store. Which was not
unlike my other psychedelic experiences, with a couple of notable
exceptions. Perhaps because the tea, which was viscous and acrid and
unexpectedly sweet, makes its alien presence felt in your stomach
and intestines, ayahuasca is a more bodily experience than some
other psychedelics. I did not get sick, but I was very much aware of
the thick brew moving through me and, as the effect of the DMT
(ayahuasca's active ingredient) came on, imagined it as a vine winding its way through the curls and convolutions of my intestines, oc-
cupying my body before slowly working its snakelike way up to and
into my head.

There followed a great many memories and images, some horrifying, others magnificent, but I want to describe one in particular
because, although I don't completely understand it, it captures something that psychedelics have taught me, something important.

Because there was still some light in the room when the ceremony
began, we were all wearing eye masks, and mine felt a little tight
around my head. Early in the journey, I became aware of the black
straps circling my skull, and these morphed into bars. My head was
caged in steel. The bars then began to multiply, moving down from
my head to encircle my torso and then my legs. I was now trapped
head to toe in a black steel cage. I pressed against the bars, but they

were unyielding. There was no way out. Panic was building when I noticed the green tip of a vine at the base of the cage. It was growing steadily upward and then turning, sinuously, to slip out between two of the bars, freeing itself and at the same time reaching toward the light. "A plant can't be caged," I heard myself thinking. "Only an animal can be caged."

I can't tell you what this means, if anything. Was the plant showing me a way out? Perhaps, but it's not as if I could actually follow it; I am an animal, after all. Yet it seemed the plant was trying to teach me *some*thing, that it was proposing a kind of visual koan for me to unpack, and I have been turning it over in my mind ever since. Maybe it was a lesson about the folly of approaching an obstacle head-on, that sometimes the answer is not the application of force but rather changing the terms of the problem in such a way that it loses its dominion without actually crumbling. It felt like some kind of jujitsu. Because the vine wasn't just escaping the confines of the cage, it was using the structure to improve its situation, climbing higher to gather more light for itself.

Or maybe the lesson was more universal, something about plants themselves and how we underestimate them. My plant teacher, as I began to think of the vine, was trying to tell me something about itself and the green kingdom it represents, a kingdom that has always figured largely in my work and my imagination. That plants are intelligent I have believed for a long time—not necessarily in the way we think of intelligence, but in a way appropriate to themselves. We can do many things plants can't, yet they can do all sorts of things we can't—escaping from steel cages, for example, or eating sunlight. If you define intelligence as the ability to solve the novel problems reality throws at the living, plants surely have it. They also possess agency, an awareness of their environment, and a kind of subjectivity—a set of interests they pursue and so a point of view. But though these are all ideas I have long believed and am happy to defend, never before have I *felt* them to be true, to be as deeply rooted as I did after my psychedelic journeys.

The un-cageable vine reminded me of that first psilocybin trip, when I felt the leaves and plants in the garden returning my gaze. One of the gifts of psychedelics is the way they reanimate the world, as if they were distributing the blessings of consciousness more widely and evenly over the landscape, in the process breaking the human monopoly on subjectivity that we moderns take as a given. To us, we are the world's only conscious subjects, with the rest of creation made up of objects; to the more egotistical among us, even other people count as objects. Psychedelic consciousness overturns that view, by granting us a wider, more generous lens through which we can glimpse the subject-hood—the spirit!—of everything, animal, vegetable, even mineral, all of it now somehow returning our gaze. Spirits, it seems, are everywhere. New rays of relation appear between us and all the world's Others.

Even in the case of the minerals, modern physics (forget psychedelics!) gives us reason to wonder if perhaps some form of consciousness might not figure in the construction of reality. Quantum mechanics holds that matter may not be as innocent of mind as the materialist would have us believe. For example, a subatomic particle can exist simultaneously in multiple locations, is pure possibility, until it is measured—that is, perceived by a mind. Only then and not a moment sooner does it drop into reality as we know it: acquire fixed coordinates in time and space. The implication here is that matter might not exist as such in the absence of a perceiving subject. Needless to say, this raises some tricky questions for a materialist understanding of consciousness. The ground underfoot may be much less solid than we think.

This is the view of quantum physics, not some psychonaut—

though it *is* a very psychedelic theory. I mention it only because it lends some of the authority of science to speculations that would otherwise sound utterly lunatic. I still tend to think that consciousness *must* be confined to brains, but I am less certain of this belief now than I was before I embarked on this journey. Maybe it too has slipped out from between the bars of that cage. Mysteries abide. But this I can say with certainty: the mind is vaster, and the world ever so much more alive, than I knew when I began.

Glossary

active placebo: A type of placebo used in drug trials to fool the volunteer into thinking he has received the psychoactive drug being tested. In the psilocybin trials, researchers have used niacin, which produces a tingling sensation, and methylphenidate (Ritalin), which is a stimulant.

ayahuasca: A psychedelic tea made from a combination of plants native to the Amazon basin, typically *Banisteriopsis caapi* and *Psychotria viridis* (or *chacruna*), and used sacramentally by indigenous peoples of South America. The *chacruna* plant contains the psychedelic compound DMT (N,N-dimethyltryptamine), but it is deactivated by digestive enzymes unless it is ingested with a monoamine oxidase inhibitor such as *Banisteriopsis*. In 2006, the U.S. Supreme Court affirmed the right of the Brazil-based UDV Church to use ayahuasca as a sacrament.

Beckley Foundation: The organization established by Amanda Feilding in England in 1998 to support research into psychedelics and advocate internationally for the reform of drug laws. The organization is named for Feilding's ancestral estate in Oxfordshire (BeckleyFoundation.org).

Council on Spiritual Practices (CSP): A nonprofit organization established by Bob Jesse in 1993 and "dedicated to making direct experience of the sacred more available to more people." CSP helped organize and fund the first experiments in psychedelic research at Johns Hopkins; CSP also supported the suit that resulted in the 2006 Supreme Court decision recognizing ayahuasca as a sacrament in the UDV Church. In 1995, CSP developed and published the "Code of Ethics for Spiritual Guides" that many underground psychedelic guides have adopted (csp.org).

default mode network (DMN): A set of interacting brain structures first described in 2001 by the Washington University neuroscientist Marcus Raichle. The default mode network, called that because it is most active when the brain is in a resting state, links parts of the cerebral cortex with deeper and evolutionarily older structures of the brain involved in emotion and memory. (Its key structures include, and link, the posterior cingulate cortex, the medial prefrontal cortex, and the hippocampus.) Neuroimaging studies suggest that the DMN is involved in such higher-order "metacognitive" activities as self-reflection, mental projection, time travel, and theory of mind—the ability to attribute mental states to others. Activity in the DMN falls during the psychedelic experiences, and when it falls most precipitously volunteers often report a dissolution of their sense of self.

DMT (or N,N-dimethyltryptamine): A rapid-onset, intense, and short-acting psychedelic compound sometimes referred to as "the businessman's trip." This tryptamine molecule is found in many plants and animals for reasons not well understood.

empathogen: A psychoactive drug that produces a heightened sense of connectedness, emotional openness, and compassion. MDMA, or Ecstasy, is such a drug. Also sometimes called an entactogen.

entheogen: From the Greek, "generating the divine within." A psychoactive substance that produces or facilitates a spiritual experience. Entheogens have been used by many cultures for thousands of years, whether by shamans or as part of religious or spiritual practices. However, the term was not coined until the 1970s, by a group of scholars that included R. Gordon Wasson, Richard Evans Schultes, Jonathan Ott, and Carl Ruck. The word was intended to help rehabilitate psychedelics by distinguishing their ancient spiritual role from the recreational uses to which they were often put beginning in the 1960s.

Esalen, or the Esalen Institute: A retreat center in Big Sur, California, founded in 1962 to explore the various methods for expanding consciousness that often go under the umbrella of the human potential movement. Esalen was closely identified with the psychedelic movement before the drugs were banned; in the years afterward, a series of meetings took place at Esalen, where strategies to rehabilitate and restart research into psychedelics were developed. Many psychedelic guides now working underground received their training at Esalen.

5-HT$_{2A}$ receptor: One of several types of receptors in the brain that respond to the neurotransmitter serotonin. Psychedelic compounds also bind to this receptor, precipitating a cascade of (poorly understood) events that produce the psychedelic experience. Because of its distinctive molecular shape, LSD binds particularly well to the 5-HT$_{2A}$ receptor. In addition, a portion of the receptor folds over the LSD molecule and holds it inside the receptor, which might explain its intensity and long duration of action.

5-MeO-DMT (5-methoxy-N,N-dimethyltryptamine): A powerful, short-acting psychedelic compound found in certain South American plants and in the venom of the Sonoran desert toad (*Incilius alvarius*). The toad venom is typically vaporized and smoked; 5-MeO-DMT obtained from plants is usually made into a snuff. The compound has been used sacramentally in South America for many years; it was first synthesized in 1936 and was not made illegal until 2011.

hallucinogen: The class of psychoactive drugs that induce hallucinations, including the psychedelics, the dissociatives, and the deliriants. The term is often used as a synonym for psychedelics, even though psychedelics don't necessarily produce full-fledged hallucinations.

Harvard Psilocybin Project: The psychological research program established by Timothy Leary and Richard Alpert (later Ram Dass) in the Department of Social Relations at Harvard in 1960. The researchers (who included Ralph

Metzner, a graduate student) administered psilocybin to hundreds of volunteers "in a naturalistic setting"; they also conducted experiments with prisoners at Concord State Prison and with theology students at Boston University's Marsh Chapel. Later, the group began working with LSD. The project was engulfed in controversy in 1962 and closed down after it had been reported that Alpert had given psilocybin to an undergraduate, in violation of its agreement with Harvard. Leary and Alpert established a successor organization in Cambridge but outside Harvard, called the International Federation for Internal Freedom.

Heffter Research Institute: A nonprofit established in 1993 by David E. Nichols, a chemist and pharmacologist at Purdue University, with several colleagues, to support scientific research into psychedelic compounds. The institute was named for Arthur Heffter, the German chemist, pharmacologist, and physician who first identified mescaline as the psychoactive component of the peyote cactus in the late 1890s. Established at a time when psychedelic research had been dormant for two decades, the Heffter Institute has played a pivotal, but quiet, role in the revival of that research, helping to fund most of the psilocybin trials done in America since the late 1990s, including the work at Hopkins and NYU (Heffter.org).

Holotropic Breathwork: A breathing exercise developed in the mid 1970s by the psychedelic therapist Stanislav Grof, and his wife, Christina, after LSD was made illegal. By breathing rapidly and exhaling deeply, nearly to the point of hyperventilation, subjects enter an altered state of consciousness without the use of a drug. This trancelike state can give access to subconscious material. "Holotropic" means "moving toward wholeness."

LSD (lysergic acid diethylamide): Also known as acid, this psychedelic compound was first synthesized in 1938 by Albert Hofmann, a Swiss chemist at Sandoz who was searching for a drug to stimulate circulation. LSD was the twenty-fifth molecule that Hofmann had derived from the alkaloids produced by ergot, a fungus that infects grain. Hofmann shelved the compound when it proved ineffective as a medicine, but five years later a premonition led him to resynthesize it. After accidentally ingesting a small quantity of LSD, he discovered its powerful psychoactive properties. In 1947, Sandoz began marketing LSD as a psychiatric drug under the name Delysid. It was withdrawn from circulation in 1966 after the drug appeared on the black market.

MAPS (Multidisciplinary Association for Psychedelic Studies): The nonprofit membership organization founded in 1986 by Rick Doblin to increase public understanding of psychedelics and support scientific research into their therapeutic applications. Based in Santa Cruz, California, MAPS has focused its efforts on MDMA, or Ecstasy, as a therapeutic intervention for people suffering from PTSD. In 2016, it won FDA approval to conduct phase 3 trials of MDMA in the treatment of PTSD; in 2017, the FDA designated MDMA as a "breakthrough therapy" for PTSD, clearing the way for an expedited review. Doblin, and MAPS, have played a central role in the revival of psychedelic research. MAPS also sponsors Psychedelic Science, the international conference on psychedelic research that takes place in Northern California every few years.

MDMA (3,4-methylenedioxymethamphetamine): A psychoactive compound first synthesized by Merck in 1912 but never marketed. After the compound was resynthesized by the Bay Area chemist Alexander "Sasha" Shulgin in the 1970s, it became a popular adjunct to psychotherapy, because its "empathogenic" qualities helped patients form a strong bond of trust with their therapists. In the 1980s, the drug showed up in the rave scene, where it was sold under the name of Ecstasy (or E or later Molly); in 1986, the U.S. government put MDMA on schedule I, declaring it a drug of abuse with no accepted medical use. However, recent drug trials sponsored by MAPS have demonstrated MDMA's value in treating PTSD. MDMA is not considered a "classical psychedelic," because it appears to operate on different brain pathways from LSD or psilocybin.

mescaline: A psychedelic compound derived from several cacti, including peyote and San Pedro. The compound was first identified and named by the German chemist Arthur Heffter in 1897. *The Doors of Perception* is a first-person account of Aldous Huxley's first mescaline experience.

microdosing: The practice of ingesting a small, "subperceptual" dose of a psychedelic, usually LSD or psilocybin, every few days as an aid to mental health or mental performance. A common protocol is to take ten micrograms of LSD (a tenth of a medium dose) every fourth day. The practice is fairly new, and as yet the evidence for its effectiveness is anecdotal. Several trials are under way.

MK-Ultra: The code name for an undercover research program on psychedelic drugs conducted by the CIA beginning in 1953; it was closed down in 1963 or 1964. At various times, the CIA sought to determine whether LSD and related compounds could be used as a means of mind control; an interrogation tool (or truth serum); a biological weapon (added to a population's water supply); or a political tool (by dosing adversaries to get them to do foolish things). As part of the research program, which at times involved forty-four universities and colleges, civilians and military personnel were dosed without their knowledge, sometimes with disastrous consequences. The public first learned about MK-Ultra during the Church Committee hearings on the CIA held in 1975; further hearings on the program were held in 1977. However, most of the agency's documents on the program had been destroyed in 1973 on orders from director Richard Helms.

Mystical Experience Questionnaire: The psychological survey, developed by Walter Pahnke and William Richards in the 1960s, used to assess whether a volunteer in a trial of a psychedelic drug has undergone a mystical-type experience. It seeks to measure, on a scale of one to five, seven attributes of a mystical experience: internal unity; external unity; transcendence of time and space; ineffability and paradoxicality; a sense of sacredness; the noetic quality; and a deeply felt positive mood. Several revised versions of the MEQ have since been developed.

noetic quality: A term introduced by William James, an American psychologist, to denote the fact that the mystical state registers not only as a feeling but as a state of knowledge. People emerge with the enduring conviction that important truths have been revealed to them. The noetic quality was, for James, one of the four marks of the mystical experience, along with ineffability, transiency, and passivity.

phenethylamines: A class of organic molecule, and the name for one of the two principal types of psychedelic compounds;

the other is the tryptamines. Mescaline and MDMA are examples of phenethylamines.

psilocin: One of the two principal psychoactive compounds found in psilocybin mushrooms. The other is psilocybin, which breaks down to psilocin under certain conditions. Both compounds were isolated (from mushrooms provided by R. Gordon Wasson) and named by Albert Hofmann in 1958. Psilocin is what gives psilocybin mushrooms their bluish tint when bruised.

Psilocybe: A genus of approximately two hundred gilled mushrooms, roughly half of which produce psychoactive compounds such as psilocybin and psilocin. *Psilocybes* are distributed throughout the world. Their possession is illegal in most jurisdictions. The best-known members of the genus are *Psilocybe cubensis, Psilocybe cyanescens, Psilocybe semilanceata*, and *Psilocybe azurescens*.

psilocybin: The main psychoactive compound found in psilocybin mushrooms and a shorthand for the class of mushrooms that contain it.

psychedelic: From the Greek for "mind manifesting." The term was coined in 1956 by Humphry Osmond to describe drugs like LSD and psilocybin that produce radical changes in consciousness.

psycholytic: A term coined in the 1960s for a drug, or dose of a drug, that loosens constraints on the mind, allowing subconscious material to enter one's awareness. Also the name for a form of psychotherapy that uses low doses of psychedelics to relax the patient's ego without obliterating it.

psychotomimetic: The name for a drug that produces effects resembling psychosis. This was a common term for LSD and drugs like it when they were first introduced to psychiatry in the 1950s; researchers believed they produced temporary psychoses that would yield insights into the nature of mental illness and give therapists the opportunity to experience madness firsthand.

reducing valve: The term used by Aldous Huxley in *The Doors of Perception* for the mental filter that admits to our awareness only a "measly trickle of the kind of consciousness" we need to survive. In his view, the value of psychedelics was to open the reducing valve, giving us access to the fullness of experience and the universal "Mind at Large."

set and setting: The inner and outer environments in which a drug experience takes place; "set" is a term for the mind-set and expectations the person brings to the experience, and "setting" is the outward circumstances in which it takes place. Set and setting are particularly influential in the case of psychedelics. The terms are usually credited to Timothy Leary, but the concept was recognized and made use of by earlier researchers such as Al Hubbard.

tryptamine: A class of organic molecule common in nature, and the name for one of the two principal types of psychedelic compounds; the other is the phenethylamines. LSD, psilocybin, and DMT are tryptamines. The neurotransmitter serotonin is also a tryptamine.

Acknowledgments

CHANGING ONE'S MIND, or one's subject as a writer, is never easy, and this book would never have been ventured, much less completed, if not for the support and encouragement of the people around me. Ann Godoff, my book editor for going on four decades now, didn't blink or blanch when I told her I wanted to write a book about psychedelics; her enthusiasm and sure-footed editorial guidance through this, our eighth book together, has been a blessing. Amanda Urban, too, abetted this adventure in so many ways; my career-long debt to her is incalculable. Thanks, too, to the superb teams in their respective offices: Sarah Hutson, Casey Denis, and Karen Mayer, at Penguin; and, at ICM, Liz Farrell, Maris Dyer, Daisy Meyrick, Molly Atlas, and Ron Bernstein.

The best thing about being a journalist is getting paid to learn whole new subjects as an adult. Yet the pursuit of such a continuing education would be impossible without the forbearance of the people we ask to be our teachers. I'm grateful to everyone—the scientists, the volunteers, the patients, the therapists, and the advocates—who endured the multiple, lengthy interviews and all the dumb questions. Special thanks to Bob Jesse, Roland Griffiths, Matthew Johnson, Mary Cosimano, Bill Richards, Katherine MacLean, Rick Doblin, Paul Stamets, James Fadiman, Stephen Ross, Tony Bossis, Jeffrey Guss, George Goldsmith, Ekaterina Malievskaia, Charles Grob, Teri Krebs, Robin Carhart-Harris, David Nutt, David Nichols, George Sarlo, Vicky Dulai, Judson Brewer, Bia Labate, Gabor Maté, Lisa Callaghan, and Andrew Weil. Though not everyone I interviewed is quoted here by name, all were excellent teachers, and I am deeply grateful for your patience with my questions and generosity with your answers. Several people took substantial risks in sharing their stories with me; although I can't thank them publicly, I owe a tremendous debt to the many underground guides who gave so freely of their time, their experience, and their wisdom. It is a shame that at least for now their healing practice depends on acts of civil disobedience.

I spent a productive and pleasurable year as a fellow of the Radcliffe Institute for Advanced Study at Harvard, which gave me the opportunity to research and write the history of psychedelic research in the city where an important chapter of it took place. The institute offered the perfect environment for pursuing a project that touches on so many different disciplines: I only had to walk down the hall to consult a brain scientist, a biologist, an anthropologist, and an investigative reporter. While at Radcliffe, I was blessed to work with a dogged undergraduate research assistant who helped me navigate the Harvard archives and turned up one hidden gem after another: thank you, Teddy Delwiche. I also owe a debt to Ed Wasserman, my dean at the Graduate School of Journalism at Berkeley, for granting me time off from teaching so that I could go to Cambridge and, later, complete the book.

Back in Berkeley, Bridget Huber did brilliant work, first as a research assistant and then as a fact-checker; that this is the most thoroughly sourced of my books owes entirely to her diligence and skill. Several of my colleagues at Berkeley contributed hugely to my education in neuroscience and psychology: David Presti, Dacher Keltner, and Alison Gopnik enriched this book in more ways than they realize and, in the case of David and his partner, Kristi Panik, who read a draft of the neuroscience chapter, saved me from errors large and small. (Though they bear no responsibility for any errors that may remain.) Mark Edmundson supplied some crucial early advice that helped shape the narrative, and Mark Danner was, as ever, an invaluable sounding board on our walks at Inspiration Point. I count myself especially lucky to be close friends with an editor as astute and generous as Gerry Marzorati; his comments on the manuscript were invaluable and saved you, dear reader, from having to read several thousand unnecessary words.

My first foray into the subject of psychedelics came in a 2015 piece in the *New Yorker*, "The Trip Treatment"; thanks to Alan Burdick, the gifted editor who assigned it, and David Remnick, for seeing it fit to publish; the piece opened all sorts of doors.

For crucial research assistance along the way, as well as their indispensable online library, I'm deeply grateful to Earth and Fire, the proprietors of Erowid, which is the single most important resource on psychedelics there is. Check it out.

For their wise, helpful, and reassuring legal counsel, I'm grateful to my dear friend Howard Sobel and his colleague Marvin Putnam at Latham & Watkins. I sleep much better knowing they have my back.

A long book project has a way of inflecting the emotional weather in a family, this one perhaps more than most. Isaac, it has meant the world to me to be able to talk through my journeys with you; I always come away from our conversations with something smart, useful, and unexpected. Your support, curiosity, and encouragement have made all the difference.

When I embarked on this long, strange trip, Judith wondered what it might mean for our thirty-year-plus collaboration. Would I return somehow changed? Never would I have imagined that after all that time anything could bring us closer together, but there it is. Thank you for pushing me to attempt something new, for the searching questions and insights along the way, for the close editing of every chapter—and, most of all, for going with me on the journey.

Notes

PROLOGUE **A NEW DOOR**

1 **The first of these molecules:** Hofmann, *LSD, My Problem Child*, 40–47.

2 **The second molecule:** Wasson and Wasson, *Mushrooms, Russia, and History*, vol. 2.

2 **a fifteen-page account:** Wasson, "Seeking the Magic Mushroom."

5 **LSD scrambled your chromosomes:** Cohen, Hirschhorn, and Frosch, "In Vivo and In Vitro Chromosomal Damage Induced by LSD-25."

7 **In the spring of 2010:** Tierney, "Hallucinogens Have Doctors Tuning In Again."

10 **For a peer-reviewed scientific paper:** Griffiths et al., "Psilocybin Can Occasion Mystical-Type Experiences Having Substantial and Sustained Personal Meaning and Spiritual Significance."

14 **emergency room admissions involving psychedelics:** Johansen and Krebs, "Psychedelics Not Linked to Mental Health Problems or Suicidal Behavior."

14 **nearly a thousand volunteers:** Personal correspondence with Matthew W. Johnson, PhD.

18 **the term "psychedelics":** Dyck, *Psychedelic Psychiatry*, 1–2.

CHAPTER ONE **A RENAISSANCE**

21 **Entering his second century:** Langlitz, *Neuropsychedelia*, 24–26.

22 **"the only joyous invention":** Hofmann, *LSD, My Problem Child*, 184–85.

22 **As a young chemist:** Ibid., 36–45.

23 **And there it remained for five years:** Ibid., 46–47.

24 **Now unfolds the world's first *bad* acid trip:** Ibid., 48–49.

24 **"My ego was suspended":** Quoted in Nichols, "LSD."

25 **"everything glistened and sparkled":** Hofmann, *LSD, My Problem Child*, 51.

25 **"in the edifice of materialist rationality":** Jonathan Ott in translator's preface to ibid., 25.

26 **"the feeling of co-creatureliness":** Langlitz, *Neuropsychedelia*, 25–26.

27 **The second watershed event of 2006:** *Gonzales v. O Centro Espirita Beneficente Uniao do Vegetal.*

29 **"major therapeutic possibilities":** Kleber, "Commentary On: Psilocybin Can Occasion Mystical-Type Experiences," 292.

30 **"hope that this landmark paper":** Schuster, "Commentary On: Psilocybin Can Occasion Mystical-Type Experiences," 289.

30 **"that, when used appropriately":** Nichols, "Commentary On: Psilocybin Can Occasion Mystical-Type Experiences," 284.

30 **"free oneself of the bounds":** Wit, "Towards a Science of Spiritual Experience."

41 **the noetic quality:** James, *Varieties of Religious Experience*, 370.

41 **"Dreams cannot stand this test":** Ibid., 389.

44 **more than a thousand scientific papers:** See, for example, Grinspoon and Bakalar, *Psychedelic Drugs Reconsidered*, 192.

45 **a PhD dissertation at Harvard:** Walter Pahnke's thesis, "Drugs and Mysticism: An Analysis of the Relationship Between Psychedelic Drugs and the Mystical Consciousness," is available in PDF form at http://www.maps.org/images/pdf/books/pahnke/walter_pahnke_drugs_and_mysticism.pdf.

45 **"Until the Good Friday Experiment":** Huston Smith, *Huston Smith Reader*, 73.

45 **a follow-up study of the Good Friday Experiment:** Doblin, "Pahnke's 'Good Friday Experiment.'"

46 **a second review:** Doblin, "Dr. Leary's Concord Prison Experiment."

47 **"would be for psychiatry":** Quoted in Nutt, "Brave New World for Psychology?," 658.

48 **the first modern trial of psilocybin:** Grob et al., "Pilot Study of Psilocybin Treatment for Anxiety in Patients with Advanced-Stage Cancer."

59 **An internal memo:** A cache of declassified CIA files related to Project Artichoke is available at http://www.paperlessarchives.com/FreeTitles/ARTICHOKECIAFiles.pdf.

69 **"my own constitution shuts me out":** James, *Varieties of Religious Experience*, 369.

69 **"The subject of it immediately says":** Ibid., 370.

69 **"Mystical states seem to those who experience them":** Ibid.

70 **"that deepened sense of the significance":** Ibid., 372.

72 "and from one recurrence to another": Ibid., 371.

72 "The mystic feels as if his own will": Ibid.

74 led to lasting changes in their personalities: MacLean et al., "Mystical Experiences Occasioned by the Hallucinogen Psilocybin Lead to Increases in the Personality Domain of Openness."

76 "Doctors encounter this strange": McHugh, review of *The Harvard Psychedelic Club*, by Don Lattin.

76 "authoritative over the individuals": James, *Varieties of Religious Experience*, 415.

77 "The existence of mystical states": Ibid., 419.

77 "might, in spite of all the perplexity": Ibid., 420.

77 "ascend[s] to a more enveloping point of view": Ibid.

77 "It is as if the opposites of the world": Ibid., 378.

78 a pilot study in smoking cessation: Johnson et al., "Pilot Study of the 5-HT$_{2AR}$ Agonist Psilocybin in the Treatment of Tobacco Addiction."

CHAPTER TWO NATURAL HISTORY: BEMUSHROOMED

91 The mycelia in a forest: Simard et al., "Net Transfer of Carbon Between Ectomycorrhizal Tree Species in the Field."

93 Humans have been using psilocybin mushrooms: Stamets, *Psilocybin Mushrooms of the World*, 11.

93 "Psilocybe mushrooms and civilization": Ibid., 16.

94 "Mistakes in mushroom identification can be lethal": Ibid., 30–32.

95 "The Stametsian Rule": Ibid., 53.

104 had personal knowledge of psychedelic drugs: Lee and Shlain, *Acid Dreams*, 71.

104 "through the eyes of a happy and gifted child": Siff, *Acid Hype*, 93.

104 *Life* gave him a generous contract: Ibid., 80.

105 "description of your own sensations": Ibid., 73.

105 a circulation of 5.7 million: Ibid.

105 "Seeking the Magic Mushroom": All quotations appear in Wasson, "Seeking the Magic Mushroom."

108 "These they ate before dawn": Wasson and Wasson, *Mushrooms, Russia, and History*, 223.

109 "the devil that they worshipped": Davis, *One River*, 95.

109 "an act of superstition condemned": Siff, *Acid Hype*, 69.

109 "carry you there where god is": Wasson, Hofmann, and Ruck, *Road to Eleusis*, 33.

110 On the night of June 29–30, 1955: Wasson, "Seeking the Magic Mushroom."

112 "Before Wasson nobody took the mushrooms": Estrada, *María Sabina*, 73.

112 "To find God, Sabina": Letcher, *Shroom*, 104.

113 *Person to Person*: Siff, *Acid Hype*, 80.

113 several other magazines: Ibid., 83.

113 **An exhibition on magic mushrooms:** Ibid., 74.

113 **Hofmann isolated and named:** Hofmann, *LSD, My Problem Child*, 128.

113 **"Thirty minutes after my taking":** Ibid., 126.

113 **In 1962, Hofmann joined Wasson:** Ibid., 139–52.

114 **"unleash[ing] on lovely Huautla":** Wasson, "Drugs," 21.

114 **"From the moment the foreigners arrived":** Estrada, *María Sabina*, 90–91.

115 **you can find him on YouTube:** The video, *The Stoned Ape Theory*, by Terence McKenna, is at https://www.youtube.com/watch?v=hOtLJwK7kdk.

115 **"access to realms of supernatural power":** McKenna, *Food of the Gods*, 26.

115 **"catalyzed the emergence of human self-reflection":** Ibid., 24.

115 **"brought us out of the animal mind":** See McKenna's talk on YouTube: https://www.youtube.com/watch?v=hOtLJwK7kdk.

124 **Samorini calls this a "depatterning factor":** Samorini, *Animals and Psychedelics*, 84–88.

127 **"Nature everywhere speaks to man":** Wulf, *Invention of Nature*, 54.

127 **"I myself am identical with nature":** Ibid., 128.

128 **"Everything," Humboldt said, "is interaction and reciprocal":** Ibid., 59.

136 **"Nature always wears the colors":** Emerson, *Nature*, 14.

136 **another form of consciousness "parted from [us]":** James, *Varieties of Religious Experience*, 377.

136 **a spiritually "realized being":** Huston Smith, *Cleansing the Doors of Perception*, 76.

137 **"forbid[s] a premature closing":** James, *Varieties of Religious Experience*, 378.

CHAPTER THREE HISTORY: THE FIRST WAVE

138 **When the federal authorities:** Leary, *Flashbacks*, 232–42.

139 **Leary was called before a committee:** Greenfield, *Timothy Leary*, 267–72.

139 **"Dreary Senate hearing and courtrooms":** Leary, *Flashbacks*, 251–52.

145 **"a tantalizing sense of portentousness":** Novak, "LSD Before Leary," 91.

146 **"enter the illness and see with a madman's eyes":** Osmond, "On Being Mad."

146 **In the years following World War II:** Dyck, *Psychedelic Psychiatry*, 17.

146 **the two researchers began to explore:** Ibid.

147 **But it was a productive hypothesis:** For an excellent overview of how this research contributed to the rise of neurochemistry, see Nichols, "Psychedelics," 267.

147 **The Saskatchewan Mental Hospital:** Weyburn would soon become the world's most important hub of research into psychedelics. Dyck, *Psychedelic Psychiatry*, 26–28.

148 **"My 12 Hours as a Madman":** For a discussion of the article, see ibid., 31–33.

148 **Their focus on LSD:** Ibid., 40–42.

149 **"seemed so bizarre that we laughed uproariously":** Ibid., 58–59.

149 **"From the first":** Ibid., 59.

151 **Based on this success:** Ibid., 71.

151 **they seemed too good to be true:** Ibid., 73.

152 **The idea that a drug could occasion:** See Novak, "LSD Before Leary," 97, and the anonymously published *"Pass It On,"* Kindle location 5372.

152 **Beginning in 1956, Bill W. had several LSD sessions:** Eisner, "Remembrances of LSD Therapy Past," 14, 26–45; Novak, "LSD Before Leary," 97.

153 **Born in 1910 in New York City:** Novak, "LSD Before Leary," 88–89.

153 **"was taken by surprise":** Ibid., 92.

153 **"the problems and strivings":** Ibid.

154 **Cohen came to think of it:** Betty Grover Eisner, draft of "Sidney Cohen, M.D.: A Remembrance," box 7, folder 3, Betty Grover Eisner Papers, Stanford University Department of Special Collections and University Archives.

154 **"psycholytic" means "mind loosening":** Grinspoon and Bakalar, *Psychedelic Drugs Reconsidered*, 7.

155 **Stanislav Grof, who trained as a psychoanalyst:** For a detailed account of this work, see Grof, *LSD*.

156 **A 1967 review article:** Grinspoon and Bakalar, *Psychedelic Drugs Reconsidered*, 208.

156 **Anaïs Nin, Jack Nicholson:** Lee and Shlain, *Acid Dreams*, 62.

157 **the most famous of these patients was Cary Grant:** Siff, *Acid Hype*, 100.

157 **declared himself "born again":** Stevens, *Storming Heaven*, 64.

157 **"All the sadness and vanities":** Siff, *Acid Hype*, 100.

157 **"I'm no longer lonely":** Ibid.

157 **"Young women have never before":** Novak, "LSD Before Leary," 103.

157 **a surge in demand for LSD therapy:** Ibid.

158 **"LSD became for us an intellectual fun drug":** Ibid., 99.

158 **Cohen was made uncomfortable:** Ibid., 99–101.

158 **He remained deeply ambivalent:** Ibid., 100.

158 **"under LSD the fondest theories":** Cohen, *Beyond Within*, 182.

159 **"any explanation of the patient's problems":** Ibid.

159 **"therapy by self-transcendence":** Cohen, "LSD and the Anguish of Dying," 71.

160 **"relish the possibility":** Dyck, *Psychedelic Psychiatry*, 1.

160 **"It was without question":** Huxley, *Moksha*, 42.

161 **"the folds of my gray flannel trousers":** Huxley, *Doors of Perception*, 33.

161 **"what Adam had seen on the morning":** Ibid., 17.

161 **"Words like 'grace' and 'transfiguration'":** Ibid., 18.

161 **"a measly trickle":** Ibid., 23.

161 **"shining with their own inner light":** Ibid., 17.

162 **a common core of mystical experience:** Huxley, *Perennial Philosophy*.

162 **"99 percent Aldous Huxley":** Novak, "LSD Before Leary," 93.

162 **"It will give that elixir a bad name":** Ibid., 95.

162 **Clearly a new name for this class:** Dyck, *Psychedelic Psychiatry*, 1–2.

163 **"had no particular connotation of madness":** Ibid., 2.

163 **"uncontaminated by other associations":** Osmond, "Review of the Clinical Effects of Psychotomimetic Agents," 429.

163 **The goal was to create the conditions:** Grinspoon and Bakalar, *Psychedelic Drugs Reconsidered*, 194–95.

164 **his FBI file:** Hubbard's FBI file is available at the Internet Archive: https://archive.org/details/AlHubbard.

165 **the best account we have of his life:** Fahey, "Original Captain Trips."

165 **the trail of Hubbard's life:** These facts, and their contradictions, are drawn from Lee and Shlain, *Acid Dreams*, and Fahey, "Original Captain Trips."

166 **We know the government kept close tabs:** Lee and Shlain, *Acid Dreams*, 45.

167 **"It was the deepest mystical thing":** Ibid.

167 **"a catalytic agent":** Ibid., 52.

168 **"if he could give the psychedelic experience":** Fahey, "Original Captain Trips."

168 **"convinced that [Al Hubbard] was the man":** Ibid.

169 **Osmond abandoned the psychotomimetic model:** Lee and Shlain, *Acid Dreams*, 54.

169 **Hubbard was the first researcher to grasp:** Dyck, *Psychedelic Psychiatry*, 93.

170 **"He said, 'Now hate them'":** R.C., "B.C.'s Acid Flashback."

171 **"We waited for him like the little old lady":** Lee and Shlain, *Acid Dreams*, 51.

171 **impressive rates of success:** Stevens, *Storming Heaven*, 175.

172 **"The CIA work stinks":** Lee and Shlain, *Acid Dreams*, 52.

172 **"I tried to tell them how to use it":** Ibid.

173 **"What came through the closed door":** Stevens, *Storming Heaven*, 56.

173 **"What Babes in the Woods":** Ibid., 54.

174 **"who, having once come to the realization":** Ibid., 57.

174 **Commission for the Study of Creative Imagination:** Eisner, "Remembrances of LSD Therapy Past," 10.

174 **"Explorers have not always been the most scientific":** Ibid., 57.

175 **"My regard for science":** Dyck, *Psychedelic Psychiatry*, 97–98.

175 **Steve Jobs often told people:** Markoff, *What the Dormouse Said*, xix.

175 **"He'd be a broader guy":** Isaacson, *Steve Jobs*, 172–73.

176 **"That was a remarkable opening":** Goldsmith, "Conversation with George Greer and Myron Stolaroff."

176 **"After that first LSD experience":** Fahey, "Original Captain Trips."

177 **"The greatest thing in the world":** Markoff, *What the Dormouse Said*, 58.

178 **Seventy-eight percent of clients:** Stevens, *Storming Heaven*, 178.

179 **"We were amazed":** Fadiman, *Psychedelic Explorer's Guide*, 185.

180 **"Our investigations of some of the current social movements":** Lee and Shlain, *Acid Dreams*, 198.

181 **"to provide the [LSD] experience":** Fahey, "Original Captain Trips."

181 **"Al never did anything resembling security work":** Ibid.

186 **his first shattering experience:** Leary, *Flashbacks*, 29–33.

187 **"In four hours by the swimming pool"**: Ibid., 33.

188 *Listen! Wake up! You are God!:* Leary, *High Priest*, 285.

188 **Experimental Expansion of Consciousness:** This course description is in the New York Public Library's collection of Leary's papers. http://archives.nypl.org/mss/18400#detailed.

189 **"We were on our own"**: Stevens, *Storming Heaven*, 135.

191 **Leary reported eye-popping results:** Lee and Shlain, *Acid Dreams*, 75.

191 **Rick Doblin at MAPS meticulously reconstructed:** Doblin, "Dr. Leary's Concord Prison Experiment."

191 **"it was the sort of research"**: Cohen, *Beyond Within*, 224.

192 **"If we learned one thing"**: Lattin, *Harvard Psychedelic Club*, 74.

192 **"We were thinking far-out history thoughts"**: Leary et al., *Neuropolitics*, 3.

193 **"We're going to teach people"**: Lee and Shlain, *Acid Dreams*, 77.

194 **"Psychedelic drugs opened to mass tourism"**: Grinspoon and Bakalar, *Psychedelic Drugs Reconsidered*, 86.

195 **A 1961 memo from David McClelland:** "Some Social Reactions to the Psilocybin Research Project," Oct. 8, 1961.

195 **"analyz[e] your data objectively"**: Memo from McClelland to Metzner, Dec. 19, 1962.

195 **"I wish I could treat this"**: Lattin, *Harvard Psychedelic Club*, 89.

196 **The next day's *Crimson:*** Robert Ellis Smith, "Psychologists Disagree on Psilocybin Research."

196 **"Hallucination Drug Fought at Harvard"**: Lattin, *Harvard Psychedelic Club*, 91.

197 **"Psychedelic drugs cause panic"**: Grinspoon and Bakalar, *Psychedelic Drugs Reconsidered*, 66.

197 **"these materials are too powerful"**: Leary and Alpert, "Letter from Alpert, Leary."

197 **"For the first time in American history"**: Ibid.

197 **"We're through playing the science game"**: Stevens, *Storming Heaven*, 189.

198 **"had talked such nonsense"**: Ibid., 190.

198 **"powerful chemicals [as] harmless toys"**: Eisner, "Remembrances of LSD Therapy Past," 145.

199 **Osmond tried once again to coin a new one:** Dyck, *Psychedelic Psychiatry*, 132.

199 **"You must face these objections"**: Ibid., 108.

199 **"wreak havoc on all of us"**: Stevens, *Storming Heaven*, 191.

199 **Leary was happy to state it:** Leary, *High Priest*, 132.

200 **"He blew in with that uniform"**: Fahey, "Original Captain Trips."

200 **"I liked Tim when we first met"**: Lee and Shlain, *Acid Dreams*, 88.

200 **"Al got greatly preoccupied"**: Fahey, "Original Captain Trips."

201 **"I suppose there is little hope"**: Stevens, *Storming Heaven*, 191.

201 **"using hallucinogens for seductions"**: Weil, "Strange Case of the Harvard Drug Scandal."

202 **"Yes, sir, I did":** Lattin, *Harvard Psychedelic Club*, 94.

202 **Alpert and Leary appear to be:** Lee and Shlain, *Acid Dreams.*

202 **"an undergraduate group":** Weil, "Strange Case of the Harvard Drug Scandal."

204 **"given to him" by Marshall McLuhan:** Strauss, *Everyone Loves You When You're Dead,* location 352.

205 **"The kids who take LSD":** This quotation appears in a video made by Retro Report, available here: https://www.retroreport.org/video/the-long-strange-trip-of-lsd/.

206 **With Ken Kesey, the CIA had turned on:** Lee and Shlain, *Acid Dreams,* 124.

207 **"by blurring the boundaries":** Grob, "Psychiatric Research with Hallucinogens."

208 **"the drugs to themselves":** Grinker, "Lysergic Acid Diethylamide."

208 **"rendering their conclusions biased":** Grinker, "Bootlegged Ecstasy."

208 **"aura of magic":** Cole and Katz, "Psychotomimetic Drugs," 758.

208 **"the transcendental into psychiatry":** Eisner, "Remembrances of LSD Therapy Past," 112.

209 **But when the study was later discredited:** Presti and Beck, "Strychnine and Other Enduring Myths," 130–31.

210 **For his first study:** Cohen, "Lysergic Acid Diethylamide."

211 **"the dangers of suicide":** Cohen and Ditman, "Complications Associated with Lysergic Acid Diethylamide (LSD-25)," 162.

211 **In another paper published:** Cohen and Ditman, "Prolonged Adverse Reactions to Lysergic Acid Diethylamide."

211 **A fourth article:** Cohen, "Classification of LSD Complications."

211 **feverish cover story:** Moore and Schiller, "Exploding Threat of the Mind Drug That Got out of Control."

212 **"LSD *has* been your Frankenstein":** Novak, "LSD Before Leary," 109.

217 **"Why if [these projects] were worthwhile":** Lee and Shlain, *Acid Dreams,* 93.

217 **"four men lay, their minds literally expanding":** Fadiman, *Psychedelic Explorer's Guide,* 186.

219 **Someone made a videotape of the event:** And it's available on YouTube: https://www.youtube.com/watch?v=rjylxvQqm0U.

219 **he's traveled from Casa Grande:** Fahey, "Original Captain Trips."

CHAPTER FOUR TRAVELOGUE: JOURNEYING UNDERGROUND

223 **there are three things human beings are afraid of:** Quoted in Epstein, *Thoughts Without a Thinker,* 119.

225 **three thousand patients and trained 150 guides:** Stolaroff, *Secret Chief Revealed,* 28, 59.

226 **"laid the Torah across my chest":** Ibid., 36.

226 **"Many times I'd be in much agony":** Ibid., 61.

227 **"*Just leave 'em alone!*":** Ibid., 50.

227 **surveying their musical practices:** Barrett et al., "Qualitative and Quantitative Features of Music Reported to Support Peak Mystical Experiences During Psychedelic Therapy Sessions."

244 **"forms of consciousness entirely different":** James, *Varieties of Religious Experience*, 377.

253 **"For the moment that interfering neurotic":** Huxley, *Doors of Perception*, 53.

264 **"the totality of the awareness belonging to Mind at Large":** Ibid., 24.

277 **"of being overwhelmed, of disintegrating":** Ibid., 55.

281 **"If one always saw like this":** Ibid., 34–35.

286 **"Standing on the bare ground":** Emerson, *Nature*, 13.

286 **"Swiftly arose and spread around me":** Whitman, *Leaves of Grass*, 29.

287 **"All at once, as it were out of the intensity":** Tennysons, "Luminous Sleep."

289 **"I saw that the universe":** Quoted in James, *Varieties of Religious Experience*, 391.

CHAPTER FIVE NEUROSCIENCE: YOUR BRAIN ON PSYCHEDELICS

292 **One candidate for that chemical:** For more detail, see David Nichols's talk "DMT and the Pineal Gland: Facts vs. Fantasy," available at https://www.youtube.com/watch?v=YeeqHUiC8Io.

293 **psychedelics like LSD and psilocybin worked:** Vollenweider et al., "Psilocybin Induces Schizophrenia-Like Psychosis in Humans via a Serotonin-2 Agonist Action."

294 **"there is nothing of which we are more certain":** Freud, *Civilization and Its Discontents*, 12.

294 **The classic thought experiment:** Nagel, "What Is It Like to Be a Bat?"

295 **consciousness may pervade the universe:** Frank, "Minding Matter."

301 **a landmark paper:** Raichle et al., "Default Mode of Brain Function."

303 **"Chaos is averted":** Raichle, "Brain's Dark Energy."

304 **It also lights up when we receive "likes":** Brewer, *Craving Mind*, 46.

304 **In an often-cited paper:** Killingsworth and Gilbert, "Wandering Mind Is an Unhappy Mind."

305 **Shortly after Carhart-Harris published:** Carhart-Harris et al., "Neural Correlates of the Psychedelic State as Determined by fMRI Studies with Psilocybin."

309 **The bee perceives a substantially different spectrum:** Srinivasan, "Honey Bees as a Model for Vision, Perception, and Cognition"; Dyer et al., "Seeing in Colour."

309 **the sense that allows bees to register:** Sutton et al., "Mechanosensory Hairs in Bumblebees (*Bombus terrestris*) Detect Weak Electric Fields."

310 **a dimension of music that conveys emotion:** Kaelen, "Psychological and Human Brain Effects of Music in Combination with Psychedelic Drugs."

311 **"serves to promote realism":** Carhart-Harris et al., "Entropic Brain."

316 **"Distinct networks became less distinct":** Carhart-Harris, Kaelen, and Nutt, "How Do Hallucinogens Work on the Brain?"

316 **the usual lines of communications:** Petri et al., "Homological Scaffolds of Brain Functional Networks."

323 **her superb book:** Gopnik, *Philosophical Baby.*

328 **"Adults have congealed in their beliefs":** Lucas et al., "When Children Are Better (or at Least More Open-Minded) Learners Than Adults."

CHAPTER SIX THE TRIP TREATMENT: PSYCHEDELICS IN PSYCHOTHERAPY

334 **"For me that is not a medical concept":** Kupferschmidt, "High Hopes," 23.

334 **"If we are to develop optimal research designs":** Grob, "Psychiatric Research with Hallucinogens."

335 **only about half of the people who take their lives:** Beacon Health Options, "We Need to Talk About Suicide," 10.

335 **"psychiatry has gone from being brainless":** Solomon, *Noonday Demon,* 102.

339 **"alter[] the experience of dying":** Cohen, "LSD and the Anguish of Dying."

339 **"of cosmic unity":** Richards et al., "LSD-Assisted Psychotherapy and the Human Encounter with Death."

347 **"I am the luckiest man on earth":** Grob, Bossis, and Griffiths, "Use of the Classic Hallucinogen Psilocybin for Treatment of Existential Distress Associated with Cancer," 303.

349 **In December 2016, a front-page story:** Hoffman, "Dose of a Hallucinogen from a 'Magic Mushroom,' and Then Lasting Peace."

351 **In a follow-up study to the NYU trial:** Belser et al., "Patient Experiences of Psilocybin-Assisted Psychotherapy: An Interpretative Phenomenological Analysis."

355 **"is to make your interests gradually wider":** Bertrand Russell, "How to Grow Old."

359 **"And suddenly I realized that the molecules":** Hertzberg, "Moon Shots (3 of 3)."

361 **80 percent of the volunteers were confirmed as abstinent:** Johnson et al., "Pilot Study of the 5-HT_{2AR} Agonist Psilocybin in the Treatment of Tobacco Addiction."

365 **This suggests that the ability:** Personal communication with the neuroscientist Draulio Araujo.

369 **The record was a complete muddle:** Krebs and Johansen, "Lysergic Acid Diethylamide (LSD) for Alcoholism."

369 **"Given the evidence for a beneficial effect":** Ibid.

369 **a 2015 pilot study:** Bogenschutz et al., "Psilocybin-Assisted Treatment for Alcohol Dependence."

374 **volunteers spent a minute looking:** Piff et al., "Awe, the Small Self, and Prosocial Behavior."

374 **the after-awe self-portraits:** Bai et al., "Awe, the Diminished Self, and Collective Engagement."

376 **researchers gave psilocybin to six men:** Carhart-Harris et al., "Psilocybin with Psychological Support for Treatment-Resistant Depression."

377 **Watts's interviews uncovered two "master" themes:** Watts et al., "Patients' Accounts of Increased 'Connectedness' and 'Acceptance' After Psilocybin for Treatment-Resistant Depression."

378 **"It was like a holiday":** Ibid.

380 **"The sheen and shine that life and existence":** For Roullier's full account, see http://inandthrough.blogspot.com/2016/08/psilocybin-trial-diary-one-year-on.html.

383 **obsessive-compulsive disorder:** Moreno et al., "Safety, Tolerability, and Efficacy of Psilocybin in 9 Patients with Obsessive-Compulsive Disorder."

383 **"Depression is a response to past loss":** Solomon, *Noonday Demon*, 65.

384 **"What started as a pleasure becomes a need":** Kessler, *Capture*, 8–9.

384 **psychedelics enhance neuroplasticity:** Vollenweider and Kometer, "Neurobiology of Psychedelic Drugs."

388 **In a college commencement address:** Reproduced, in part, at Brain Pickings: https://www.brainpickings.org/2012/09/12/this-is-water-david-foster-wallace/.

391 **"how we *relate* to our thoughts and feelings":** Brewer, *Craving Mind*, 115.

EPILOGUE IN PRAISE OF NEURAL DIVERSITY

397 **"We are not the counterculture":** Schwartz, "Molly at the Marriott."

398 **mentioned the plenary panel:** A video of the talk is at https://www.youtube.com/watch?v=_oZ_v3QFQDE.

403 **a videotaped interview with Ram Dass:** Available at https://www.youtube.com/watch?v=NhlTrDIOcrQ&feature=share.

Bibliography

Bai, Yang, Laura A. Maruskin, Serena Chen, Amie M. Gordon, Jennifer E. Stellar, Galen D. McNeil, Kaiping Peng, and Dacher Keltner. "Awe, the Diminished Self, and Collective Engagement: Universals and Cultural Variations in the Small Self." *Journal of Personality and Social Psychology* 113, no. 2 (2017): 185–209. doi:10.1037/pspa0000087.

Barrett, Frederick S., Hollis Robbins, David Smooke, Jenine L. Brown, and Roland R. Griffiths. "Qualitative and Quantitative Features of Music Reported to Support Peak Mystical Experiences During Psychedelic Therapy Sessions." *Frontiers in Physiology* 8 (July 2017): 1–12. doi:10.3389/fpsyg.2017.01238.

Beacon Health Options. "We Need to Talk About Suicide." 2017.

Belser, Alexander B., Gabrielle Agin-Liebes, T. Cody Swift, Sara Terrana, Neşe Devenot, Harris L. Friedman, Jeffrey Guss, Anthony Bossis, and Stephen Ross. "Patient Experiences of Psilocybin-Assisted Psychotherapy: An Interpretative Phenomenological Analysis." *Journal of Humanistic Psychology* 57, no. 4 (2017): 354–88. doi:10.1177/0022167817706884.

Bogenschutz, Michael P., Alyssa A. Forcehimes, Jessica A. Pommy, Claire E. Wilcox, P. C. R. Barbosa, and Rick J. Strassman. "Psilocybin-Assisted Treatment for Alcohol Dependence: A Proof-of-Concept Study." *Journal of Psychopharmacology* 29, no. 3 (2015): 289–99. doi:10.1177/0269881114565144.

Brewer, Judson. *The Craving Mind: From Cigarettes to Smartphones to Love—Why We Get Hooked and How We Can Break Bad Habits.* New Haven, Conn.: Yale University Press, 2017.

Buckner, Randy L., Jessica R. Andrews-Hanna, and Daniel L. Schacter. "The Brain's Default Network: Anatomy, Function, and Relevance to Disease." *Annals of the New York Academy of Sciences* 1124, no. 1 (2008): 1–38. doi:10.1196/annals.1440.011.

Carbonaro, Theresa M., Matthew P. Bradstreet, Frederick S. Barrett, Katherine A. MacLean, Robert Jesse, Matthew W. Johnson, and Roland R. Griffiths. "Survey Study of Challenging Experiences After Ingesting Psilocybin Mushrooms: Acute and Enduring Positive and Negative Consequences." *Journal of Psychopharmacology* 30, no. 12 (2016): 1268–78.

Carhart-Harris, Robin L., et al. "Neural Correlates of the Psychedelic State as Determined by fMRI Studies with Psilocybin." *Proceedings of the National Academy of Sciences of the United States of America* 109, no. 6 (2012): 2138–43. doi:10.1073/pnas.1119598109.

———. "Psilocybin with Psychological Support for Treatment-Resistant Depression: An Open-Label Feasibility Study." *Lancet Psychiatry* 3, no. 7 (2016): 619–27. doi:10.1016/ S2215-0366(16)30065-7.

Carhart-Harris, Robin L., Mendel Kaelen, and David J. Nutt. "How Do Hallucinogens Work on the Brain?" *Psychologist* 27, no. 9 (2014): 662–65.

Carhart-Harris, Robin L., Robert Leech, Peter J. Hellyer, Murray Shanahan, Amanda Feilding, Enzo Tagliazucchi, Dante R. Chialvo, and David Nutt. "The Entropic Brain: A Theory of Conscious States Informed by Neuroimaging Research with Psychedelic Drugs." *Frontiers in Human Neuroscience* 8 (Feb. 2014): 20. doi:10.3389/fnhum.2014.00020.

Cohen, Maimon M., Kurt Hirschhorn, and William A. Frosch. "In Vivo and In Vitro Chromosomal Damage Induced by LSD-25." *New England Journal of Medicine* 277, no. 20 (1967): 1043–49. doi:10.1056/NEJM197107222850421.

Cohen, Sidney. *The Beyond Within: The LSD Story.* New York: Atheneum, 1964.

———. "A Classification of LSD Complications." *Psychosomatics* 7, no. 3 (1966): 182–86.

———. "LSD and the Anguish of Dying." *Harper's Magazine*, Sept. 1965, 69–78.

———. "Lysergic Acid Diethylamide: Side Effects and Complications." *Journal of Nervous and Mental Disease* 130, no. 1 (1960): 30–40.

Cohen, Sidney, and Keith S. Ditman. "Complications Associated with Lysergic Acid Diethylamide (LSD-25)." *Journal of the American Medical Association* 181, no. 2 (1962): 161–62.

———. "Prolonged Adverse Reactions to Lysergic Acid Diethylamide." *Archives of General Psychiatry* 8, no. 5 (1963): 475–80.

Cole, Jonathan O., and Martin M. Katz. "The Psychotomimetic Drugs: An Overview." *Journal of the American Medical Association* 187, no. 10 (1964): 758–61.

Davis, Wade. *One River: Explorations and Discoveries in the Amazon Rain Forest.* New York: Simon & Schuster, 1996.

Doblin, Rick. "Dr. Leary's Concord Prison Experiment: A 34-Year Follow-Up Study." *Journal of Psychoactive Drugs* 30, no. 4 (1998): 419–26. doi:10.1080/02791072.1998.10399715.

———. "Pahnke's 'Good Friday Experiment': A Long-Term Follow-Up and Methodological Critique." *Journal of Transpersonal Psychology* 23, no. 1 (1991): 1–28. doi:10.1177/0269881108094300.

Dyck, Erika. *Psychedelic Psychiatry: LSD from Clinic to Campus.* Baltimore: Johns Hopkins University Press, 2008.

Dyer, Adrian G., Jair E. Garcia, Mani Shrestha, and Klaus Lunau. "Seeing in Colour: A Hundred Years of Studies on Bee Vision Since the Work of the Nobel Laureate Karl von Frisch." *Proceedings of the Royal Society of Victoria* 127 (July 2015): 66–72. doi:10.1071/ RS15006.

Eisner, Betty Grover. "Remembrances of LSD Therapy Past." 2002. http://www.maps.org/ images/pdf/books/remembrances.pdf.

Emerson, Ralph Waldo. *Nature*. Boston: James Munroe, 1836.

Epstein, Mark. *Thoughts Without a Thinker: Psychotherapy from a Buddhist Perspective*. New York: Basic Books, 1995.

Estrada, Alvaro. *María Sabina: Her Life and Chants*. Santa Barbara, Calif.: Ross-Erikson, 1981.

Fadiman, James. *The Psychedelic Explorer's Guide: Safe, Therapeutic and Sacred Journeys*. Rochester, Vt.: Park Street Press, 2011.

Fahey, Todd Brendan. "The Original Captain Trips." *High Times*, Nov. 1991.

Frank, Adam. "Minding Matter." *Aeon*, March 2017.

Freud, Sigmund. *Civilization and Its Discontents*. New York: Norton, 1961.

Goldsmith, Neal. "A Conversation with George Greer and Myron Stolaroff." 2013. https:// erowid.org/culture/characters/stolaroff_myron/stolaroff_myron_interview1.shtml.

Gonzales v. O Centro Espirita Beneficente Uniao do Vegetal, 546 U.S. 418 (2006).

Gopnik, Alison. *The Philosophical Baby: What Children's Minds Tell Us About Truth, Love, and the Meaning of Life*. New York: Farrar, Straus and Giroux, 2009.

Greenfield, Robert. *Timothy Leary: A Biography*. Orlando, Fla.: Harcourt, 2006.

Griffiths, R. R., W. A. Richards, U. McCann, and R. Jesse. "Psilocybin Can Occasion Mystical-Type Experiences Having Substantial and Sustained Personal Meaning and Spiritual Significance." *Psychopharmacology* 187, no. 3 (2006): 268–83. doi:10.1007/ s00213-006-0457-5.

Grinker, Roy R. "Bootlegged Ecstasy." *Journal of the American Medical Association* 187, no. 10 (1964): 768.

———. "Lysergic Acid Diethylamide." *Archives of General Psychiatry* 8, no. 5 (1963): 425. doi:10.1056/NEJM196802222780806.

Grinspoon, Lester, and James B. Bakalar. *Psychedelic Drugs Reconsidered*. New York: Basic Books, 1979.

Grob, Charles S. "Psychiatric Research with Hallucinogens: What Have We Learned?" *Yearbook for Ethnomedicine and the Study of Consciousness*, no. 3 (1994): 91–112.

Grob, Charles S., Anthony P. Bossis, and Roland R. Griffiths. "Use of the Classic Hallucinogen Psilocybin for Treatment of Existential Distress Associated with Cancer." In *Psychological Aspects of Cancer: A Guide to Emotional and Psychological Consequences of Cancer, Their Causes and Their Management*, edited by Brian I. Carr and Jennifer Steel, 291–308. New York: Springer, 2013. doi:10.1007/978-1-4614-4866-2.

Grob, Charles S., Alicia L. Danforth, Gurpreet S. Chopra, Marycie Hagerty, Charles R. McKay, Adam L. Halberstadt, and George R. Greer. "Pilot Study of Psilocybin Treatment for Anxiety in Patients with Advanced-Stage Cancer." *Archives of General Psychiatry* 68, no. 1 (2011): 71–8. doi:10.1001/archgenpsychiatry.2010.116.

Grof, Stanislav. *LSD: Doorway to the Numinous: The Groundbreaking Psychedelic Research into Realms of the Human Unconscious*. Rochester, Vt.: Park Street Press, 2009.

Hertzberg, Hendrik. "Moon Shots (3 of 3): Lunar Epiphanies." *New Yorker*, Aug. 2008.

Hoffman, Jan. "A Dose of a Hallucinogen from a 'Magic Mushroom,' and Then Lasting Peace." *New York Times*, Dec. 1, 2016.

Hofmann, Albert. *LSD, My Problem Child*. Santa Cruz, Calif.: Multidisciplinary Association for Psychedelic Studies, 2009.

Huxley, Aldous. *The Doors of Perception, and Heaven and Hell*. New York: Harper & Row, 1963.

————. *Moksha: Writings on Psychedelics and the Visionary Experience (1931–1963).* Edited by Michael Horowitz and Cynthia Palmer. New York: Stonehill, 1977.

————. *The Perennial Philosophy.* London: Chatto & Windus, 1947. doi:10.1017/S0031819100023330.

Isaacson, Walter. *Steve Jobs.* New York: Simon & Schuster, 2011.

James, William. *The Varieties of Religious Experience.* EBook. Project Gutenberg, 2014.

Johansen, Pål-Ørjan, and Teri Suzanne Krebs. "Psychedelics Not Linked to Mental Health Problems or Suicidal Behavior: A Population Study." *Journal of Psychopharmacology* 29, no. 3 (2015): 270–79. doi:10.1177/0269881114568039.

Johnson, Matthew W., Albert Garcia-Romeu, Mary P. Cosimano, and Roland R. Griffiths. "Pilot Study of the 5-HT$_{2A}$R Agonist Psilocybin in the Treatment of Tobacco Addiction." *Journal of Psychopharmacology* 28, no. 11 (2014): 983–92. doi:10.1177/0269881114548296.

Kaelen, Mendel. "The Psychological and Human Brain Effects of Music in Combination with Psychedelic Drugs." PhD diss., Imperial College London, 2017.

Kessler, David A. *Capture: Unraveling the Mystery of Mental Suffering.* New York: Harper Wave, 2016.

Killingsworth, Matthew A., and Daniel T. Gilbert. "A Wandering Mind Is an Unhappy Mind." *Science* 330, no. 6006 (2010): 932. doi:10.1126/science.1192439.

Kleber, Herbert D. "Commentary On: Psilocybin Can Occasion Mystical-Type Experiences Having Substantial and Sustained Personal Meaning and Spiritual Significance." *Psychopharmacology* 187 (2006): 291–92.

Krebs, Teri S., and Pål-Ørjan Johansen. "Lysergic Acid Diethylamide (LSD) for Alcoholism: Meta-analysis of Randomized Controlled Trials." *Journal of Psychopharmacology* 26, no. 7 (2012): 994–1002. doi:10.1177/0269881112439253.

Kupferschmidt, Kai. "High Hopes." *Science* 345, no. 6192 (2014).

Langlitz, Nicolas. *Neuropsychedelia: The Revival of Hallucinogen Research Since the Decade of the Brain.* Berkeley: University of California Press, 2013.

Lattin, Don. *The Harvard Psychedelic Club: How Timothy Leary, Ram Dass, Huston Smith, and Andrew Weil Killed the Fifties and Ushered in a New Age for America.* New York: HarperCollins, 2010.

Leary, Timothy. *Flashbacks: A Personal and Cultural History of an Era: An Autobiography.* New York: G. P. Putnam's Sons, 1990.

————. *High Priest.* Berkeley, Calif.: Ronin, 1995.

Leary, Timothy, and Richard Alpert. "Letter from Alpert, Leary." *Harvard Crimson*, 1962.

Leary, Timothy, and James Penner. *Timothy Leary, The Harvard Years: Early Writings on LSD and Psilocybin with Richard Alpert, Huston Smith, Ralph Metzler, and Others.* Rochester, Vt.: Park Street Press, 2014.

Leary, Timothy, Robert Anton Wilson, George A. Koopman, and Daniel Gilbertson. *Neuropolitics: The Sociobiology of Human Metamorphosis.* Los Angeles: Starseed/Peace Press, 1977.

Lee, Martin A., and Bruce Shlain. *Acid Dreams: The Complete Social History of LSD: The CIA, the Sixties, and Beyond.* New York: Grove Press, 1992.

Lieberman, Jeffrey A. *Shrinks: The Untold Story of Psychiatry.* New York: Little, Brown, 2015.

Lucas, Christopher G., Sophie Bridgers, Thomas L. Griffiths, and Alison Gopnik. "When

Children Are Better (or at Least More Open-Minded) Learners Than Adults: Developmental Differences in Learning the Forms of Causal Relationships." *Cognition* 131, no. 2 (2014): 284–99. doi:10.1016/j.cognition.2013.12.010.

MacLean, Katherine A., Matthew W. Johnson, and Roland R. Griffiths. "Mystical Experiences Occasioned by the Hallucinogen Psilocybin Lead to Increases in the Personality Domain of Openness." *Journal of Psychopharmacology* 25, no. 11 (2011): 1453–61. doi:10.1177/0269881111420188.

McHugh, Paul. Review of *The Harvard Psychedelic Club*, by Don Lattin. *Commentary*, April 2010.

McKenna, Terence. *Food of the Gods: The Search for the Original Tree of Knowledge*. New York: Bantam Books, 1992.

Markoff, John. *What the Dormouse Said: How the Sixties Counterculture Shaped the Personal Computer Industry*. New York: Penguin, 2005.

Moore, Gerald, and Larry Schiller. "The Exploding Threat of the Mind Drug That Got out of Control." *Life*, March 25, 1966.

Moreno, Francisco A., Christopher B. Wiegand, E. Keolani Taitano, and Pedro L. Delgado. "Safety, Tolerability, and Efficacy of Psilocybin in 9 Patients with Obsessive-Compulsive Disorder." *Journal of Clinical Psychiatry* 67, no. 11 (2006): 1735–40. doi:10.4088/JCP.v67n1110.

Nagel, Thomas. "What Is It Like to Be a Bat?" *Philosophical Review* 83, no. 4 (1974): 435–50. doi:10.2307/2183914.

Nichols, David E. "Commentary On: Psilocybin Can Occasion Mystical-Type Experiences Having Substantial and Sustained Personal Meaning and Spiritual Significance." *Psychopharmacology* 187, no. 3 (2006): 284–86. doi:10.1007/s00213-006-0457-5.

———. "LSD: Cultural Revolution and Medical Advances." *Chemistry World* 3, no. 1 (2006): 30–34.

———. "Psychedelics." *Pharmacological Reviews* 68, no. 2 (2016): 264–355.

Nour, Matthew M., Lisa Evans, and Robin L. Carhar-Harris. "Psychedelics, Personality and Political Perspectives." *Journal of Psychoactive Drugs* (2017): 1–10.

Novak, Steven J. "LSD Before Leary: Sidney Cohen's Critique of 1950s Psychedelic Drug Research." *History of Science Society* 88, no. 1 (1997): 87–110.

Nutt, David. "A Brave New World for Psychology?" *Psychologist* 27, no. 9 (2014): 658–60. doi:10.1097/NMD.0000000000000113.

———. *Drugs Without the Hot Air: Minimising the Harms of Legal and Illegal Drugs*. Cambridge, England: UIT Cambridge, 2012.

Osmond, Humphry. "On Being Mad." *Saskatchewan Psychiatric Services Journal* 1, no. 2 (1952).

———. "A Review of the Clinical Effects of Psychotomimetic Agents." *Annals of the New York Academy of Sciences* 66, no. 1 (1957): 418–34.

Pahnke, Walter, "The Psychedelic Mystical Experience in the Human Encounter with Death." *Harvard Theological Review* 62, no. 1 (1969): 1–22.

"Pass It On": The Story of Bill Wilson and How the A.A. Message Reached the World. New York: Alcoholics Anonymous World Services, 1984.

Petri, G., P. Expert, F. Turkheimer, R. Carhart-Harris, D. Nutt, P. J. Hellyer, and F. Vaccarino. "Homological Scaffolds of Brain Functional Networks." *Journal of the Royal Society Interface* 11, no. 101 (2014).

Piff, Paul K., Pia Dietze, Matthew Feinberg, Daniel M. Stancato, and Dacher Keltner. "Awe, the Small Self, and Prosocial Behavior." *Journal of Personality and Social Psychology* 108, no. 6 (2015): 883–99. doi:10.1037/pspi0000018.

Pollan, Michael. "The Trip Treatment." *New Yorker*, Feb. 9, 2015.

Preller, Katrin H., Marcus Herdener, Thomas Pokorny, Amanda Planzer, Rainer Krahenmann, Philipp Stämpfli, Matthias E. Liechti, Erich Seifritz, and Franz X. Vollenweider. "The Fabric of Meaning and Subjective Effects in LSD-induced States Depend on Serotonin 2A Receptor Activation." *Current Biology* 27, no. 3 (2017): 451–57.

Presti, David, and Jerome Beck. "Strychnine and Other Enduring Myths: Expert and User Folklore Surrounding LSD." In *Psychoactive Sacramentals: Essays on Entheogens and Religion*, edited by Thomas B. Roberts, 125–35. San Francisco: Council on Spiritual Practices, 2001.

Raichle, Marcus E. "The Brain's Dark Energy." *Scientific American* 302, no. 3 (2010): 44–49. doi:10.1038/scientificamerican0310-44.

Raichle, Marcus E., Ann Mary MacLeod, Abraham Z. Snyder, William J. Powers, Debra A. Gusnard, and Gordon L. Shulman. "A Default Mode of Brain Function." *Proceedings of the National Academy of Sciences* 98, no. 2 (2001): 676–82. doi:10.1073/pnas.98.2.676.

R.C. "B.C.'s Acid Flashback." *Vancouver Sun*, Dec. 8, 2001.

Richards, William A. *Sacred Knowledge: Psychedelics and Religious Experiences.* New York: Columbia University Press, 2015.

Richards, William, Stanislav Grof, Louis Goodman, and Albert Kurland. "LSD-Assisted Psychotherapy and the Human Encounter with Death." *Journal of Transpersonal Psychology* 4, no. 2 (1972): 121–50.

Samorini, Giorgio. *Animals and Psychedelics: The Natural World and the Instinct to Alter Consciousness.* Rochester, Vt.: Park Street Press, 2002.

Schuster, Charles R. "Commentary On: Psilocybin Can Occasion Mystical-Type Experiences Having Substantial and Sustained Personal Meaning and Spiritual Significance." *Psychopharmacology* 187, no. 3 (2006): 289–90. doi:10.1007/s00213-006-0457-5.

Schwartz, Casey. "Molly at the Marriott: Inside America's Premier Psychedelics Conference." *New York Times*, May 6, 2017.

Siff, Stephen. *Acid Hype: American News Media and the Psychedelic Experience.* Urbana: University of Illinois Press, 2015.

Simard, Suzanne W., David A. Perry, Melanie D. Jones, David D. Myrold, Daniel M. Durall, and Randy Molina. "Net Transfer of Carbon Between Ectomycorrhizal Tree Species in the Field." *Nature* 388 (1997): 579–82.

Smith, Huston. *Cleansing the Doors of Perception: The Religious Significance of Entheogenic Plants and Chemicals.* New York: Jeremy P. Tarcher/Putnam, 2000.

———. *The Huston Smith Reader.* Edited by Jeffery Paine. Berkeley: University of California Press, 2012.

Smith, Robert Ellis. "Psychologists Disagree on Psilocybin Research." *Harvard Crimson*, March 15, 1962.

Solomon, Andrew. *The Noonday Demon: An Atlas of Depression.* New York: Scribner, 2015.

Srinivasan, Mandyam V. "Honey Bees as a Model for Vision, Perception, and Cognition." *Annual Review of Entomology* 55, no. 1 (2010): 267–84. doi:10.1146/annurev. ento.010908.164537.

Stamets, Paul. *Psilocybin Mushrooms of the World.* Berkeley, Calif.: Ten Speed Press, 1996.

Stevens, Jay. *Storming Heaven: LSD and the American Dream*. New York: Grove Press, 1987.

Stolaroff, Myron J. *The Secret Chief Revealed*. Sarasota, Fla.: Multidisciplinary Association for Psychedelic Studies, 2004.

Strauss, Neil. *Everyone Loves You When You're Dead: Journeys into Fame and Madness*. E-Book, 2011.

Sullivan, Walter. "The Einstein Papers. A Man of Many Parts," *New York Times*, March 29, 1972.

Sutton, Gregory P., Dominic Clarke, Erica L. Morley, and Daniel Robert. "Mechanosensory Hairs in Bumblebees (*Bombus terrestris*) Detect Weak Electric Fields." *Proceedings of the National Academy of Sciences* 113, no. 26 (2016): 7261–65. doi:10.1073/pnas.1601624113.

Tennyson, Alfred. "Luminous Sleep." *The Spectator*, Aug. 1, 1903.

Tierney, John. "Hallucinogens Have Doctors Tuning In Again." *New York Times*, April 12, 2010.

U.S. Congress Senate Subcommittee on Executive Reorganization of the Committee on Government Operations: Hearing on the Organization and Coordination of Federal Drug Research and Regulatory Programs: LSD. 89th Cong., 2nd sess., May 24–26, 1966.

Vollenweider, Franz X., and Michael Kometer. "The Neurobiology of Psychedelic Drugs: Implications for the Treatment of Mood Disorders." *Nature Reviews Neuroscience* 11, no. 9 (2010): 642–51. doi:10.1038/nrn2884.

Vollenweider, Franz X., Margreet F. I. Vollenweider-Scherpenhuyzen, Andreas Bäbler, Helen Vogel, and Daniel Hell. "Psilocybin Induces Schizophrenia-Like Psychosis in Humans via a Serotonin-2 Agonist Action." *NeuroReport* 9, no. 17 (1998): 3897–902. doi:10.1097/00001756-199812010-00024.

Wasson, R. Gordon. "Drugs: The Sacred Mushroom." *New York Times*, Sept. 26, 1970.

———. "Seeking the Magic Mushroom." *Life*, May 13, 1957, 100–120.

Wasson, R. Gordon, Albert Hofmann, and Carl A. P. Ruck. *The Road to Eleusis: Unveiling the Secret of the Mysteries*. Berkeley, Calif.: North Atlantic Books, 2008.

Wasson, Valentina Pavlovna, and R. Gordon Wasson. *Mushrooms, Russia, and History*. Vol. 2. New York: Pantheon Books, 1957.

Watts, Rosalind, Camilla Day, Jacob Krzanowski, David Nutt, and Robin Carhart-Harris. "Patients' Accounts of Increased 'Connectedness' and 'Acceptance' After Psilocybin for Treatment-Resistant Depression." *Journal of Humanistic Psychology* 57, no. 5 (2017): 520–64. doi:10.1177/0022167817709585.

Weil, Andrew T. "The Strange Case of the Harvard Drug Scandal." *Look*, Nov. 1963.

Whitman, Walt. *Leaves of Grass: The First (1855) Edition*. New York: Penguin, 1986.

Wit, Harriet de. "Towards a Science of Spiritual Experience." *Psychopharmacology* 187, no. 3 (2006): 267. doi:10.1007/s00213-006-0462-8.

Wulf, Andrea. *The Invention of Nature: Alexander von Humboldt's New World*. New York: Alfred A. Knopf, 2015.

Index

Note: Page numbers in *italics* refer to illustrations.

Timothy Leary 58
hyphae 90
2·4 miles wide hairy fungus 90
Description of Psilocybe 94/5
Tchotchkes. 231
Holotropic Breathwork 260
252/3
261 Mask
268 Music
269. evanescent
285 Gratuitous / INEFFABLE

Trepanation (Amanda Feilding 298
353.NB 355 Atheist 359 NB 373 383 35
389

ALLEN LANE
an imprint of
PENGUIN BOOKS

Also Published

David Brooks, *The Second Mountain*

Roberto Calasso, *The Unnamable Present*

Lee Smolin, *Einstein's Unfinished Revolution: The Search for What Lies Beyond the Quantum*

Clare Carlisle, *Philosopher of the Heart: The Restless Life of Søren Kierkegaard*

Nicci Gerrard, *What Dementia Teaches Us About Love*

Edward O. Wilson, *Genesis: On the Deep Origin of Societies*

John Barton, *A History of the Bible: The Book and its Faiths*

Carolyn Forché, *What You Have Heard is True: A Memoir of Witness and Resistance*

Elizabeth-Jane Burnett, *The Grassling*

Kate Brown, *Manual for Survival: A Chernobyl Guide to the Future*

Roderick Beaton, *Greece: Biography of a Modern Nation*

Matt Parker, *Humble Pi: A Comedy of Maths Errors*

Ruchir Sharma, *Democracy on the Road*

145

David Wallace-Wells, *The Uninhabitable Earth: A Story of the Future*

Randolph M. Nesse, *Good Reasons for Bad Feelings: Insights from the Frontier of Evolutionary Psychiatry*

Anand Giridharadas, *Winners Take All: The Elite Charade of Changing the World*

Richard Bassett, *Last Days in Old Europe: Triste '79, Vienna '85, Prague '89*

Paul Davies, *The Demon in the Machine: How Hidden Webs of Information Are Finally Solving the Mystery of Life*

Toby Green, *A Fistful of Shells: West Africa from the Rise of the Slave Trade to the Age of Revolution*

Paul Dolan, *Happy Ever After: Escaping the Myth of The Perfect Life*

Sunil Amrith, *Unruly Waters: How Mountain Rivers and Monsoons Have Shaped South Asia's History*

Christopher Harding, *Japan Story: In Search of a Nation, 1850 to the Present*

Timothy Day, *I Saw Eternity the Other Night: King's College, Cambridge, and an English Singing Style*

Richard Abels, *Aethelred the Unready: The Failed King*

Eric Kaufmann, *Whiteshift: Populism, Immigration and the Future of White Majorities*

Alan Greenspan and Adrian Wooldridge, *Capitalism in America: A History*

Philip Hensher, *The Penguin Book of the Contemporary British Short Story*

Paul Collier, *The Future of Capitalism: Facing the New Anxieties*

Andrew Roberts, *Churchill: Walking With Destiny*